THE LIBRARY
ST. MARY'S COLLEGE OF MARYLAND
ST. MARY'S CITY, MARYLAND 20686

D1557171

Chromatographic Detectors

CHROMATOGRAPHIC SCIENCE SERIES

A Series of Monographs

Editor: JACK CAZES
Cherry Hill, New Jersey

1. Dynamics of Chromatography, *J. Calvin Giddings*
2. Gas Chromatographic Analysis of Drugs and Pesticides, *Benjamin J. Gudzinowicz*
3. Principles of Adsorption Chromatography: The Separation of Nonionic Organic Compounds, *Lloyd R. Snyder*
4. Multicomponent Chromatography: Theory of Interference, *Friedrich Helfferich and Gerhard Klein*
5. Quantitative Analysis by Gas Chromatography, *Josef Novák*
6. High-Speed Liquid Chromatography, *Peter M. Rajcsanyi and Elisabeth Rajcsanyi*
7. Fundamentals of Integrated GC-MS (in three parts), *Benjamin J. Gudzinowicz, Michael J. Gudzinowicz, and Horace F. Martin*
8. Liquid Chromatography of Polymers and Related Materials, *Jack Cazes*
9. GLC and HPLC Determination of Therapeutic Agents (in three parts), *Part 1 edited by Kiyoshi Tsuji and Walter Morozowich, Parts 2 and 3 edited by Kiyoshi Tsuji*
10. Biological/Biomedical Applications of Liquid Chromatography, *edited by Gerald L. Hawk*
11. Chromatography in Petroleum Analysis, *edited by Klaus H. Altgelt and T. H. Gouw*
12. Biological/Biomedical Applications of Liquid Chromatography II, *edited by Gerald L. Hawk*
13. Liquid Chromatography of Polymers and Related Materials II, *edited by Jack Cazes and Xavier Delamare*
14. Introduction to Analytical Gas Chromatography: History, Principles, and Practice, *John A. Perry*
15. Applications of Glass Capillary Gas Chromatography, *edited by Walter G. Jennings*
16. Steroid Analysis by HPLC: Recent Applications, *edited by Marie P. Kautsky*
17. Thin-Layer Chromatography: Techniques and Applications, *Bernard Fried and Joseph Sherma*
18. Biological/Biomedical Applications of Liquid Chromatography III, *edited by Gerald L. Hawk*
19. Liquid Chromatography of Polymers and Related Materials III, *edited by Jack Cazes*
20. Biological/Biomedical Applications of Liquid Chromatography, *edited by Gerald L. Hawk*
21. Chromatographic Separation and Extraction with Foamed Plastics and Rubbers, *G. J. Moody and J. D. R. Thomas*

22. Analytical Pyrolysis: A Comprehensive Guide, *William J. Irwin*
23. Liquid Chromatography Detectors, *edited by Thomas M. Vickrey*
24. High-Performance Liquid Chromatography in Forensic Chemistry, *edited by Ira S. Lurie and John D. Wittwer, Jr.*
25. Steric Exclusion Liquid Chromatography of Polymers, *edited by Josef Janča*
26. HPLC Analysis of Biological Compounds: A Laboratory Guide, *William S. Hancock and James T. Sparrow*
27. Affinity Chromatography: Template Chromatography of Nucleic Acids and Proteins, *Herbert Schott*
28. HPLC in Nucleic Acid Research: Methods and Applications, *edited by Phyllis R. Brown*
29. Pyrolysis and GC in Polymer Analysis, *edited by S. A. Liebman and E. J. Levy*
30. Modern Chromatographic Analysis of the Vitamins, *edited by André P. De Leenheer, Willy E. Lambert, and Marcel G. M. De Ruyter*
31. Ion-Pair Chromatography, *edited by Milton T. W. Hearn*
32. Therapeutic Drug Monitoring and Toxicology by Liquid Chromatography, *edited by Steven H. Y. Wong*
33. Affinity Chromatography: Practical and Theoretical Aspects, *Peter Mohr and Klaus Pommerening*
34. Reaction Detection in Liquid Chromatography, *edited by Ira S. Krull*
35. Thin-Layer Chromatography: Techniques and Applications. Second Edition, Revised and Expanded, *Bernard Fried and Joseph Sherma*
36. Quantitative Thin-Layer Chromatography and Its Industrial Applications, *edited by Laszlo R. Treiber*
37. Ion Chromatography, *edited by James G. Tarter*
38. Chromatographic Theory and Basic Principles, *edited by Jan Åke Jönsson*
39. Field-Flow Fractionation: Analysis of Macromolecules and Particles, *Josef Janča*
40. Chromatographic Chiral Separations, *edited by Morris Zief and Laura J. Crane*
41. Quantitative Analysis by Gas Chromatography, Second Edition, Revised and Expanded, *Josef Novák*
42. Flow Perturbation Gas Chromatography, *N. A. Katsanos*
43. Ion-Exchange Chromatography of Proteins, *Shuichi Yamamoto, Kazuhiro Nakanishi, and Ryuichi Matsuno*
44. Countercurrent Chromatography: Theory and Practice, *edited by N. Bhushan Mandava and Yoichiro Ito*
45. Microbore Column Chromatography: A Unified Approach to Chromatography, *edited by Frank J. Yang*
46. Preparative-Scale Chromatography, *edited by Eli Grushka*
47. Packings and Stationary Phases in Chromatographic Techniques, *edited by Klaus K. Unger*
48. Detection-Oriented Derivatization Techniques in Liquid Chromatography, *edited by Henk Lingeman and Willy J. M. Underberg*
49. Chromatographic Analysis of Pharmaceuticals, *edited by John A. Adamovics*
50. Multidimensional Chromatography: Techniques and Applications, *edited by Hernan Cortes*

51. HPLC of Biological Macromolecules: Methods and Applications, *edited by Karen M. Gooding and Fred E. Regnier*
52. Modern Thin-Layer Chromatography, *edited by Nelu Grinberg*
53. Chromatographic Analysis of Alkaloids, *Milan Popl, Jan Fähnrich, and Vlastimil Tatar*
54. HPLC in Clinical Chemistry, *I. N. Papadoyannis*
55. Handbook of Thin-Layer Chromatography, *edited by Joseph Sherma and Bernard Fried*
56. Gas–Liquid–Solid Chromatography, *V. G. Berezkin*
57. Complexation Chromatography, *edited by D. Cagniant*
58. Liquid Chromatography–Mass Spectrometry, *W. M. A. Niessen and Jan van der Greef*
59. Trace Analysis with Microcolumn Liquid Chromatography, *Miloš Krejcí*
60. Modern Chromatographic Analysis of Vitamins: Second Edition, *edited by André P. De Leenheer, Willy E. Lambert, and Hans J. Nelis*
61. Preparative and Production Scale Chromatography, *edited by G. Ganetsos and P. E. Barker*
62. Diode Array Detection in HPLC, *edited by Ludwig Huber and Stephan A. George*
63. Handbook of Affinity Chromatography, *edited by Toni Kline*
64. Capillary Electrophoresis Technology, *edited by Norberto A. Guzman*
65. Lipid Chromatographic Analysis, *edited by Takayuki Shibamoto*
66. Thin-Layer Chromatography: Techniques and Applications, Third Edition, Revised and Expanded, *Bernard Fried and Joseph Sherma*
67. Liquid Chromatography for the Analyst, *Raymond P. W. Scott*
68. Centrifugal Partition Chromatography, *edited by Alain P. Foucault*
69. Handbook of Size Exclusion Chromatography, *edited by Chi-san Wu*
70. Techniques and Practice of Chromatography, *Raymond P. W. Scott*
71. Handbook of Thin-Layer Chromatography: Second Edition, Revised and Expanded, *edited by Joseph Sherma and Bernard Fried*
72. Liquid Chromatography of Oligomers, *Constantin V. Uglea*
73. Chromatographic Detectors: Design, Function, and Operation, *Raymond P. W. Scott*

ADDITIONAL VOLUMES IN PREPARATION

Chromatographic Analysis of Pharmaceuticals: Second Edition, *edited by John A. Adamovics*

Chromatographic Detectors

Design, Function, and Operation

Raymond P. W. Scott

Georgetown University
Washington, D.C.

Birkbeck College, University of London
London, United Kingdom

Marcel Dekker, Inc. New York•Basel•Hong Kong

Library of Congress Cataloging-in-Publication Data

Scott, Raymond P. W. (Raymond Peter William).
  Chromatographic detectors: design, function, and operation / Raymond P. W. Scott.
    p. cm. — (Chromatographic science series; v. 70)
  Includes bibliographical references (p. -) and index.
  ISBN 0-8247-9779-5 (alk. paper)
  1. Chromatographic detectors. I. Title. II. Series: Chromatographic science; v. 70.
  QD79.C4S383   1996
  681'.754—dc20   96-18799
                                                                   CIP

The publisher offers discounts on this book when ordered in bulk quantities. For more information, write to Special Sales/Professional Marketing at the address below.

This book is printed on acid-free paper.

Copyright © 1996 by MARCEL DEKKER, INC. All Rights Reserved.

Neither this book nor any part may be reproduced or transmitted in any form or by any means, electronic or mechanical, including photocopying, microfilming, and recording, or by any information storage and retrieval system, without permission in writing from the publisher.

MARCEL DEKKER, INC.
270 Madison Avenue, New York, New York 10016

Current printing (last digit):
10 9 8 7 6 5 4 3 2 1

PRINTED IN THE UNITED STATES OF AMERICA

Preface

The development of detectors for gas chromatography, liquid chromatography and thin layer chromatography has now reached a "steady state" and, consequently, a book that covers all three detection methods is timely and appropriate. This book is written in a form similar to that of my previous book on liquid chromatography detectors, but has been brought up-to-date and now includes gas chromatography detectors and some thin layer detection methods. Not only are the well known detectors and detecting systems discussed but, to give a comprehensive and balanced presentation, the equally fascinating but more obscure detection devices are also included. I have again emphasized the importance of employing rational methods to specify detector performance so that the selection of a detector is based on sound and pertinent technical information. A modern trend, that resulting from the impressive developments in column technology, has been the association of the chromatograph with spectroscopic devices in the form of tandem systems. Consequently, tandem systems are also discussed here in some considerable detail. Advice is given on many practical aspects of detector operation, derivatization procedures and the use of the detector in accurate quantitative analysis.

As in with my previous books in this series, I have adhered to what I believe will be the 21st century style of technical writing. The book is written to facilitate the rapid assimilation of information and is virtually devoid of technical jargon. References have been kept to a minimum and only those acknowledging the original inventor or having some particular pertinence are included. In addition, for rapid comprehension, the use of acronyms has been restricted and any that are used are sparsely employed and unambiguously defined.

I would like to take this opportunity to thank the staff of Marcel Dekker Inc. for its much appreciated help and advice in the production of this book.

Raymond P. W. Scott

Acknowledgments

I would like to thank the Royal Society of Chemistry for permission to reproduce figures 6, 7 and 8 in chapter 6.

Figure 1 in chapter 1 is reprinted from *J. Chromatogr.*, **253**(1982)159-178; figure 10 in chapter 2 is reprinted from *J. Chromatogr.*, **218**(1981)97-122; figure 15 in chapter 2 is reprinted from *J. Chromatogr.*, **186**(1979)475-487 and figure 3 in Chapter 8 is reprinted from *J. Chromatogr.*, **169**(1979)51-72 with kind permission of Elsevier Science-NL, Sara Bugerhartstraat 25, 1055 KV Amsterdam, The Netherlands.

Many instrument manufacturers have kindly provided detailed information on their products including photographs and their help is greatly appreciated. Among the many, special thanks are due to ATI Unicam Ltd., Bruker Inc., CAMAG Scientific Inc., the ΔPACKARD Company, ESA Inc., the GOW-MAC Instrument Co., the Hewlett-Packard Corporation, JM Science Inc., LabConections Inc., LDC Analytical, Nicolet Analytical, the Perkin Elmer Corporation, Polymer Laboratories Inc., Supelco Inc., VALCO Instruments Inc., Varian Instruments Inc., VG Organic Inc., and Whatman Inc. for permission to reproduce instrument details and application examples from their technical literature.

Contents

Preface iii

PART 1 Detector Properties and Specifications 1

Chapter 1 An Introduction to Chromatography Detectors 3
Detector Classification 6
Chromatography Nomenclature 9
Extracolumn Dispersion 14
References 16

Chapter 2 Detector Specifications 17
The Detector Output 18
Units Employed in Detector Specifications 21
The Dynamic Range of the Detector 23
Detector Linearity 24
The Linear Dynamic Range of a Detector 31
Detector Response 31
Detector Noise 32
Detector Sensitivity or the Minimum Detectable Concentration 36
The Mass Sensitivity of a Chromatographic System 37
The Concentration Sensitivity of a Chromatographic System 39
The Maximum Capacity Factor of an Eluted Peak 39
Peak Dispersion in the Chromatographic System 42
Pressure Sensitivity 60
Flow Sensitivity 61
Temperature Sensitivity 62
Summary of Detector Criteria 63
References 65

Chapter 3   Data Acquisition and Processing 67
The Acquisition of Chromatography Data 68
Transmission of the Data to the Computer. 72
Data Processing and Reporting 75
Data Processing 76
Chromatographic Control 78
Reference 80

Part 2 Gas Chromatography Detectors 81

**Chapter 4   Gas Chromatography Detectors: Their
Evolution and General Properties** 83
The Evolution of GC Detectors 83
The General Properties of GC Detectors 95
References 97

**Chapter 5 The Flame Ionization Detector and Its
Extensions** 99
The Design of the FID 100
The Response Mechanism of the FID 103
The Operation of the FID 107
Typical Applications of the FID 107
The Nitrogen Phosphorus Detector (NPD) 110
The Emissivity or Photometric Detector 114
References 118

Chapter 6 The Argon Ionization Family of Detectors 119
Introduction 119
The Simple or Macro Argon Detector Sensor 121
The Micro Argon Detector 124
The Triode Detector. 127
The Thermal Argon Detector. 129
The Helium Detector 132
The Pulsed Helium Discharge Detector 135
The Electron Capture Detector. 137
The Pulsed Discharge Electron Capture Detector 143
References 147

**Chapter 7 The Katherometer and Some of the Less
Well Known Detectors** 149
The Simple Gas Density Balance 155
The Radioactivity Detector 157
Some Less Common GC Detectors 159

Part 3 Liquid Chromatography Detectors 175

Chapter 8 Introduction to LC Detectors and the UV Detectors 177
The UV Absorption Detectors 180
References 198

Chapter 9 The Fluorescence and Other Light Processing Detectors 199
The Fluorescence Detectors 201
The Evaporative Light Scattering Detector 211
Liquid Light Scattering Detectors 215
References 222

Chapter 10 The Electrical Conductivity Detector and the Electrochemical Detector 223
The Electrical Conductivity Detector 224
The Electrochemical Detector 233
The Multi Electrode Array Detector 240
References 246

Chapter 11 The Refractive Index and Associated Detectors 247
The Refractive Index Detector 247
The Christiansen Effect Detector 253
The Interferometer Detector 255
Applications of the Refractive Index Detector 259
Detectors Associated with Refractive Index Measurement 264
References 272

Chapter 12 Multifunctional Detectors and Transport Detectors 273
Multifunctional Detectors 273
Transport Detectors 284
References 296

Chapter 13 Chiral Detectors 297
The Production and Properties of Polarized Light 301
Piezo-Optical Modulators 306
Practical Chiral Detectors 308
References 314

Chapter 14 The Radioactivity Detector and Some Lesser Known Detectors 315
The Radioactivity Detector 315
Some Lesser Known Detectors 327
References 360

Chapter 15 Detection in Thin Layer Chromatography 363
Spot Detection in Thin Layer Chromatography 363
General Spot Derivatizing Methods 364
Specific Derivatizing Methods 366
Fluorescence Detection 367
Scanning Densitometry 368
References 374

Part 4 General Detector Techniques 375

Chapter 16 Spectroscopic Detectors and Tandem Systems 377
Gas Chromatography Tandem Systems 380
The Atomic Emission Detector 393
Liquid Chromatography Tandem Systems 395
Thin Layer Chromatography Tandem Systems 433
References 436

Chapter 17 Practical Detector Techniques 439
General Detector Operation 440
Differential and Integral Detector Operation 452
Vacancy Chromatography 460
Derivatization 464
References 472

Chapter 18 Quantitative Analysis 475
Chromatographic Resolution 475
Peak Deconvolution 476
The Detector Response 480
The Quantitative Evaluation of the Chromatogram 481
Peak Area Measurements 482
Peak Height Measurements 487
Quantitative Analytical Methods for GC and LC 490
Quantitative Analysis by TLC 499

Appendix I	503
Appendix II	505
Appendix III	507
Index	509

Chromatographic Detectors

Part 1

Detector Properties and Specifications

CHAPTER 1

AN INTRODUCTION TO CHROMATOGRAPHY DETECTORS

A device that monitors the presence of a solute as it leaves the chromatographic system is an essential adjunct to all chromatography instruments. In fact, without it, chromatography would have a very limited performance and a very restricted field of application. The monitoring device has been given the general term *detector*, which encompasses all types of mobile phase monitoring instruments ranging from the relatively simple flame ionization detector (FID) used in gas chromatography (GC), to very sophisticated spectrometric equipment such as the mass spectrometer.

The slow development of the technique in the early part of the twentieth century was largely due to the lack of a sensitive, in–line detector and it was not until the first GC detectors were developed in the early 1950s that chromatography could finally evolve into the highly sensitive and accurate analytical procedure that is known today. Since the early 1950s there has been a continuous synergistic interaction between improved detector performance and improved column performance; each advance being mutually dependent on the other. Initially, high sensitivity detectors permitted a precise column theory to be developed which resulted in the production of columns having much higher efficiencies. The improved efficiency however produced peaks of very small volume, small that is compared with the dispersion that occurred in the connecting tubes and sensing volume of the contemporary detectors themselves. It followed that the ultimate efficiency obtainable from the column was now determined by the geometry of the detector, not by its sensitivity. As a result the design of the detector was changed; the dimensions and geometric form of the connecting tubes were modified and the volume of the sensing cell was greatly reduced. In liquid chromatography (LC) this allowed the full potential of columns packed with very small particles to be exploited, the advantages of which were predicted by Martin as long ago as 1941 [1]. The small particles provided much higher efficiencies and even smaller peak volumes [2] provoking further modifications in detector design. In GC the introduction of the capillary or open tubular column by Golay [3] produced the same effect. Solute bands eluted from open tubular column are contained in a few microliters of mobile phase and thus stipulate the use of detectors that have small volume sensors and connecting tubes with minimal peak dispersion. Both the small diameter open tubular columns in GC and the smaller particle packings in LC provided columns that could produce extremely fast separations Examples of high speed LC separations are afforded by the work of DeCesere *et al.* [4], and Scott and Katz [5] who separated solute mixtures of moderate complexity in a few seconds.

In such separations, individual peaks were only a few milliseconds in width and thus, the standard deviation of such peaks would be of the same order of magnitude as the response time constant of the sensor and electronics. Similar speeds were achieved by Desty [6] (using small diameter glass capillary columns in GC) evoking the redesign of both GC and LC detector sensors and supporting electronics to provide adequately fast responses. An example of the high speed LC separation of a simple mixture by Katz and Scott (5) is shown in figure 1. It is seen that the first peak is only about 125 milliseconds wide.

The introduction of small bore packed columns in LC [7] reduced the peak volume still further and placed an even greater strain on LC detector design. Due to the relatively lower sensitivity of LC detectors compared with that of GC detectors, the LC detector sensor volume was forced down to a level which, for present day technology, may well be the practical limit for many types of LC detectors. This interaction between detector design and column design continues to this day and probably will do so for many years to come.

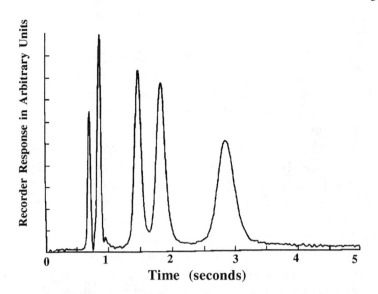

Packing : Hypersil 3-m, Column I.D. 0.26 cm, length 2.5 cm. Mobile Phase: 2.2% v/v methyl acetate in n-pentane. Linear velocity: 3.3 cm/sec. 1, p-xylene; 2, anisole; 3, nitrobenzene; 4, acetophenone; 5, dipropyl phthalate.

Figure 1 High-Speed Separation of a Five Component Mixture

Current attempts to utilize capillary columns in LC require extremely small sensor volumes that, even for selected detector systems, are exceedingly difficult to realize in practice. Successes that have so far been achieved have been more apparent than real. Due to unsuitable detector specifications, relatively large charges must be placed on the column and so the column performance is impaired and the high resolution expected from such columns far from realized. To date, nothing like the separations obtained from open tubular columns in GC have been realized in LC, particularly for muticomponent mixtures. Furthermore, it is unlikely that the specifications of existing LC detectors will permit such separations to be realized. Nevertheless, comparable technical challenges that have fostered scientific progress in the past and are still active today, so hopefully with the advent of a more sensitive LC detectors high efficiency open tubular columns will become a reality in the future.

Detector Classification

The detector can be defined as a device that locates, in the dimensions of space or time, the positions of the components of a mixture after they have been subjected to a chromatographic process and thus permits the senses to appreciate the nature of the separation that has been obtained. This definition, by necessity, must be broad as it needs to encompass all detecting systems ranging from the elaborate electronic devices presently available, to the human eye or even the sense of smell. Tswett in his pioneering chromatographic separation of plant pigments used the eye to determine the nature of the separation that he obtained. Even today the eye is still frequently used as a means of detection in thin layer chromatography (TLC) although more sophisticated methods are now available. As a result of international dialogue between the respective organizations involved with chromatography, detector specifications have been clearly and unambiguously defined. In contrast, the classification of detectors has not been the subject of debate, nor has it been considered by IUPAC, and as a result the present accepted classification has arisen largely by default. In fact, detectors can be classified in a number of different ways with little or no logical relationship between them.

Detectors are primarily classified with respect to the technique with which they are used, *viz*. GC detectors, LC detectors and TLC detectors, and this type of classification is generally accepted. There is more debate regarding the secondary classification of detectors and, unfortunately, there are three methods in common use. Each of these methods of classification will be considered in detail.

Bulk Property and Solute Property Detectors

One secondary method of classification, and one that is probably the most frequently used, is based on a more rational differentiation and defines a detector as either a 'bulk property' detector or a 'solute property' detector. This secondary classification applies to GC, LC and TLC detectors.

Bulk Property Detectors

Bulk property detectors function by measuring some bulk physical property of the mobile phase, *e.g.*, thermal conductivity in the case of a GC detector or refractive index in the case of an LC detector. For obvious reasons bulk property detectors are not commonly used for TLC detection. As a bulk property is being measured, they are very susceptible to changes in mobile phase composition or temperature and can not be used for gradient elution in LC. They are also very sensitive to changes in pressure and flow rate and consequently the operating conditions of the chromatograph need to be very carefully controlled. It follows that bulk property detectors used in GC must be very carefully thermostatted in a separate oven if temperature programming is to be employed.

Solute Property Detectors

Solute property detectors function by measuring some property of the solute that the mobile phase does not possess, or has to a very reduced extent. An example of a solute property detector would be the phosphorus-nitrogen detector (PND) in GC, which responds only to nitrogen or phosphorus, or the UV detector in LC, which detects only those substances that absorb in the UV. This classification, however, is rarely precise, particularly in LC, as it is often difficult to employ a phase system to which the detector has no response relative to that of the solute. For example, the UV detector may be employed with a mobile phase that contains small quantities of a polar solvent such as ethyl acetate. The ethyl acetate at the higher wavelengths in the UV range will exhibit little or no absorption and the detector may well behave as a true solute property detector. However, if the wavelength is reduced (*e.g.*, to detect amino acids), then the ethyl acetate will absorb quite strongly and the system will begin to resemble that of a bulk property detector. Solute property detectors are used effectively in GC, LC and TLC. However, it is evident that this type of classification is not clear cut and can be confusing, as a given detector can exhibit both characteristics depending on the conditions under which it is operated.

Mass Sensitive and Concentration Sensitive Detectors

Some detectors respond to changes in solute concentration while others respond to the change in mass passing through the sensor per unit time. This difference has also been used as a basis with which to classify detectors.

Concentration Sensitive Detectors

Concentration sensitive detectors provide an output that is directly related to the concentration of solute in the mobile phase passing through it. In GC the katherometer would be an example of this type of detector whereas in LC the UV absorption detector would be typical of a concentration sensitive detector.

Mass Sensitive Detectors

The mass sensitive detector responds to the mass of solute passing through it per unit time and is thus independent of the volume flow of mobile phase. A typical example of a mass sensitive GC detector is the FID. The function of the FID will be described later but it should be noted here that the eluent from a capillary column (at a very low flow rate) is mixed with a hydrogen flow and then passed into the FID. As the mass entering the detector is determined solely by the rate of elution from the column, the response of the detector is entirely independent of the actual flow of hydrogen *although* it will significantly *reduce* the solute *concentration* entering the detector.

This classification is satisfactory for GC detectors but as there is only one LC detector that is mass sensitive [8] (the transport detector), which at the time of writing this book is also not commercially available, this manner of classification is of little use for LC detectors.

Specific and Non–Specific Detectors

Detectors have also been classified on the basis of the nature of their response whether it is specific or non-specific.

Specific Detectors

Specific detectors respond to a particular type of compound or a particular chemical group. An example of a specific GC detector would be NPD that responds specifically to compounds containing nitrogen and phosphorus. In LC the fluorescence detector would be a typical specific detector that responds only to those substances that fluoresce.

Non-Specific Detectors

A non-specific detector responds to all solutes present in the mobile phase and its catholic performance makes it a very useful and popular type of detector. Unfortunately, non-specific detectors in LC tend to be relatively insensitive. The FID is a nonspecific detector in that (with very few exceptions) it responds to all solutes that contain carbon. As an added advantage the FID also has a very high sensitivity. In LC the refractive index detector is probably the most non-specific detector but as already mentioned it also has the least sensitivity of the commonly used detectors.

In general, the classification based on bulk property and solute property response seems to be most useful and consequently this is the classification that will be employed throughout this book.

Chromatography Nomenclature

The design and performance of any detector depends heavily on the column and chromatographic system with which it is associated. Throughout this book it will be necessary to relate the detector design to the properties of the column and chromatography apparatus. As a consequence, the characteristics of a separation need to be defined and in due course some basic column chromatography column theory will need to be referred to in order to identify detector specifications. A simple chromatogram of a mixture, appropriately labeled, is shown in Figure 2. The chromatogram could be from a GC or an LC analysis and the terminology employed would be identical for both techniques.

The elution point of an unretained substance occurs at the *dead time* (t_O) and the volume of mobile phase that has passed through the column between the injection point and the dead point is the *dead volume* (V_O). The dead volume is given by ($Q\, t_O$), where (Q) is the flow rate of mobile phase through the column. The volume of mobile phase that passes through the column between the injection point and the peak maximum is called the *retention volume* (V_r), which is given by ($Q\, t_r$) where (t_r) is the time that has elapsed between the injection point and the peak maximum. The difference between the retention volume and the dead time ($V_r - V_O$) is called the *corrected retention volume* (V'_r) which is also equal to the product of the *corrected retention time* (t'_r) and the flow rate (Q).

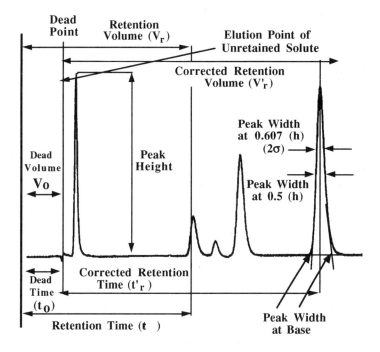

Figure 2 The Characteristics of a Chromatogram

The distance between the baseline produced beneath the peak and the peak maximum is called the *peak height* (h). The width of the peak at

(0.607 h), the position of the points of inflection of the Gaussian curve, is called the *peak width* and is equivalent to 2 standard deviations of the Gaussian curve. The peak width measured at (0.5 h) is the *peak width at half height*. The distance between the points of intersection of the tangents drawn to the points of inflection, and the base line produced beneath the peak, is the *base width* and is equivalent to 4 standard deviations of the Gaussian curve.

In fact, elution curves from both GC and LC columns are often not perfectly Gaussian but tend to be slightly asymmetric. There are a number of reasons for this but their discussion is not germane to the subject of this book. As far as detector design is concerned any imperfections in the shape of the elution curves can be ignored and a Gaussian form can be safely assumed for all calculations. The characteristics of the thin layer plate employ an entirely different nomenclature although many of the concepts are common to both. A diagram of a thin layer plate after development and appropriately labeled is shown in figure 3.

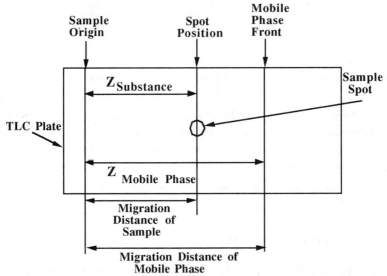

Figure 3   The Characteristics of a Thin Layer Plate

The migration distance of the spot (denoted as $Z_{substance}$) is the distance between the sample origin and the spot position after development. The Migration distance of the mobile phase (denoted as $Z_{Mobile\ Phase}$) is the distance between the sample origin and the solvent front at the completion of the development

The basic parameter used in TLC for solute identification is the *Retardation Factor* (R_F), which is given by

$$R_F = \frac{Z_{Solute}}{Z_{Mobile\ Phase}} = \frac{Z_s}{Z_f} \tag{1}$$

If, by analogy, it is tentatively assumed that (Z_f) is equivalent to the retention volume (V_r), and (Z_s) is equivalent to the dead volume (V_m), then the equivalent value for the capacity ratio (k') is given by,

$$k' = \frac{V_r - V_m}{V_m} = \frac{Z_f - Z_s}{Z_s} = \frac{1 - \frac{Z_s}{Z_f}}{\frac{Z_s}{Z_f}} = \frac{1 - R_f}{R_f} \tag{2}$$

(The assumptions that (Z_f) is equivalent to the retention volume (V_r), and (Z_s) is equivalent to the dead volume (V_m) in LC or GC are, in the author's opinion not valid. Furthermore, in practice the development occurs under gradient elution conditions and thus the distribution coefficient (K) is not constant throughout the development, and retention will depend on the homogeneity of the stationary phase layer. Nevertheless, the capacity factor calculated in this way is a parameter that is used in TLC to help identify a solute. The assumption, right or wrong, has little impact on TLC detector design or function.)

The Summation of Variances

In this introduction, one further important basic concept needs to be discussed as it is used extensively in detector design and in particular, the design of detector connections and detector sensor geometry and that is the principle of the *summation of variances*.

The width of the band of an eluted solute relative to the distance between it and its nearest neighbor determines whether two solutes are resolved or not. The ultimate band width of a solute peak, as sensed by the detector, is the result of a number of individual dispersion processes taking place in the chromatographic system, some of which take place in the column itself and some in the sample valve, connecting tubes and detector. In order to determine the ultimate dispersion of the solute band and thus identify the true column performance the final peak variance must be calculated. This is achieved by taking into account all the individual dispersion processes that take place in the total chromatographic system. It is not possible to sum the band widths resulting from each individual dispersion process to obtain the final band width, but it is possible to sum all the respective variances. However, the summation of all the variances resulting from each process is only possible if each process is non-interacting and random in nature. That is to say, the extent to which one dispersion process progresses is independent of the development and progress of any other dispersion process. In both GC and LC this condition of non-interactive dispersion is generally true. Thus, assuming there are (n) non-interacting, random dispersion processes occurring in the chromatographic system, then any one process, (p), acting alone will produce a Gaussian curve having a variance σ_p^2,

$$\text{Hence,} \quad \sigma_1^2 + \sigma_2^2 + \sigma_3^2 + \ldots + \sigma_n^2 = \sigma^2 \quad (3)$$

where σ^2 is the variance of the solute band as sensed by the detector.

The above equation is the algebraic description of the principle of the summation of variances and is fundamentally important. If the individual dispersion processes that are taking place in a column can be identified, and the variance that results from each dispersion determined, then the variance of the final band can be calculated from the sum of all the individual variances. An example of the use of this principle is afforded by the calculation of the maximum extracolumn dispersion that can be tolerated for a particular column. This calculation is pertinent to both GC and LC columns but not thin layer plates.

Extracolumn Dispersion

Extracolumn dispersion is that contribution to the total band dispersion that arises from spreading processes taking place *outside* the column and for convenience will be taken as (σ_E^2).

Now from the summation of variances,

$$\sigma^2 = \sigma_C^2 + \sigma_E^2 \qquad (4)$$

where (σ^2) is the variance of the peak entering the detector,
and (σ_C^2) is the variance of the peak leaving the column.

Now the maximum increase in band width that can be tolerated before loss of resolution becomes unacceptable was determined by Klinkenberg [9] to be 10% of the column variance; thus,

$$1.1\sigma_C^2 = \sigma_C^2 + \sigma_E^2 \quad \text{or} \quad \sigma_E^2 = 0.1\sigma_C^2 \qquad (5)$$

Now, from the Plate Theory [10], it can be shown

$$\sigma_C^2 = \frac{V_r^2}{n} \qquad (6)$$

where (V_r) is the retention volume of the pertinent solute,
and (n) is the efficiency of the column.

Thus substituting for (σ_C^2) from (6) in (5)

$$\sigma_E^2 = 0.1\frac{V_r^2}{n} \quad \text{or} \quad \sigma_E = 0.32\frac{V_r}{\sqrt{n}} \qquad (7)$$

In practice equation (7) is usually applied to the dead volume peak as all peaks have equal importance and the dead volume peak is the most narrow and thus tolerates the least extracolumn dispersion.

$$\sigma_E = 0.32 \frac{V_0}{\sqrt{n}} \quad (8)$$

where (V_0) is the dead volume of the column.

As an example of the use of equation (8) in detector design, let us calculate the standard deviation of the extracolumn dispersion that can be tolerated by an LC column 10 cm long 4.6 mm I.D packed with particles 10 µ in diameter.

Now, the height of the theoretical plate (HETP) of a column packed with particles having a diameter (dp) will be approximately (2dp), and thus the efficiency (n) for a column of length (l) will be

$$n = \frac{l}{2dp} = \frac{10}{0.002} \quad 5,000 \quad \text{theoretical plates.}$$

Now, if the column diameter is (d) the dead volume of the column will be given by

$$V_0 = 0.6\pi \left(\frac{d}{2}\right)^2 l$$

where it is assumed that 60% of the column volume is occupied by mobile phase. This of course will only be true for a *packed* column.

Thus, $\quad V_0 = 0.6\pi(0.23)^2 10 = 0.997$ ml

Thus from equation (8),

$$\sigma_E = 0.32 \frac{0.997}{\sqrt{5,000}} = 0.0045$$

It is seen that the maximum value for the standard deviation of all the processes that contribute the extracolumn dispersion must be less than 4.5 microliters and gives an indication of the difficulties involved in designing detectors that can function well with microbore columns. It is also seen that equation (8) can be extremely useful in detector design and provides the necessary data that would allow a detecting system to be constructed to suit a particular range of column sizes. It is obvious that, although the maximum value for (σ_E) is now known, it will be necessary to examine quantitatively the contribution of the various extracolumn dispersion processes to the overall value of (σ_E). These details will be discussed later in this book.

References

1. A. J. P. Martin and R. L. M. Synge, *Biochem. J.*, **35**(1941)1358.
2. K. K. Unger, W. Messer and K. F. Krebs, *J. Chromatogr.*, **149**(1978)1.
3. M. J. E. Golay, "*Gas Chromatography, 1958*" (ed. D. H. Desty) Butterworths, London (1958)36.
4. J. L. DiCesare, M. M. Dong and L. S. Ettre, *Chromatographia*, **14**(1981)257.
5. E. Katz and R. P. W. Scott, *J. Chromatogr.*, **253**(1982)159.
6. D. H. Desty, "*Gas Chromatography, 1958*" (ed. D. H. Desty) Butterworths, London (1958)126.
7. R. P. W. Scott and P. Kucera, *J. Chromatogr.*, **169**(1979)51.
8. R. P. W. Scott and J. G. Lawrence, *Anal. Chem.*, **39**(1967)830.
9. A. Klinkenberg, "*Gas Chromatography 1960* " (Ed. R.P.W. Scott), Butterworths, London,(1960)194.
10. R. P. W. Scott, "Liquid Chromatography Column Theory", John Wiley and Sons, New York, (1992)26.
11. R. P. W. Scott, "Liquid Chromatography Column Theory", John Wiley and Sons, New York, (1992)109.

CHAPTER 2

DETECTOR SPECIFICATIONS

Accurate performance criteria or specifications must be available to determine the suitability of a detector for a specific application. This is necessary, not only to compare its performance with alternatives supplied by other instrument manufactures, but also to determine the optimum chromatography system with which it must be used to achieve the maximum efficiency. The specifications should be presented in a standard form and in standard units, so that detectors that function on widely different principles can be compared. The major detector characteristics that fulfill these requirements together with the units in which they are measured are summarized in table 1.

Unfortunately, there has been considerable confusion and disagreement over the respective units that should be used to define certain detector specifications and in some cases there has been dissension over the exact definition of the specifications themselves. This confusion has arisen largely from the adoption of criteria used in other instrumental devices that have been borrowed for use in detector technology and are sometimes not precisely applicable. In addition, specifications have also been adopted by some manufacturers in order to present their products in the best possible light and which may not be easily translatable into specifications that are pertinent to the chromatographer. In the following discussions on detector specifications, the units that are used have been chosen as those most relevant to chromatography and that are directly related to the needs of the analyst, albeit for use in GC or LC.

Table 1   Detector Specifications

Specifications	Units
Dynamic Range	(D_R) g/ml (*e.g.* 3×10^{-9} to 6×10^{-5})
Response Index	(r) dimensionless
Linear Dynamic Range	(D_{LR}) g/ml (*e.g.* 1×10^{-8} to 2×10^{-5})
Detector Response	(R_C) Volts/g or (specific units of measurement/g)
Detector Noise Level	(N_D) usually in millivolts but may be in specific units (*e.g.* Refractive Index Units)
Sensitivity or minimum detectable concentration	(X_D) g/ml (*e.g.* 3×10^{-8}) but may be in specific units (*e.g.* Absorption Units)
Total Detection System Dispersion	(σ_D^2) (ml² often µl²)
Cell Dimensions	(length (l), and radius (r)), (cm)
Cell Volume	(V_D), µl.
Overall Time Constant (sensor and electronics)	(T_D), seconds (sometimes milliseconds)
Pressure Sensitivity	(D_P) usually in the USA, p.s.i in Europe MPa
Flow Rate Sensitivity	(D_Q) usually in ml/min
Temperature Range	°C

The Detector Output

Depending on the basic function of the detector the output can be normal, differential or integral. The three different types of detector response are shown in figure 1. The proportional detector, as its name implies, provides a signal that tends to be directly proportional to the concentration of the solute in the mobile phase passing through it.

Thus, $y = Ac$

where (y) is the output of the detector in appropriate units,
(c) is the concentration of solute in the mobile phase,
and (A) is a constant.

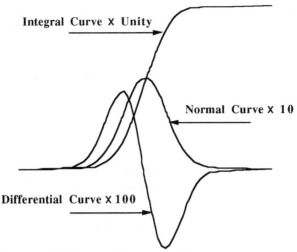

Figure 1 The Different Forms of Detector Response

All proportional detectors are designed to provide a response that is as close to linear as possible in order to facilitate calculation procedures in quantitative analysis. In many detectors the output from the detector *sensor* is *not* linearly elated to the solute concentration, in which case appropriate signal modifying circuits must be used to furnish a linear output. For example the output from a light adsorption sensor will be exponential and consequently it must be used with a logarithmic amplifier to produce an output that is linearly related to solute concentration.

In fact, there is no reason why a detector should not have a logarithmic, exponential or any other functional output as long as the function can be explicitly defined. However, chromatograms obtained from such detectors would be unfamiliar and difficult to interpret and calculations for quantitative analysis would become very involved. The concentration profile of an eluted peak closely resembles that of the Gaussian or Error function where the independent variable is the volume of mobile phase passed through the column and the dependent variable solute concentration in the mobile phase. If the flow rate through the chromatographic system is constant, then the independent variable can be replaced by time. Thus a linear detector will provide a close representation of the Gaussian concentration profile of the peak as shown in figure 1.

For a concentration sensitive detector, the concentration (C_m) of a solute in the mobile phase is given by:

$$C_m = AV$$

where (V) is the output of the detector in millivolts,
and (A) is the proportionality constant.

To provide an integral response (Y) that is equal to the mass of solute eluted,

$$Y = \int_0^v C_m dv = \int_0^v AV dv = AVv = C_m v = m$$

where (v) is the volume of mobile phase passed through the column
and (m) is the mass of solute eluted.

A detector with an integral response would give a curve consisting of a series of steps, each step height being proportional to the mass of solute in each peak. At this time, no commercially available detector provides a true integral response, although by electronically integrating the output from a detector with a normal response, an integral output can be achieved. The original mass detector [1] accumulated the eluted solutes onto a recording balance and thus did, indeed, produce a true integral response but this detector was never developed into a commercial product. The advantages of an integral detector for quantitative analysis is more apparent than real. If all the peaks in a chromatogram are completely resolved, then it could be the ideal detector response for accurate analysis. However, if resolution is not complete, which is the more common situation, the merged steps are very difficult to interpret and deconvoluting software will be less precise. Finally, the peak maximum that is shown by a detector with a normal response, and which is used to identify the solute, occurs in the center of the step of an integral curve and consequently can also be difficult to identify precisely.

The differential curve (Y') is obtained by differentiating the normal output,

$$Y' = \frac{dC_m}{dv} = A\frac{dV}{dv}$$

The value of this type of output is very limited as each peak obtained by a detector with a normal response would result in a negative and positive peak from the detector with a differential response. This makes the interpretation of a chromatogram containing unresolved peaks very difficult indeed. Furthermore, the differentiation of the signal increases the noise by a factor of ten, reducing the overall sensitivity of the detector by the same amount. An interesting point about the differential curve is that at the peak maximum the differential signal rapidly goes from a relatively large positive value to a relatively large negative value. This effect, processed either by analog circuitry or digitally with appropriate software, can be used to accurately locate the position of the peak maximum, and consequently provide a precise measure of the retention time or retention volume. There are very few detectors where the sensor gives a differential response and the only one that readily comes to mind is the heat of adsorption detector developed by Claxton [2] and Groszek [3]. This detector will be discussed later in the book; it is sufficient to say at this time that the output is approximately proportional to the first derivative of the normal error curve but the precise nature of the relationship depends strongly on the geometry of the detector sensing cell.

Units Employed in Detector Specifications

The simple detector properties such as geometry and sensor cell dimensions are given in length (cm), time (sec) or mass (g) and there is no disagreement in the literature regarding such specifications. The units employed to define detector *sensitivity*, and to lesser extent, the *dynamic* and *linear dynamic* ranges, however, have been the subject of considerable controversy. The dispute has arisen from the diverse physical properties to which different detectors respond. The FID detector (GC), for example, responds to changes in the carbon content of the mobile phase passing through it whereas the refractive index (RI) detector (LC) responds to changes in refractive index. As a consequence the RI detector sensitivity is often given in refractive index units and the FID sensitivity in g/sec of carbon passing through the detector. This method of defining sensitivity allows different FID's to be compared and also different RI detectors to be compared but the relative sensitivity of the FID and RI detector is not immediately obvious. In a similar way defining the sensitivity of the electrical conductivity detector in mho's (reciprocal ohms) permits no direct comparison with either the FID or the refractive index detector.

The most important detector specification is probably detector sensitivity as it not only defines the minimum concentration of solute that can be detected but also allows the overall mass sensitivity of the chromatographic system to be calculated. The detector sensitivity also places a limit on the maximum (k') (capacity factor) at which a solute can be eluted from a chromatographic column. In order to calculate the mass sensitivity or the maximum (k') value, the detector sensitivity must be available in concentration units, *e.g.* g/ml. Moreover, if all detector sensitivities were given in units of g/ml, then all detecting devices, functioning on quite different principles, could then be rationally compared.

The problem arises in choosing the most appropriate solute with which to measure the concentration sensitivity of all detectors. Obviously toluene could be used to define the sensitivity of the FID, UV detector and the RI detector but would be useless for measuring the sensitivity of the electrical conductivity detector or the nitrogen phosphorus detector (NPD) in GC. In a similar way, if the reference solute was chosen to have high refractive index and low UV absorption and contain no phosphorus or nitrogen then the RI detector would appear to have a higher sensitivity than the UV or NPD detectors. Further complications would ensue if the UV detector has multi-wavelength facilities whereby the sensitivity of the detecting system can be changed by selecting different wavelengths for absorption.

It is, therefore, hardly surprising that some confusion has arisen over the units in which detector sensitivity is measured. Instrument manufacturers have avoided the problem by not employing concentration as a unit of sensitivity, leaving the analyst to convert the detector sensitivity given in RI units %carbon etc. to g/ml of a given solute for calculation purposes. This might well be considered unreasonable by the analyst, who can not easily calculate the mass and concentration sensitivity of his/her equipment and often does not know the vapor thermal conductivity, the refractive index or electrical conductivity etc. of the solutes contained in the mixture being analyzed. Furthermore, in almost all analyses the *mass* concentration of the components are the quantity of real interest.

It is not unreasonable to expect the detector manufacturer to specify their products in units that are most useful to their customers. It is therefore recommended that detector sensitivities be given not only in the basic units of measurement but also in g/ml of a readily available solute. The solute chosen should be one that often occurs in mixtures with which the detector will be frequently used for analysis. Unfortunately, instrument manufacturers are not reputed to listen favorably to such simple suggestions and it is likely the analyst will need to measure the detector sensitivity experimentally. A simple procedure for measuring detector sensitivity will be given later in this chapter.

The Dynamic Range of the Detector

There are two important ranges that are specified for a detector (both GC and LC) and these are the *dynamic* range and the *linear dynamic* range. The dynamic range extends from the minimum detectable concentration (*i.e.* the sensitivity) to that concentration at which the detector no longer responds to any increase. The dynamic range is not usually pertinent to general analytical work but is important in preparative chromatography where much higher solute concentrations are used and a linear response is usually not required. In preparative chromatography, high sensitivities are often a *disadvantage* and the most popular detectors for this purpose are the least sensitive, which is in complete contrast to those used for analytical use. The dynamic range of a detector (D_R) is usually given in the form

$$D_R = 6 \times 10^5$$

The manufacturer should give the dynamic range in dimensionless units as it is a ratio of concentrations and independent of the units used. Employing the minimum detectable concentration in conjunction with the dynamic range the analyst can then calculate the *maximum* concentration detectable.

Detector Linearity

Detector linearity is probably the most important specification for any detector that is to be used for quantitative analysis. It is defined as the concentration range over which the detector response is linearly related to the concentration of solute passing through it.

That is, $$V = A C_m$$

where the symbols have the meanings previously ascribed to them.

Beacause of the imperfections in mechanical and electrical devices, true linearity is a hypothetical concept and practical detectors can only approach this ideal response. It is therefore important that the analyst has some measure of linearity that is specified in numerical terms so that comparisons can be made between detectors and the proximity of the detector to true linearity understood. Fowlis and Scott [4] proposed a method of measuring detector linearity. They assumed that for a closely linear detector the response could be described by the following power function

$$V = AC_m^r \tag{1}$$

where (r) is defined as the Response Index and the other symbols have the meanings previously ascribed to them.

It follows that for a truly linear detector, $r = 1$, and the proximity of (r) to unity will indicate the extent to which the response of the detector deviates from true linearity. The response of some detectors having different values for (r) are shown as curves relating the detector output (V) to solute concentration (C_m^r) in figure 2. It is seen that the individual curves appear as straight lines but the errors that occur in assuming linearity can be quite large.

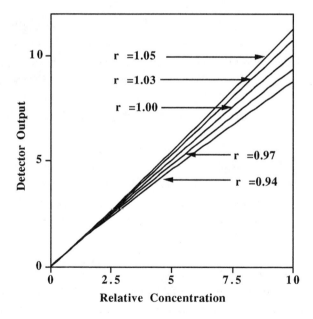

Figure 2 Graph of Detector Output against Solute Concentration for Detectors Having Different Response Indices

The errors actually involved are shown in table 2.

Table 2 The Analysis of a Binary Mixture Employing Detectors with Different Response Indices

Solute	r = 0.94	r = 0.97	r = 1.00	r = 1.03	r = 1.05
1	11.25%	10.60%	10.00%	9.42%	9.05%
2	88.75%	89.40%	90.00%	90.58%	90.95%

Examination of table 2 shows that errors in the level of the smaller component can be as much as 12.5% (1.25% absolute) for r = 0.94 and 9.5% (0.95% absolute) for r = 1.05. In general analytical work, if reasonable linearity is to be assumed, then

$$0.98 < r < 1.03$$

The basic advantage of defining linearity in this way is that if the detector is not perfectly linear, and a value for (r) is known, then a correction can be applied to accommodate the non-linearity. There are alternative methods for defining linearity which, in the authors opinion, are less precise and less useful. The recommendations of the ASTM, E19 committee [5] on linearity measurement were as follows:

the linear range of a PD (photometric detector) is that concentration range of the test substance over which the response of the detector is constant to within 5% as determined from a linearity plot – the linear range should be expressed as the ratio of the highest concentration on the linearity to the minimum detectable concentration.

This method for defining detector linearity is perfectly satisfactory and ensures a minimum linearity from the detector and consequently an acceptable quantitative accuracy. However, the specification is significantly 'looser' than that given above and there is no means of correcting for any non–linearity that may exist as there is no correction factor given that is equivalent to the response index. It is strongly advised that the response index of all detectors (GC and LC) that are to be used for quantitative analysis be determined. In most cases, (r) need only be determined once unless the detector experiences some catastrophic event, in which case (r) may need to be checked again.

The Determination of the Response Index of a Detector

There are two methods that can be used to measure the response index of a detector, the *incremental method of measurement* and the *logarithmic dilution method* of measurement. The former requires no special apparatus but the latter requires a log-dilution vessel which fortunately is relatively easy to fabricate.

The Incremental Method of Linearity Measurement

The apparatus necessary is the detector itself with its associated electronics and recorder or computer system, a mobile phase supply, pump, sample valve and virtually any kind of column. In practice the chromatograph itself which will be used for the subsequent analyses is normally employed. The solute is chosen as typical of the type of substances that will be analyzed and a mobile phase is chosen that will elute the solute from the column in a reasonable time. Initial sample concentrations are chosen to be appropriate for the detector under examination.

Duplicate samples are placed on the column, the sample solution is diluted by a factor of three and again duplicate samples are placed on the column. This procedure is repeated increasing the detector sensitivity setting where necessary until the height of the eluted peak is commensurate with the noise level. If the detector has no data acquisition and processing facilities, then the peaks from the chart recorder can be used. The width of the peak at 0.607 of the peak height is measured and from the chart speed and the mobile phase flow rate the peak volume can be calculated. Now the concentration at the peak maximum will be twice the average peak concentration, which can be calculated from the following equation

$$C_p = \frac{ms}{wQ}$$

where (C_p) is the concentration of solute in the mobile phase at the peak height in (g/ml),
(m) is the mass of solute injected,
(w) is the peak width at 0.6067 of the peak height,
(s) is the chart speed of the recorder or printer
(Q) is the flow rate in ml/min.

The log of the peak height (y) (where y is the peak height in millivolts) is then plotted against the log of the solute concentration at the peak maximum (C_p).

From equation (1),

$$\log(y) = \text{Log } A + (r) \log (C_p)$$

Thus the slope of the Log/Log curve will give the value of the response index (r). If the detector is truly linear, $r = 1$ (*i.e.* the slope of the curve will be $\sin \pi/4 = 1$). Alternatively, if suitable software is available, the data can be curved fitted to a power function and the value of (r) extracted from the results. The same data can be employed to determine the linear range as defined by the ASTM E19 committee. In this case, however, a linear plot of detector output against solute concentration at the peak maximum should be used and the point where the line deviates from 45º by 5% determines the limit of the linear dynamic range.

The Logarithmic Dilution Method of Linearity Measurement

The logarithmic dilution method for detector calibration was introduced by Lovelock [6] for the calibration of GC detectors and was modified by Fowlis and Scott [4] for the calibration of LC detectors. The system provides a continuous flow of gas or liquid through the detector that contains a solute, the concentration of which decreases exponentially with time. A diagram of the log dilution apparatus is shown in figure 3.

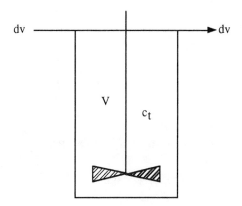

Figure 3   Diagram of a Logarithmic Dilution Vessel

A known mass of solute is introduced into a well–stirred vessel through which a flow of gas or solvent continuously passes. The mixture is thus continuously diluted and the concentration of the solute in the exit flow progressively monitored by means of the detector.

Let the vessel have a volume (V), and let the concentration of the solute in the vessel after time (t) be (C_t). Let a volume (dv) of pure solvent enter the vessel and displace a similar volume (dv) from the vessel.

The mass of solute removed (dm) is given by

$$dm = C_t dv$$

Now the change in mass (dm) in the vessel will result in a change in concentration (dC_t)

$$V dC_t + C_t dv = 0$$

and

$$\frac{dC_t}{C_t} = \frac{-dv}{V}$$

Integrating $\quad \log C_t = \dfrac{-v}{V} + k$

$$= \dfrac{-Qt}{V} + k$$

where (Q) is the flow rate,
(v) is the volume flow of mobile phase through the system after time (t)
and (k) is the integration constant.

Now when t=0, $C_t = C_0$, where C_0 is the initial concentration of solute.

Thus, $\quad k = \log C_0 \quad$ and $\quad \log C_t = \dfrac{-v}{V} + \log C_0$

or $\quad C_t = C_0 e^{\dfrac{-Qt}{V}}$

Thus if the logarithm of the detector output is plotted against time, then for a truly linear detector, a straight line will be produced having a slope (–Q/V). If the detector has a response index of (r) and the slope of the line is (φ), then

$$\varphi = \dfrac{-Qr}{V} \quad \text{or} \quad r = \dfrac{-\varphi V}{Q}$$

Thus, the response index can be easily determined. However the accuracy of the measurement will depend on the flow rate remaining constant throughout the calibration, and consequently for a GC detector a precision flow controller must be employed and for an LC detector, a good quality solvent pump. Manufacturers do not usually provide the response indices for their detectors and so it is left to the analysts to measure it for themselves.

It should be pointed out that the logarithmic dilution method should not be used if the linearity is to be measured by the method recommended by the E19 committee of the ASTM.

The Linear Dynamic Range of a Detector

The linearity of most detectors deteriorates at high concentrations and thus the *linear dynamic range* of a detector will always be less than the *dynamic range*. The symbol for the linear dynamic range is usually taken as (D_{LR}) and might be specified in the following form for the FID as an example

$$D_{LR} = 2 \times 10^{-3} \quad \text{for } 0.98 < r < 1.02$$

The lowest concentration in the linear dynamic range is usually equal to the *minimum detectable concentration* or the *sensitivity* of the detector. The largest concentration in the linear dynamic range would be that where the response factor (r) falls outside the range specified. Unfortunately many manufacturers do not differentiate between (D_R) and (D_{LR}) and do not quote a range for the response index (r). Some manufacturers do mark the least sensitive setting on a detector as N/L (non–linear), which, in effect, accepts that there is a difference between the linear dynamic range and the dynamic range.

Detector Response

Detector response can be defined in two ways. It can be taken as the voltage output for unit change in solute concentration in which case in a similar way to detector sensitivity and dynamic range, the solute used for measurement must to be specified. Alternatively, it can be taken as the voltage output that would result from unit change in the physical property that the detector measures, *e.g.* refractive index or carbon content. In the latter case the dimensions of the response will vary with the nature of the property being measured.

The detector response (R_c) can be determined by injecting a known mass of the chosen solute (m) onto the column and measuring the response from the dimensions of the peak. Again assuming the concentration of the solute at the peak maximum is twice the average peak concentration, then

$$R_c = \frac{hwQ}{sm}$$

where (h) is the peak height and the other symbols have the meaning previously attributed to them.

Inevitably the response of a detector will differ between different solutes as well as between different detectors. *Ipso eo* the response of two detectors of the same type and geometry can only be compared using the same solute and the same mobile phase. Even so, if the response is specified in units of the physical property being measured, using the same solute and the same mobile phase, accurate comparisons may still not be possible. For example, in comparing the response of two UV detectors using the same wavelength, the same solute and mobile phase, the path length of the respective cells must still be taken into account to arrive at a rational comparison of response. This problem will be discussed in detail when light absorption detectors are discussed.

Detector Noise

Detector noise is the term given to any perturbation on the detector output that is not related to an eluted solute. It is a fundamental property of the detecting system and determines the ultimate sensitivity or minimum detectable concentration that can be achieved. Detector noise has been arbitrarily divided into three types, *'short term noise'*, *'long term noise'* and *'drift'* all three of which are depicted in figure 4.

Short Term Noise

Short term noise consists of baseline perturbations that have a frequency that is significantly higher than the eluted peak. Short term detector noise is not often a serious problem as it can be easily removed by appropriate noise filters without significantly affecting the profiles of the peaks. Its source is usually electronic, originating from either the detector sensor system or the amplifier.

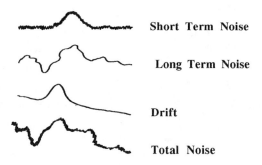

Figure 4   Different Types of Detector Noise

Long Term Noise

Long term noise consists of baseline perturbations that have a frequency that is similar to that of the eluted peak. This type of detector noise is the most significant and damaging as it is indiscernible from very small peaks in the chromatogram. Long term noise cannot be removed by electronic filtering without affecting the profiles of the eluted peaks. In figure 3, it is clear that the peak profile can easily be discerned above the high frequency noise but is lost in the long term noise. Long term noise usually arises from temperature, pressure or flow rate changes in the sensing cell. Long term noise is largely controlled by detector cell design and ultimately limits the detector *sensitivity* or the *minimum detectable concentration*.

Drift

Drift results from baseline perturbations that have a frequency that is significantly larger than that of the eluted peak. Drift is almost always due to either changes in ambient temperature, changes in mobile flow rate, or column bleed in GC; in LC drift can be due to pressure changes, flow rate changes or variations in solvent composition. As a consequence, with certain detectors baseline drift can become very significant in GC at high column temperatures and in LC at high column flow rates. Drift is easily constrained by choosing operating parameters that are within detector and column specifications.

A combination of all three sources of noise is shown by the trace at the bottom of figure 3. In general, the sensitivity of the detector should never be set above the level where the combined noise exceeds 2% of the F.S.D. (full scale deflection) of the recorder (if one is used), or appears as more than 2% F.S.D. of the computer simulation of the chromatogram.

Measurement of Detector Noise

The detector noise is defined as the maximum amplitude of the combined short- and long-term noise measured over a period of 10 minutes (the E19 committee recommends a period of 15 minutes). The detector must be connected to a column and mobile phase passed through it during measurement. The detector noise is obtained by constructing parallel lines embracing the maximum excursions of the recorder trace over the defined time period as shown in figure 5. The distance between the parallel lines measured in millivolts is taken as the measured noise (v_n), and the noise level (N_D) is calculated in the following manner.

Figure 5 The Measurement of Detector Noise

$$N_D = v_n A = \frac{v_n}{B}$$

where (v_n) is the noise measured in volts from the recorder trace,
(A) is the attenuation factor,
and (B) is the alternative amplification factor.

It should be noted that attenuation is the reciprocal of amplification and manufacturers may use either function as a control of detector sensitivity. For example the sensitivity settings on a RI detector for LC may take either of the following forms.

Minimum	(Amplification)	1	2	4	8	16	32	Maximum
Sensitivity	(Attenuation)	32	16	8	4	2	1	Sensitivity

The noise level of detectors that are particularly susceptible to variations in column pressure or flow rate (*e.g.* the katherometer and the refractive index detector) are often measured under static conditions (*i.e.* no flow of mobile phase). Such specifications are not really useful, as the analyst can never use the detector without a column flow. It could be argued that the manufacturer of the detector should not be held responsible for the precise control of the mobile phase, beitmay a gas flow controller or a solvent pump. However, all mobile phase delivery systems show some variation in flow rates (and consequently pressure) and it is the responsibility of the detector manufacturer to design devices that are as insensitive to pressure and flow changes as possible.

It should be noted that at the high sensitivity range settings of some detectors, filter circuits are automatically introduced to reduce the noise. Under such circumstances the noise level should be determined at the lowest attenuation (or highest amplification) that does not include noise-filtering devices (or at best the lowest attenuation with the fastest response time) and then corrected to an attenuation of unity.

Detector Sensitivity or the Minimum Detectable Concentration

Detector sensitivity or minimum detectable concentration (MDC) is defined as the minimum concentration of solute passing through the detector that can be unambiguously discriminated from noise. The size of the signal that will make it distinctly apparent from the noise (the signal-to-noise ratio) is a somewhat arbitrary choice. It is generally accepted that a signal can be differentiated from the noise in electronic measurements when the signal to noise ratio is *two* and this criteria has been adopted for defining detector sensitivity.

Thus for a concentration sensitive detector, the detector sensitivity (X_D) is given by

$$X_D = \frac{2N_D}{R_c} \quad (g/ml)$$

(R_c) and (N_D) being determined in the manner previously described.

The sensitivity of a detector is *not* the minimum mass that can be detected. This would be the *system mass sensitivity,* which would also depend on the characteristics of the apparatus as well as the detector and, in particular, the type of column employed. During the development of a separation the peaks become broader as the retention increases. Consequently, a given mass may be detected if eluted as a narrow peak early in the chromatogram, but if eluted later, its peak height may be reduced to such an extent that it is impossible to discern it from the noise. Thus detector sensitivity quoted as the 'minimum mass detectable' must be carefully examined and related to the chromatographic system and particularly the column with which it was used. If the data to do this are not available, then the sensitivity must be calculated from the detector response and the noise level in the manner described above.

Some manufacturers have taken the minimum detectable concentration and multiplied it by the sensor volume and defined the product as the minimum detectable mass. This way of defining detector sensitivity is particularly misleading. For example a detector having a true sensitivity of 10^{-6} g/ml and a sensor volume of 10 µl would be attributed a sensitivity of 10^{-8} g. This value is, of course, grossly incorrect as it is the peak volume that controls the mass sensitivity, not the sensor volume. Conversely, if the peak volume does approach that of the sensor, then very serious peak distortion and loss of resolution occurs, and this effect will be discussed in detail later. Detector specifications given by manufacturers are becoming more rational and helpful to the analyst, but misleading values are still given and any specifications not described in this book need to be careful examined to assess their true significance.

The Mass Sensitivity of a Chromatographic System

The mass sensitivity of a chromatographic system (including column, sample valve or injection system, connecting tubes and detector) is defined as the mass of solute (m_D) that will provide a peak with a height equivalent to twice the noise level.

Consider a chromatographic peak with height equivalent to twice the noise level, being sensed by a detector with a maximum sensitivity of (X_D) g/ml. The peak volume can be taken as ($4\sigma_c$) [7], where (σ_c) is the volume standard deviation of the peak as it is eluted from the column and the concentration at the peak maximum as twice the mean concentration of the solute in the peak.

Then, $$\frac{2m_D}{4\sigma_c} = X_D$$

or $$m_D = 2\sigma_c X_D \qquad (2)$$

Now from the Plate Theory (8),

$$\sigma_c = \frac{V_r}{\sqrt{n}}$$

where (V_r) is the retention volume of the solute,
and (n) is the column efficiency.

Thus,
$$m_D = \frac{2V_r X_D}{\sqrt{n}} \qquad (3)$$

Now, also from the Plate theory (8),

$$V_r = V_o(1 + k') \qquad (4)$$

where (V_o) is the column dead volume,
and (k') is the capacity ratio of the solute.

and
$$V_o = \varepsilon \pi r^2 l \qquad (5)$$

where (l) is the column length,
(r) is the column radius,
and (ε) is the fraction of the column volume that is considered filled with mobile phase (usually taken as 0.65)

Consequently, substituting for (V_r) in equation (3) from equation (4) and for (V_o) from equation (5),

$$m_D = \frac{2\varepsilon \pi r^2 (1+k') X_D}{\sqrt{n}} \qquad (6)$$

Thus (m_D), the mass sensitivity of the chromatographic system depends on the detector sensitivity, column dimensions, column efficiency and the capacity factor of the eluted solute. However, irrespective of the column properties, the mass sensitivity is still directly related to the detector sensitivity. It will also be seen that the column radius will depend on the extracolumn dispersion, much of which arises from the detector connecting tubes and sensor. It follows that the design of the detector and its sensitivity has a major influence on the mass sensitivity of the overall chromatographic system.

The Concentration Sensitivity of a Chromatographic System

The concentration sensitivity of a *chromatographic system* (X_C) is defined as that which will provide a peak with a height equivalent to twice the noise level and can be obtained directly from the system mass sensitivity. If the minimum detectable mass is dissolved in the maximum permissible sample volume [9] (that sample volume that will limit the increase in sample variance to 10% of the column variance), then this solute concentration will constitute the minimum detectable sample concentration.

Now the maximum sample volume (V_S) that can be used while maintaining the limited peak dispersion is $1.1\sigma_c$ and the mass sensitivity give by equation (2) is $2\sigma_c X_D$. Thus the concentration sensitivity (X_C) is given by

$$X_c = \frac{2\sigma_c X_D}{1.1\sigma_c} = 1.8 X_D \qquad (7)$$

It is seen that the system concentration sensitivity is also directly (and solely) dependent on the detector sensitivity and, in contrast to the system mass sensitivity, is independent of column or solute properties. It should be emphasized, however, that in order to realize the minimum concentration sensitivity, the maximum permissible sample volume must be used.

The Maximum Capacity Factor of an Eluted Peak

The maximum (k') at which a solute can be eluted is also determined by the detector sensitivity. As the (k') of the eluted solute increases, the peak becomes more disperse, the peak height is reduced and eventually the peak will disappear into the detector noise. In a general chromatographic analysis, if it were merely necessary to unambiguously identify the existence of a peak, then the specified signal–to–noise ratio might not be acceptable and the peak maximum would probably need to be about 5 times the noise level. This is also an arbitrary assumption based on experience but in fact any desired signal to noise level can be chosen and the mathematical arguments will remain the same.

Now if the detector sensitivity as defined is (X_D), at the maximum (k') the concentration will be 2.5 X_D. The concentration at the peak maximum can be taken as twice the average concentration so if there is a mass (m) contained in a peak of base width $4\sigma_c$ then from the Plate Theory [8],

$$4\sigma_c = 4\sqrt{n}(v_m + K v_s)$$

where (v_m) is the volume of mobile phase in a theoretical plate,
(v_s) is the volume of stationary phase in a theoretical plate,
and K is the distribution coefficient of the solute between the two phases.

Then,
$$\frac{2m}{4\sqrt{n}(v_m + K v_s)} = 2.5 X_D \qquad (8)$$

or
$$m = 5 X_D \sqrt{n}(v_m + K v_s)$$

Now as stated previously, the maximum permissible sample volume (V_s) is given by

$$V_s = 1.1 \sigma_c$$

and from the Plate Theory (7),

$$\sigma_c = \sqrt{n}(v_m + K v_s)$$

Thus,
$$V_s = 1.1 \sqrt{n}(v_m + K v_s) \qquad (9)$$

The early peaks, being the narrowest, are the most affected by sample volume. Furthermore the resolution of both early and late peaks must be given the same priority. Consequently, the sample volume should be chosen so that it does not increase the dead volume peak width by more than 5%. Thus (K) in equation (8) must be made zero and thus,

$$V_s = 1.1\sqrt{n}(v_m)$$

Consequently, if the sample concentration is (X_S),

$$m = X_s V_s = 1.1 X_s \sqrt{n}(v_m) \qquad (10)$$

Equating equations (8) and (10),

$$1.1 X_s \sqrt{n}(v_m) = 5 X_D \sqrt{n}(v_m + K v_s)$$

Now, bearing mind that from the Plate Theory (8), $k' = \dfrac{K v_s}{v_m}$

Then $\quad \dfrac{1.1 X_s}{5 X_D} = (1+k') \quad$ or $\quad k' = \dfrac{0.22 X_s}{X_D} - 1 \qquad (11)$

It is seen that the maximum (k') depends on both the detector sensitivity and the sample concentration. Consequently, the detector sensitivity determines the greatest value for (k) that can be obtained from a column.

Table 3 Limiting Values for (k') for Detectors of Different Sensitivities and a Sample Concentration of 0.1% w/v

Detector Sensitivity	Maximum (k')
10^{-5} g/ml	21
10^{-6} g/ml	219
10^{-7} g/ml	2199
10^{-8} g/ml	22,000

Providing the detector has a reasonable sensitivity the limit for (k'), however, is fairly high as shown in table 2 where the sample concentration is assumed to be 0.1 % w/v. The data given above is subject to the caveat that the sample is placed on the column in the maximum permissible volume but not overloaded.

Peak Dispersion in the Chromatographic System

Peak dispersion takes place in all parts of the chromatographic system through which the mobile phase passes from the sampling device to the detector sensor. The function of the column is to separate the solutes into individual bands and, at the same time, contain the dispersion so that they are eluted discretely. It follows that it is important that dispersion in other parts of the system is also contained so that, having separated the peaks in the column, they are not merged together again before detection. The dispersion processes outside the column can be algebraically combined and the total dispersion, not surprisingly, is called *extracolumn dispersion.*

Extracolumn Dispersion

There are four major sources of extracolumn dispersion which are measured in terms of their variance.

1/ Dispersion due to the sample volume (σ_s^2).

2/ Dispersion taking place in the valve–column and column–detector connecting tubing (σ_T^2).

3/ Dispersion in the sensor volume resulting from Newtonian flow (σ_{CF}^2).

4/ Dispersion in the sensor volume from band merging (σ_{CM}^2).

5/ Dispersion from the sensor and electronics time constant (σ_t^2).

The sum of the variances will give the overall variance for the extra-column dispersion (σ_E^2). Thus,

$$\sigma_E^2 = \sigma_s^2 + \sigma_T^2 + \sigma_{CF}^2 + \sigma_{CM}^2 + \sigma_t^2 \tag{12}$$

Equation (12) shows how the extracolumn dispersion is made up and according to Klinkenberg [9] must not exceed 10% of the column variance if the resolution of the column is to be maintained, *i.e.*,

$$\sigma_E^2 = \sigma_s^2 + \sigma_T^2 + \sigma_{CF}^2 + \sigma_{CM}^2 + \sigma_t^2 = 0.1\sigma_c^2$$

It is of considerable interest to calculate the maximum extracolumn dispersion that can be tolerated for different types of columns. This will indicate the level to which dispersion in the detector and its associated conduits must be contained to avoid abrogating the chromatographic resolution.

From the Plate Theory [8] the column variance is given by $(\frac{V_r^2}{n})$ and for a peak eluted at the dead volume the variance will be $(\frac{V_0^2}{n})$.

Thus for a capillary column of radius (r_t) and length (l_t),

$$\sigma_E^2 = \frac{0.1(\pi r_t^2 l_t)^2}{n}$$

Now from *Appendix I* the efficiency (n) of the dead volume peak from a capillary column can be approximated to

$$n = \frac{1}{0.6\, r_t}$$

Thus, $$\sigma_E^2 = 0.06\pi^2 \, r_t^5 l_t \tag{13}$$

For a packed column, of radius (r_p) and length (l_p), the permissible extracolumn variance will be much larger. Again $\sigma_c^2 = \dfrac{V_r^2}{n}$ and for the dead volume peak,

$$\sigma_c^2 = \dfrac{\left(\varepsilon \pi r^2 l\right)^2}{n} \quad \text{Thus,} \quad \sigma_E^2 = 0.1 \dfrac{\left(\varepsilon \pi r^2 l\right)^2}{n} \qquad (14)$$

where (ε) is the fraction of the coluln volume that is occupied by the mobile phase.

In *Appendix II* it is shown to a first approximation the efficiency (n) of the dead volume peak of a packed column is given by $n = \dfrac{1}{1.6 dp}$, where (dp) is the particle diameter of the packing. Thus, assuming the fraction of the column occupied by the mobile phase (ε) is 0.65, substituting for (n) in equation (14),

$$\sigma_E^2 = 0.068 \pi^2 r^4 l \, dp \qquad (15)$$

Equations (13) and (15) allow the permissible extracolumn dispersion to be calculated for a range of capillary columns and packed columns. The results are shown in table 4. The standard deviation of the extra-column dispersion is given as opposed to the variance, because it is easier to visualize from a practical point of view. The values for (σ_E) represents half the width (in volume flow of mobile phase) at 0.607 of the height of the peak that would have been caused by extracolumn dispersion *alone*. It is seen the values vary widely with the type of column that is used. (σ_E) values for GC capillary columns range from about 12 μl for a relatively short, wide, macrobore column to 1.1 μl for a long, narrow, high efficiency column.

The packed GC column has a value for (σ_E) of about 55 μl whereas the high efficiency microbore LC column only 0.23 μl. It is clear that problems of extracolumn dispersion with packed GC columns are not very severe. However, shorter GC capillary columns with small diameters will have a very poor tolerance to extracolumn dispersion.

In the same way, short microbore LC columns packed with small particles will make very stringent demands on dispersion control in LC detecting systems.

Table 4 The Permissible Extracolumn Dispersion for a Range of Different Types of Column

Capillary Columns (GC)

Dimensions	Macrobore	Standard	High Efficiency
length (l)	10 m	100 m	400 m
radius (r)	0.0265 cm	0.0125 cm	0.005 cm
(σ_E)	2.78 µl	1.34 µl	0.27 µl

Packed Column

Dimensions	Packed LC	2 m GC	Microbore LC
length (l)	100 cm	200 cm	25 cm
radius (r)	0.05 cm	0.23 cm	0.05 cm
particle diam.	0.0020 cm	0.0080 cm	0.0005 cm
(σ_E)	0.91 µl	54.8 µl	0.23 µl

It is necessary to recall that the maximum allowable dispersion will include contributions from *all* the different dispersion sources. Furthermore, the analyst may frequently be required to place a large volume of sample on the column to accommodate the specific nature of the sample. The peak spreading resulting from the use of the maximum possible sample volume is likely to reach the permissible dispersion limit. It follows that the dispersion that takes place in the connecting tubes, sensor volume and other parts of the detector must be reduced to the absolute minimum and if possible kept to less than 10% of that permissible, to allow large sample volumes to be used when necessary.

It is clear that reducing dispersion to these low limits places strict demands on detector design. The solutes are actually sensed in the sensor cell or sensing volume of the detector but must be carried to the sensor by a suitable conduit. This is usually accomplished by a short length of cylindrical connecting tubing of very small diameter. As a consequence, the dispersion in both the tubes connecting the eluent to the cell from the column and the cell itself must be considered.

Dispersion in Connecting Tubes

The dispersion in open tubes was examined by Golay [10] and Atwood and Golay [11] and experimentally by Scott and Kucera [12] and Lochmuller and Sumner [13]. The variance per unit length of an open tube (H) according to Golay is given by

$$H = \frac{2D_m}{u} + \frac{r^2 u}{24 D_m}$$

where (D_m) is the diffusivity of the solute in the mobile phase,
(u) the linear velocity of the mobile phase,
and the other symbols have the meaning previously ascribed to them.

At relatively high velocities (that is at velocities much greater than the optimum velocity of the tube, which will usually be true for all connecting tubes)

$$H = \frac{r^2 u}{24 D_m}$$

Furthermore, $$Q = \pi r^2 u$$

where (Q) is the flow rate through the tube.

Thus, $$H = \frac{Q}{24 \pi D_m}$$

Now, (H) is the variance per unit length of the tube but a more useful parameter to the analyst is the volume variance (σ_v^2). This can be derived using the relatioinship predicted by the Plate Theory.

$$\sigma_v^2 = \frac{(V_0)^2}{n} = \frac{(\pi r^2 l)^2}{n} = \frac{\pi^2 r^4 l^2}{n}$$

Now $H = \dfrac{l}{n}$, consequently $\sigma_v^2 = \pi^2 r^4 l H = \dfrac{\pi r^4 l Q}{24 D_m}$

Thus, expression for the volume standard deviation ($\sigma_{v(l)}$) for tubes of different length is

$$\sigma_v^2 = \left(\frac{\pi l Q}{24 D_m}\right)^{0.5} r^2 \qquad (16)$$

Employing equation (16) it is possible to calculate the value of ($\sigma_{v(l)}$) for a range of cylindrical connecting tubes of different radii and different lengths.

Table 5 Standard Deviation of Connecting Tubes of Different Sizes

Connecting Tubes for Gas Chromatography

Tube Radius	Standard Deviation of Tube Dispersion				
	l = 1 cm	l = 2 cm	l = 5 cm	l = 10 cm	l = 15 cm
0.001 in, 0.00254 cm	1.1 nl	1.6 nl	2.5 nl	3.5 nl	4.3 nl
0.002 in, 0.00508 cm	4.3 nl	6.1 nl	9.6 nl	13.6 nl	16.6 nl
0.003 in, 0.00762 cm	9.6 nl	13.6 nl	21.5 nl	30.4 nl	37.2 nl
0.005 in, 0.01270 cm	26.6 nl	37.6 nl	59.5 nl	84.1 vl	103.2 nl
0.010 in, 0.02540 cm	106 nl	149 nl	237 vl	0.335 µl	0.410 µl

Connecting Tubes for Liquid Chromatography

Tube Diameter	Standard Deviation of Tube Dispersion				
	l=1 cm	l=2 cm	l=5 cm	l=10 cm	l=15 cm
0.001 in, 0.00254 cm	22.3 nl	31.5 nl	49.9 nl	70.5 nl	86.4 nl
0.002 in, 0.00508 cm	47.6 nl	67.3 nl	106.4 nl	150 nl	184.4 nl
0.003 in, 0.00762 cm	107 nl	151.3 nl	239.2 nl	0.34 µl	0.41 µl
0.005 in, 0.01270 cm	298 nl	421 nl	0.67 µl	0.94 µl	1.15 µl
0.010 in, 0.02540 cm	1.19 µl	1.68 µl	2.66 µl	3.76 µl	4.61 µl

Two sets of results are calculated, one for GC where (D_m) is taken as 0.1 cm^{2}sec^{-1} and the flow rate 20 ml/min and the second set for LC, where (D_m) is taken as 2 x 10^{-5} cm^{2}sec^{-1} and the flow rate at 0.5 ml/min. All values are fairly typical for the normal operation of the chromatographic system near optimum conditions.

It is seen from table 5 that in GC, because of the relatively high value of the diffusion coefficient, the dispersion in connecting tubes is very small. It is seen from table 4 that the dispersion from capillary columns is measured in microliters so the effect of the connecting tubes, even 10 cm long, is not too serious. It must be noted, however, that connecting tubes less than 0.002 in I.D. (0.00508 cm) can easily become blocked and thus the smaller diameter tubes can only be used in an extremely clean chromatographic system. Employing 0.003 in tubing, it is seen that the connecting tube to the detector can be 2 to 5 cm long and still restrain the dispersion to about 10% of the column dispersion.

The effect of dispersion in connecting tubes used in LC is far worse because of the very low diffusivity of solutes in liquids. In table 4 it is clear that the tube dispersion needs to reduced to about 80 nl if the integrity of the Microbore LC separations is to be maintained. Again assuming that to minimize the chance of tube blocking, the limiting minimum I.D. for the connecting tube is made to be 0.003 in, then the connecting tube must be less than 1 cm long. It is clear that limiting the length of the connecting tube between column and detector is critical. If the tubing diameter is reduced further and the column is designed to have radii that provide larger column dispersion, then longer tubing lengths can possibly be employed. Alternatively, the integrity of the early peaks can be sacrificed in favor of later eluting peaks which will also allow longer connecting tubes to be used. These techniques to reduce the effect of connecting tube dispersion in LC are common with most manufacturers. The simple solution of designing the chromatographic system such that the detector sensor is situated close to the end of the column does not appear to have been adopted.

Low Dispersion Tubing

The dispersion that takes place in an open tube results from the parabolic velocity profile that occurs under conditions of Newtonian flow, *i.e.* when the velocity is significantly below that which produces turbulence. Under condition of Newtonian flow, the distribution of fluid velocity across the tube adopts a parabolic profile, the velocity at the walls being virtually zero and that at the center a maximum. This situation is depicted diagramatically in Figure 6A. Due to the relatively high velocity at the center of the tube and the very low velocity at the walls, the center of the band of solute passing down the tube will move ahead of that situated at the walls. This dispersive effect is depicted in figure 6B.

A. Newtonian Flow

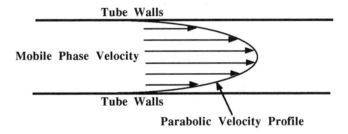

B. Band Dispersion Due to Newtonian Flow

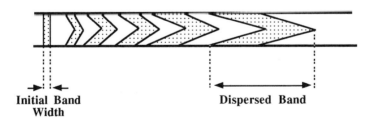

Figure 6 Dispersion Due to Newtonian Flow

The effect is much less when the fluid is a gas, where the solute diffusivity is high and the solute tends to diffuse across the tube and partially compensate for the non–linear velocity profile. However, when the fluid is a liquid, the diffusivity will be very small and the dispersion proportionally larger. It follows that to reduce dispersion, the tube must have a geometry that causes strong *radial* flow that aids in diffusion and destroys the parabolic velocity profile.

The first attempt to produce low dispersion tubing was by Halasz *et al.* [14], who crimped and bent the tube into different shapes to interrupt the Newtonian flow and introduce radial flow within the tube. His devices had limited success and the tubes had a tendency to block very easily.

Figure 7 Low Dispersion Tubing

The next attempt was by Tijssen [15], who developed a theory to describe the radial flow that was induced into coiled tubes by the continual change in direction of the fluid as it flowed round the spirals. Coiling the tubes significantly reduced dispersion, particularly at high flow rates, but they were a little clumsy to form as the radius of the coil was required to be less than 3 times the internal radius of the tube for optimum performance. A more practical system was introduced by Katz and Scott [16], who developed a serpentine form of connecting tube that met the requirement that the radius of the serpentine bends (a/2 in the diagram) was less that 3 times that of the internal radius of the tube. A diagram of a serpentine tube is shown in figure 7. This system had the advantage that the mobile phase flow direction was changed by 180º as it passed from one serpentine bend to another and resulted in extensive radial flow which greatly reduced the dispersion. This is shown by the curves relating the variance against flow rate for straight and serpentine tubes shown in figure 8. It is seen that the dispersion is reduced by over an order of magnitude at the higher flow rates. Low dispersion serpentine tubing, although extremely effective, appears to have been employed in only one commercial LC detector.

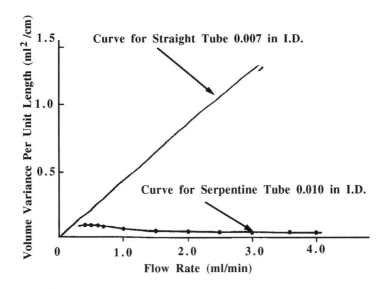

Figure 8 Graphs of Peak Variance against Flow Rate for Straight and Serpentine Tubes

Any conduit system that has low dispersion will also provide very fast heat transfer rates. Serpentine tubing has been also used in commercial column ovens to heat the mobile phase rapidly to the column oven temperature before it enters the column. The serpentine tubing allows effective heat exchange with a minimum of heat exchanger volume to distort the concentration profile of the solvent gradient.

The different forms of dispersion profiles that are obtained from various types of connecting tubes used in LC are shown in figure 9. These dispersion curves were obtained using a low dispersion UV detector (cell volume, 1.4 µl) in conjunction with a sample valve with a 1 µl internal loop. All tubes were of the same length and a flow rate of 2 ml/min was employed. The peaks were recorded on a high speed recorder. The peak from the serpentine tubing is seen to be symmetrical and has the smallest width. The peak from the coiled tube, although still very symmetrical is the widest at the points of inflexion of all four peaks. The peak from the straight tube 0.25 mm I.D. is grossly asymmetrical and has an extremely wide base.

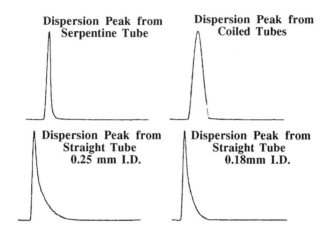

Figure 9 Dispersion Profiles from Different Types of Tube

The width and asymmetry is reduced using a tube with an I.D. of 0.18 mm but serious asymmetry remains. It is clear that under conditions where the tube length between the end of the column and the detector sensor can not be made sufficiently short to restrain dispersion, the use of low dispersion serpentine tubing may be a satisfactory alternative.

Dispersion in the Detector Sensor Volume

The sensor volume of a detector can cause dispersion and contribute to the peak variance in two ways. Firstly there will be dispersion resulting from the viscous flow of fluid through the cell sensor volume, which will furnish a variance similar in form to that from cylindrical connecting tubes. Secondly, there will be a peak spreading which results from the finite volume of the sensor. If the sensor has a significant volume, it will not measure the instantaneous concentration at each point on the elution curve, but it will measure the average concentration of a slice of the peak equivalent to the sensor volume. Thus, the true profile of the peak will not be monitored. The net effect will first (at sensor volumes still significantly smaller than the peak volume) give the peak an apparent dispersion. Then (as the sensor volume becomes of the same order of magnitude as the peak volume) it will distort the profile and eventually seriously impair resolution. In the worst case two peaks could coexist in the sensor at one time and only a single peak will be represented. This apparent peak spreading is not strictly a dispersion process but in practice will have a similar effect on chromatographic resolution. The effect of viscous flow on dispersion will first be considered.

Dispersion in Detector Sensors Resulting from Newtonian Flow

Most sensor volumes whether in LC (*e.g.*, a UV absorption cell) or in GC (*e.g.*, a katherometer cell) are cylindrical in shape, are relatively short in length and have a small length-to-diameter ratio.

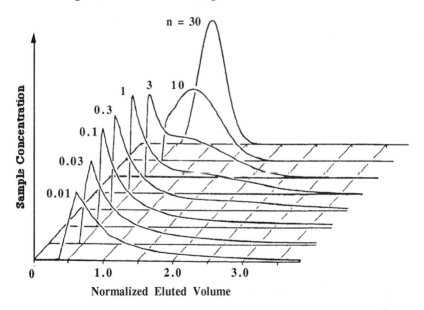

Figure 10 Elution Curves Presented as a Function of the Normalized Tube Length

The small length-to-diameter ratio is in conflict with the premises adopted in the development of the Golay equation for dispersion in an open tube and consequently its conclusions are not pertinent to detector sensors. Atwood and Golay [17] extended the theory of dispersion in open tubes to tubes of small length-to-diameter ratio. The theory developed is not pertinent here as it will be seen that dispersion from viscous sources is negligible. Nevertheless, the effect of the cell on solute profiles is shown in figure 10. Fortunately, this situation rarely arises in practice as the profile is further modified by the manner of entrance and exit of the mobile phase.

The conduits are designed to produce secondary flow and break up the parabolic velocity profile causing peak distortion as shown in figure 11. The Newtonian flow through the detector is broken up by the conformation of the inlet and outlet conduits to and from the cell. Mobile phase enters the cell at an angle that is directed at the cell window. As a consequence, the flow has to virtually reverse its direction to pass through the cell producing a strong radial flow and disrupting the Newtonian flow. The same situation is arranged to occur at the exit end of the cell.

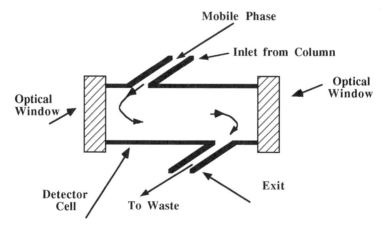

Figure 11 The Design of a Modern Absorption Cell

The flow along the axis of the cell must reverse its direction to pass out of the port that is also set at an angle to accomplish the same effect. Employing this type of cell geometry dispersion resulting from viscous flow is practically eliminated

Apparent Dispersion from Detector Sensor Volume

The detector responds to an average value of the total amount of solute in the sensor cell. In the extreme, the sensor volume or cell could be large enough to hold two closely eluted peaks and thus give a response that would appear as though only a single solute had been eluted, albeit very distorted in shape. This extreme condition rarely happens but serious peak distortion and loss of resolution can still result. This is particularly so if the sensor volume is of the same order of magnitude as the peak volume. The problem can be particularly severe when open tubular columns and columns of small diameter are being used.

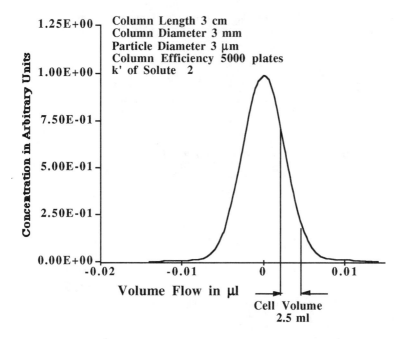

Figure 12 Effect of Sensor Volume on Detector Output

The situation is depicted in figure 12. It is the elution profile of a peak eluted from a column 3 cm long, 3 mm I.D. packed with particles 3 µ in diameter. The peak is considered to be eluted at a (k') of 2 and it is seen that the peak width at the base is about 14 µl wide. The sensor cell volume is 2.5 µl and the portion of the peak in the cell is included in the figure. It is clear the detector will respond to the mean concentration of the slice contained in the 2.5 µl sensor volume. It is also clear that, as the sensor volume is increased, the greater part of the peak will be contained in the cell and the output will be an average value of an even larger portion of the peak resulting in serious peak distortion. The effect of a finite sensor volume can be easily simulated with a relatively simple computer program and the output from such a program is shown in figure 13.

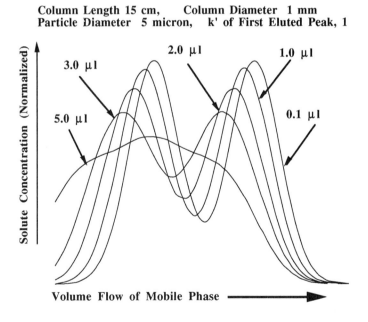

Figure 13 The Effect of Detector Sensor Volume on the Resolution of Two Solutes

The example given, although not the worst case scenario, is a condition where the sensing volume of the detector can have a very serious effect on the peak profile and, consequently, the resolution. The column is a small bore column and thus the eluted peaks have a relatively small peak volume which is commensurate with that of the sensing cell. It is seen that even a sensor volume of 1 µl has a significant effect on the peak width and it is clear that if the maximum resolution is to be obtained from the column, then the sensor cell volume should be no greater than 2 µl. It should also be noted that the results from the use of a sensor cell having a volume of 5 µl are virtually useless and that many commercially available detectors do, indeed, have sensor volumes as great as, if not greater than, this. It follows that if small bore columns are to be employed, such sensor volumes must be studiously avoided.

Dispersion Resulting from the Overall Detector Time Constant

In addition to the sources of dispersion so far discussed, the peak can appear to be further dispersed by the time constant of the sensor and its associated electronics.

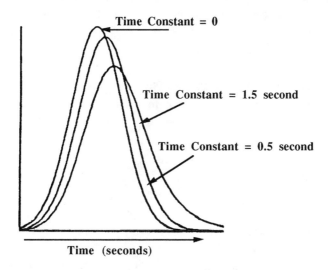

Figure 14 Peak Profiles Demonstrating Distortion Resulting from Detector Time Constant

The term *appear* is used as the solvent profile itself is not actually changed, only the profile as presented on the recorder or printer. The effect of the detector time constant can be calculated and the results from such a calculation are shown in figure 14. The undistorted peak, that would be monitored by a detector with a zero time constant, is about 4 seconds wide. Thus, for a GC packed column operating at 20 ml/min this would represent a peak having a volume of about 1.3 ml.

For an LC column operating at a flow rate of 1 ml/min, a peak with a base width of 4 seconds would represent a peak volume of about 67 µl. Consequently, the peaks depicted would represent those eluted fairly late in the chromatogram. It is seen that despite the late elution, the distortion is still quite severe. In order to elute early peaks without distortion the time constant would need to be at least an order of magnitude less.

Scott *et al.* [18] examined the time constants of two photocells and their results are shown in figure 15. The output of the photocells to fast transient changes in highlight intensity was monitored with a high speed recorder. The curves for the cadmium sulfide photocell, Figure 15 (chosen as an old type, sensor with a very slow response) is shown in the top part of the diagram. From the slope of the log curve the time constant can be calculated to be about 2.5 seconds. Such a slow response would be impossible to use with modern chromatographic systems, as two or more peaks could be eluted within the period of the time constant. Because of the cadmium sulfide sensor's slow response, the peaks would be blended together into a single distorted peak.

The performance of the photomultiplier (representative of a very fast responding sensor even in terms of modern solid state devices) is shown in the lower curves and its performance is in complete contrast to that of the cadmium sulfide cell. The time constant, determined again from the slope of the log curve, was found to be only 40 milliseconds. Such a response time is generally acceptable for most GC and LC separations. Nevertheless in both fast GC and fast LC solutes can be eluted in less than 100 milliseconds in which case an even faster response might be necessary.

The effect of the overall time constant of the sensor and associated electronics have been discussed by Vandenheuval [20] Schmauch [21] and Sternberg [22]. For those readers interested in pursuing the subject further, the discussion of Sternberg is recommended which, although it is nearly 30 years old, considers the subject in a manner that is still pertinent to contemporary chromatography.

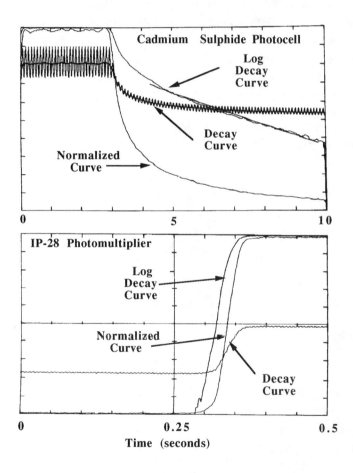

Figure 15 The Response Curves of Two Photocells

Modern sensors and electronic systems depend heavily on fast solid state sensors and solid state electronic components. Thus the majority of detector systems commercially available are sufficiently fast for the vast majority of chromatography applications. In general, the overall time constant of the detecting system should be less than 50 milliseconds. For special applications involving very fast separations, this value may need to be reduced to around 15 milliseconds. Sensors and electronics, with very small time constants, unfortunately, will also readily respond to high frequency noise. Consequently the whole chromatographic system must be carefully designed to reduce short term noise, which, as already stated, is not normally a problem in general chromatographic analysis. This may involve careful magnetic screening to reduce the effect of stray, low-frequency electromagnetic fields from nearby power supplies and any high energy consuming laboratory equipment.

Pressure Sensitivity

The pressure sensitivity of a detector is extremely important as it is one of the detector parameters that determines both the long term noise and the drift. As it influences long term noise, it will also have a direct impact on detector sensitivity or minimum detectable concentration together with those other characteristics that depend on detector sensitivity. Certain detectors are more sensitive to changes in pressure than others. The katherometer detector, which is used frequently for the detection of permanent gases in GC, can be very pressure sensitive as can the LC refractive index detector. Careful design can minimize the effect of pressure but all bulk property detectors will tend to be pressure sensitive.

The pressure sensitivity (D_P) should be given as the output in millivolts for unit pressure change in the detector *e.g.* as mV/p.s.i or mV/kg/m^2. The pressure sensitivity can be used to calculate the pressure change (N_P) that would provide a signal equivalent to the detector noise (N_D),

$$\text{i.e.} \quad N_P = \frac{N_D}{D_P}$$

A knowledge of (N_P) can be utilized in a number of ways but in particular allows the effect of pressure fluctuations on detector noise to be estimated.

Finally, in LC multidimensional analysis, where the chromatographic separation is developed through two or more columns in series is now becoming more popular. The procedure usually requires that a detector is situated between two of the columns and consequently, is at high pressure. It follows that the sensor must be capable of operating at elevated pressures and thus, to determine the suitability of a detector for such purposes the maximum working pressure (P_D) must be known. For normal operation in LC, a satisfactory value for (P_D) should range from 30 to 100 p.s.i. For multidimensional analysis the working pressure may need to be as great as 2,000 or 3,000 p.s.i.. In GC much lower pressures are employed. For normal operation in GC, a satisfactory value for (P_D) should range from 2 to 5 p.s.i. and in multidimensional analysis the working pressure may need never exceed 200 p.s.i. However, in GC, many of the detectors employed are not suitable for operation above atmospheric pressure (e.g. the FID) unless extremely complicated modifications are carried out.

Flow Sensitivity

Flow sensitivity is the second detector property that can have a significant effect on long term noise and, consequently, also on detector sensitivity. The same bulk property detectors are again the most likely to have high flow sensitivities. For example the katherometer is usually fitted with a reference cell through which a reference flow of mobile passes. As will be seen later, the two sensors for the column eluent and the reference flow are placed in the arms of a Wheatstone bridge so that any changes in flow rate are to a large extent compensated. The same procedure is used in LC with the RI detector which has both an eluent sensor and a reference cell sensor that are also situated in the arms of a Wheatstone bridge. Again any response that is identical in both cells (*i.e.* a change in flow rate) are balanced by the bridge and only differential signals are monitored.

The flow sensitivity (D_Q) should be defined in a similar manner to pressure sensitivity, *i.e.* mV/ml/min. The flow sensitivity can be used to calculate the flow change (N_Q) that would provide a signal equivalent to the detector noise (N_D),

i.e. $$N_Q = \frac{N_D}{D_Q}$$

A knowledge of (N_Q) can also be utilized in different ways but in particular allows the effect of flow fluctuations on detector noise to be estimated.

Temperature Sensitivity

The sensitivity of a detector to temperature varies greatly from one detector to another. The FID used in GC is virtually insensitive to temperature changes but this may not necessarily be true for the associated electronics. In contrast the katherometer detector is extremely sensitive to temperature changes (the reason for this will be clear when the katherometer detector is discussed) and must be thermostatted in a separate oven. Many LC detectors are temperature sensitive but, as most LC columns are not operated at very high temperatures, reasonable thermal insulation is sufficient to prevent drift. Temperature changes together with changes in mobile phase composition are the two main sources of *drift* in LC detectors.

The overall temperature sensitivity of the detector system (D_T) is defined as the change in output in millivolts for one degree change in temperature (C^o). Some detectors have a limited temperature range over which they can operate satisfactorily and thus the maximum and minimum operating temperatures should also be available. The temperature sensitivity can be used to calculate the temperature change (N_T) that would provide a signal equivalent to the detector noise (N_D),

$$N_T = \frac{N_D}{D_T}$$

A knowledge of (N_T) can be used in the same way as (N_P) and (N_Q)

and allows the effect of temperature fluctuations on detector noise to be estimated.

Summary of Detector Criteria

1. Dynamic Range – (R_D) – The dynamic range of a detector is that concentration range over which it will give a concentration dependent output. It is given as the ratio of the concentration at which the detector fails to respond to further concentration changes to the minimum detectable concentration or sensitivity. Consequently, the units are dimensionless. A knowledge of the dynamic range of a detector is useful in preparative chromatography.

2. The Response Index – (r) – The response index of detector is a measure of its linearity and for a truly linear detector would take the value of unity. In practice the value of (r) should lie between 0.98 and 1.02. If (r) is known, quantitative results can be corrected for any nonlinearity.

3. Linear Dynamic Range – (D_L) – The linear dynamic range of a detector is that concentration range over which the detector response is linear within defined *response index* limits. It is also dimensionless and is taken as the ratio of the concentration at which the response index falls outside its defined limits, to the minimum detectable concentration or sensitivity. The linear dynamic range is important when the components of a mixture being analyzed cover a wide concentration range.

4. Detector Response – (R_C) – The detector response can be defined in two ways. Firstly, as the detector output per unit change in concentration (*e.g.* volts/g/ml) or secondly, as the detector output per unit change of physical property being measured (*e.g.* for the FID, volts/gram of carbon/ml). In conjunction with the detector noise level it allows the sensitivity or minimum detectable concentration to be measured.

5. Noise Level – (N_D) – The noise level of a detector is taken as the maximum amplitude of the combined short and long term noise taken over a period of 10 minutes; it is usually measured in volts.

*It must be emphasized that detectors can **not** be compared on the basis of their noise or response. They can only be compared on the basis of their sensitivity or signal-to-noise ratio at a specific solute concentration.*

6. Detector Sensitivity – Minimum Detectable Concentration – (X_D). The detector sensitivity can also be defined in two ways. Firstly, as that concentration that will produce a signal equivalent to twice the noise. Secondly, it can be defined as that change in the physical property being measured that will provide a signal equivalent to twice the noise. The sensitivity defined in concentration units is more useful to the analyst as it allows the system *concentration sensitivity* and *mass sensitivity* to be calculated.

7. Detector Dispersion – (σ_d^2) – The total detector dispersion, in conjunction with other extracolumn sources of dispersion, determines the optimum column radius [19] to provide maximum mass sensitivity and minimum solvent consumption. It is usually given units of (μl^2). The length and internal diameter of the fluid conduits of the detector should also be provided together with the geometry, dimensions and volume of the sensor cell.

8. Detector Time Constant – (D_t) – The overall time constant of the sensor and electronics and is usually given in milliseconds. It is mainly of interest in high speed chromatography.

9. Maximum Working Pressure – (P_D) – The maximum working pressure of the detector determines the maximum impedance that can be permitted in post–detector conduits, valves or columns and is largely used in multidimensional analysis and superfluid chromatography. It is specified either in p.s.i or kg/m^2

10. Pressure Sensitivity - (D_P) – The pressure sensitivity of a detector is the output that results from unit change in pressure. It is usually specified in V/p.s.i. or V/kg/m^2.

11. Flow Sensitivity – (D_Q) – The flow sensitivity is the output that results from unit change in flow rate. It is specified in V/ml/min. The flow sensitivity determines the required performance from the gas flow controller in GC and pump in LC.

12. Temperature Sensitivity – (DT) – The temperature sensitivity is defined as the output that results from 1°C change in temperature. It is given in V/^{o}C.

References

1. R. P. W. Scott and J. G. Lawrence, *Anal. Chem.*, **39**(1967)830.
2. G. Claxton, *J. Chromatogr.*, **2**(1959)136.
3. A. J. Groszek, *Nature*, **182**(1958)1152.
4. I. A. Fowlis and R. P. W. Scott, *J. Chromatogr.*, **11**(1963)1.
5. C. G. Scott, ASTM E19 No. E689-79.
6. J. E. Lovelock, *"Gas Chromatography 1960"* (Ed. R. P. W. Scott), Butterworths, London, (1960)26.
7. R. P. W. Scott, *"Liquid Chromatography Column Theory"*, John Wiley and Sons, New York, (1992)46.
8. R. P. W. Scott, *"Liquid Chromatography Column Theory"*, John Wiley and Sons, New York, (1992)15.
9. A. Klinkenberg, *""Gas Chromatography 1960"* (Ed. R. P. W. Scott), Butterworths, London, (1960)194.
10. M. J. E. Golay, *"Gas Chromatography 1958",* (Ed. D. H. Desty), Butterworths, London (1958)36.
11. J. G. Atwood and M. J. E. Golay, *J. Chromatogr.*, **218**(1981)97.
12. R. P. W. Scott and P. Kucera, *J. Chromatogr. Sci.*,**18**(1971)641.
13. C. H. Lochmuller and M. Sumner, *J. Chromatogr. Sci.*,**18**(1980)159.
14. I.Halasz,H.O.Gerlach, K.F.Gutlich and P. Walkling, US Patent **3,820,660**, (1974).
15. R. Tijssen, *Separ, Sci. Technol.*, **13**(1978)681.
16. E. D. Katz and R. P. W. Scott, *J. Chromatogr.*, **268**(1983)169.
17. J. G. Atwood and M. J. E. Golay, *J. Chromatogr.*, **218**(1981)97.
18. R. P. W. Scott, P Kucera and M. Munroe, *J. Chromatogr.*, **186**(1979)475.
19. R. P. W. Scott, *"Liquid Chromatography Column Theory"*, John Wiley and Sons, New York, (1992)196.
20. F. A. Vandenheuvel, *Anal. Chem.*, **35**(1963)1193.
21. L. J. Schmauch, *Anal. Chem.*, **31**(1959)225.
22. J. C. Sternberg, *Advances in Chromatography*, (Ed. J. C. Giddings and R. A. Keller), Marcel Dekker, New York, **2**(1966)205

CHAPTER 3

DATA ACQUISITION AND PROCESSING

The vast majority of chromatographs purchased today are provided with data acquisition and processing equipment. This will usually include a dedicated computer which will also control the operation of the chromatograph. In large laboratories the data acquisition and processing requirements for a number of different chromatographs are sometimes handled by a central computer. Central computers can time share with five, ten or even twenty different chromatographs and while data processing they can also provide laboratory housekeeping functions, such as monitoring the progress of samples through the laboratory from admission to the final report. The analyst or chromatographer does not need to know the details of the necessary electronic circuitry, nor the format of the software that is used in data processing. Notwithstanding, some perception of the logic behind the acquisition process and the mathematical algorithms that are used will help to identify the accuracy limits and the precision that can be expected from analysis.

In the early days of chromatography, data was processed using a simple integrator but today it has become far more complicated and at the same time more efficient. Data handling can be logically divided into two stages. Firstly the analog data is converted into digital form, acquired and stored in an appropriate memory. Secondly, the digital data is recalled, processed and the results reported or presented in a chosen format. The overall system can reside entirely in the computer or the data acquisition procedure can be partly contained in the detector and thus be part of the detector electronics. As already stated, there is no particular advantage to either as far as efficiency, accuracy and precision is concerned. However, if the data acquisition electronics are included in the detector the user is free to choose any convenient computer providing appropriate data processing software is available.

The Acquisition of Chromatography Data

Data systems vary slightly between different manufacturers but the scheme outlined below is generally representative of the majority of those presently used for handling chromatographic data. A diagram showing the individual steps involved in acquiring and processing chromatography data is shown in Figure 1.

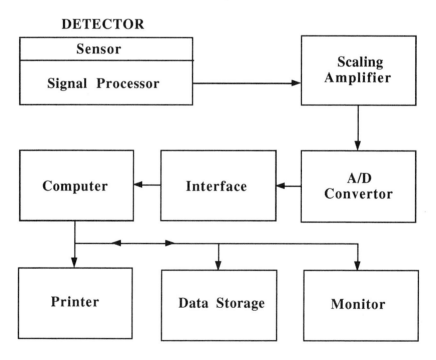

Figure 1 Block Diagram of a Typical Chromatography Data Acquisition System

After processing the output from the sensor, the detector provides an analog signal that passes to the scaling amplifier and then to an analog-to-digital converter (A/D converter). The signal from the A/D converter, now the digital equivalent of the analog signal, then passes via an appropriate interface to the computer and is stored in memory.

Either in real time or at the end of the analysis the computer recalls the data, carries out the necessary calculations and displays the result on a monitor and/or printer.

The Scaling Amplifier

The output from most detectors ranges from 0 to 10 mV? whereas the input to most A/D converters is considerably greater *e.g.* 0 to 1.0 V. Thus, the instantaneous measurement of 2 mV assumed from the detector must be scaled up to 0.2 volt which is carried out by a simple linear scaling amplifier. This is achieved by a simple linear amplifier with an appropriate gain.

The A/D Converter

After the signal has been scaled to an appropriate value it must be converted to digital form. There are a number of ways to digitize an analog signal and among others these will include sample and hold successive approximation conversion, the dual slope integrating converter, the single slope integrating converter and the voltage to frequency (V/F) converter. A detailed discussion of the advantages and disadvantages of different A/D converters is not germane to the subject of this book and consequently only the simple V/F converter will be described as an example. Readers wishing to know more about the subject of A/D converters are referred to the *Data Acquisition and Conversion Handbook* [1]. A diagram showing the operating principle of an V/F type A/D converter is shown in figure 2. The converter consists of an integrator that can be constructed from an operational amplifier with a feedback capacitor. The capacitor is charged by the voltage from the scaling amplifier through the operational amplifier.

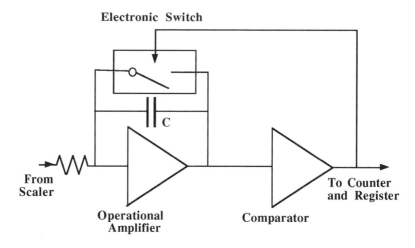

Figure 2   The Basic V/F Analog–to–Digital Converter

The output from the integrator is sensed by a comparator which activates the electronic switch when the potential across the capacitor reaches a preset voltage. The activation of the comparator also causes a pulse to be passed to a counter and at the same time the capacitor is discharged by the electronic switch. The process then starts again. The time taken to charge the capacitor to the prescribed voltage will be inversely proportional to the applied voltage and consequently the frequency of the pulses from the comparator will be directly proportional to the applied voltage. The frequency of the pulses generated by the voltage controlled oscillator is sampled at regular intervals by a counter which then transfers the count in binary form to a register. The overall system is shown diagramatically in figure 3.

As already stated the output from most detectors ranges from zero to ten millivolts and the input range of many A/D converters is usually from zero to one volt. Thus, the instantaneous measurement of 0.2 mV from the detector must be scaled up by a factor of 100 to 0.2 volts, which is carried out by the scaling amplifier. Now the A/D converter changes the analog voltage to a digital number, the magnitude of which is determined by the number of "bits" that the computer employs in its calculations.

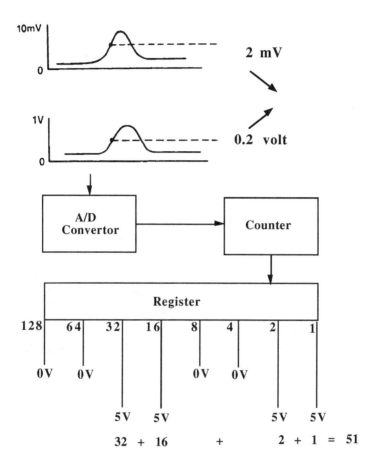

Figure 3 Stages of Data Acquisition

If, for example, eight bits are used, the largest decimal number will be 255. The digital data shown in figure 3 can be processed backward to demonstrate A/D procedure. It is seen that the third and fourth most significant "bits" (which are counted from the far left) and the two least significant "bits" (which are counted from the far right) are at the five volt level (high), which as shown in figure 3 is equivalent to 51 in decimal notation (32+16+2+1). It follows that the voltage that was converted must be $\dfrac{51}{255}$ x 1 volt = 0.2 volt. It should also be noted that because of the limitation of 8 "bits", the minimum discrimination that can be made between any two numbers is $\frac{1}{255}$ x 100 ≈ 0.4%. It follows that 8 bit systems are rarely used today and contemporary A/D converters usually have at least 12 or 18 bit outputs.

Transmission of the Data to the Computer

After the analog signal has been converted to digital form the next step is to make the digital data available to the computer. There are two major modes of signal transmission: serial transmission and parallel transmission.

Serial Transmission

In the serial mode, the digital word (number) is sent to the computer one bit at a time. Now a binary counter provides a parallel output since each of the output bits has its own data output channel and the value of each output bit is simultaneously available. To use a serial transmission scheme, this parallel output must be put into serial form. One way to accomplish this is to use a Universal Asynchronous Receiver Transmitter (UART). The detailed operation of the UART will not be given here as it is not germane to the subject of this book. It is sufficient to say that the heart of the UART is a shift register and the shift register is strobed by a signal from the computer that displaces the binary number, bit by bit, sequentially from the register to the computer.

The serial data transmission mode finds its greatest use in multiple detector/converter systems where data must be sent over moderate to long distances to the computer. The system is easy to implement, has a good noise immunity, and is reliable. The main disadvantage to the serial system is its moderate speed of transmission, which is about 3000 bytes/sec. This relatively low speed of transfer may limit its use in some very high-speed applications *e.g.* data acquisition of mass spectrometric data. However, its speed would be satisfactory for the vast majority of chromatography applications.

Parallel Transmission

In parallel transmission the outputs of the counter are connected directly to a peripheral interface adapter (PIA) and thence to the data bus of the computer. The computer data bus is a parallel system of conductors by which the binary data is transferred between the central processor, memory and peripheral circuits. As the data bus is used for all data transfer, and each transfer involves different data levels, the data bus can not be continuously connected to the register of the A/D output. The isolation is achieved by means of a series of dual input and-gates with tri-state outputs, one gate for each bit. The and–gate only allows the data on the input to appear at the output on reception of a signal from the computer.

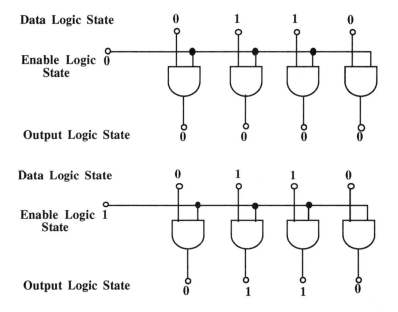

Figure 4 The Function of the And-Gates in the PIA

An and–gate is a device, the output of which will be identical with the logic state of the input on applying a logic 1 voltage to the second input (i.e. the output of an and–gate will only go high when both inputs are high). The operation of the and-gate system is shown in figure 4. In the upper diagram the digital output from the A/D Converter appears on one input of each and-gate and, as the second inputs are inactive, the output of all gates are logical zero. On activation of the second input of each gate with a logical 1 voltage from the computer, the logic state of the output of the gates are exactly the same as their first inputs. As soon as the data has been read into the computer, the activating voltage (the enabling pulse) is removed. The output of the gates is zero and the digital output from the A/D Converter is again isolated from the computer. The data now located in the accumulator of the computer is then transferred either to the computer random access memory (RAM), to a buffer store or to disc. The parallel data transfer mode is simple to use and allows for greater flexibility than the serial mode. However, it requires numerous connections between the PIA and the A/D Converter, and is, therefore limited to those cases where the converter and computer are very close. This type of transmission is very popular in chromatographs which have a built-in dedicated computer.

When the separation is complete, all the chromatographic information is stored in the computer memory as a series of binary words. Each binary word represents the detector output at a given time and the period between each consecutive word represents the same interval of time. Consequently, the computer has a record of the concentration of solute passing through the detector taken at regular intervals of time throughout the separation. Thus, the chromatogram can be reconstructed by simply plotting detector reading against time (which in fact is the reconstructed chromatogram). If the chromatogram is registered on a potentiometric recorder, as opposed to a printer, the output must be passed through a D/A recorder to recover the original analog signals. Typically, one chromatographic data point will occupy 2 words in memory (which for an 8 bit machine would require 16 bits). Consequently, a chromatogram 20 minutes long acquired at 5 data points a second would occupy 12,000 words. Part or all of the chromatogram can be reconstructed or, if so desired, certain portions can be expanded to permit more detailed examination of those areas of particular interest. Once the data is in memory, many different types of calculations can be carried out to help identify the components or carry out quantitative analysis. The details of the calculation carried out will be discussed in the chapters on quantitative and qualitative analysis but some general points will be made here.

Data Processing and Reporting

Data from the chromatograph can be handled in two ways. The data can either be processed in real time as it arrives at the computer and only the results of the calculations, such as retention time, peak height and peak area etc. are stored, the raw data being discarded. This is called "on the fly" processing and has the dubious advantage of providing results immediately after the peak is eluted. This procedure requires much less memory and, in the days when memory was limited and expensive, it was an attractive alternative. Today when many megabytes of RAM are available together with hundreds of megabytes of disc storage, this procedure is rarely used. The big disadvantage is that the deletion of the raw data prevents further or alternative data processing if, subsequently, it is shown that the results require it. The only application where this procedure might still be useful is in process control where results are required rapidly in order to control a production process to ensure optimum yield and maximum purity.

The alternative, and far more advantageous, form of data processing is to store each detector output value as it is received (usually on a hard disc) and then process the results after the analysis has been completed. This method give complete flexibility in data reduction, allows the chromatogram to be reconstructed in a number of alternative forms and different algorithms selected for base line correction or peak area measurement, etc. It also allows specific portions of the chromatogram to be amplified to show the presence of trace materials.

Data Processing

The first function the software must achieve is peak detection. The start of the peak can be identified by a significant change in detector output (the level of the change being defined by the user) or by the rate of change of the detector output (the rate also being defined prior to the analysis). If peak areas are being measured, the start of the peak defines the point where integration must commence. In the simplest form integration involves summing the detector output from the peak start to the peak end.

The end of the peak is identified in the same way, but if the baseline is sloping then the computer will extrapolate the baseline across the peak and calculate the level at which the peak can be assumed to have reached the baseline. This procedure will be discussed shortly. The next important parameter to identify is the peak maximum. There are a number of ways of doing this but the simplest is to subtract consecutive values at (n) and (n–1) between the peak start and the peak finish and determine when,

$$S_{(n)} - S_{(n-1)} \quad \text{becomes negative.}$$

The is a very simple example and would only work if there was no noise on the signal, which in practice is rarely the case. Often, some smoothing routine is carried out on the points proximate to the peak maximum and then peak identification is repeated. In this way an accurate measure of the position of the peak maximum can be obtained.

The certain identity of a component depends on the accuracy of the retention time measurement and thus the correct identification of the position of the peak maximum is critical. The peak height is taken as the difference between the signal at the point of the peak maximum and that directly beneath the peak maximum on the baseline produced under the peak. The projected baseline under the peak is calculated using a procedure similar to that described below.

Thus the data processing has provided the position of the start of the peak, the position of the peak maximum, the position of the end of the peak, the peak area and the peak height. One method for constructing the baseline under a peak when the peak is eluted on a sloping baseline is as follows. In figure 5 a peak is depicted eluted on a sloping baseline.

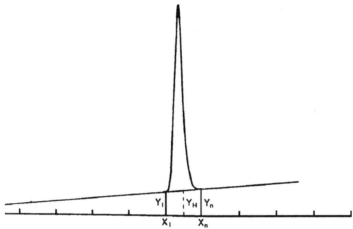

Figure 5   Baseline Correction

The average value of the baseline (Y_H) is obtained by the following simple relationship,

$$Y_H = \frac{Y_1 + Y_n}{2}$$

where (Y_1) is the detector signal at the start of the peak,
and (Y_n) is the detector signal at the end of the peak.

In general, the slope is usually considered linear under the peak and, in fact, if it is not, the chromatographic operating conditions probably need adjustment. Under some circumstances (*e.g.* at the end of a high-temperature program) the slope may not be linear and adjusting the operating conditions will not help. Thus, the curve of the baseline under the peak can be approximated to a second order polynomial function. If the slope of the baseline under the peak is linear, then the extrapolated baseline under the peak will obviously be given by

$$Y_n' = Y_1 + \frac{Y_n - Y_1}{n_n - n_1}(n_{n'} - n_1)$$

where (X_1) is the data point at the peak start,
  (n) is the data point at the peak end,
  (n') is the data point between the peak start and peak end,
  $Y_{(n')}$ is the baseline at the data point (n').

Consequently the peak area (A) will be calculated as

$$A = \sum_{p=1}^{p=n'} Y_p - Y_1 + \frac{Y_n - Y_1}{n_n - n_1}(n_p - n_1)$$

The data processing techniques so far discussed are basic and relatively simple. A more detailed treatment is not appropriate here but the subject of peak skimming and deconvolution will be discussed in the chapters on qualitative and quantitative analysis.

However, there is one important point that should be made. In whatever way the software may manipulate the data, the results can never be more precise or more accurate than the basic data that has been produced by the chromatograph. Furthermore, clever algorithms and ingenious software are no substitute for poor chromatography.

Chromatographic Control

Although not directly pertinent to detectors, the computer that handles the output from the detector should also provide other information in the analytical report. Today, most chromatographic systems, gas and liquid, have a dedicated computer associated with them which, as well as processing the information provided by the detector, will also control and record the operating conditions of the chromatograph. Temperatures, flow rates programs, etc., will be entered via the keyboard of the computer and the information stored for reporting purposes when required.

Software Version	:4.1(build 143)	
Date : 6/7/95	12:44 PM	
Sample Name	Tabasco	
Data File	C/TC4/Data/Pepper/Tobasd. Raw	Date 6/4/95 08:54 PM
Sequence File	C/TC4/Data/Pepper/Capsai.Seq.	Cycle: 7 Channel :A
Instrument	LC140 – 4:A Rack File 0/10	Operator
Sample Amount	1.000 Dilution Factor 1.00	

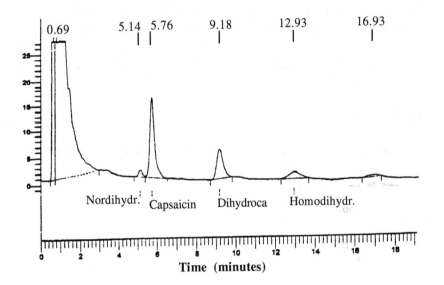

SCOVILLE "HEAT" IN PEPPER REPORT

Peak	Time (min)	Area (µVs)	Component Name	Raw NG	λ max	Peak Purity	Capsaicin	Scov val.	Scov. Heat
1	5.138	17342	Nordihydrocapsaicin	3.374	279	1.3	0.001	9.3	78
2	5.759	257942	Capsaicin	50.183	280	1.2	0.013	16.1	2020
3	9.179	126620	Dihydrocapsaicin	24.440	280	1.2	0.006	16.1	984
4	12.931	51739	Homohydrocapsaicin	10.066	280	1.5	0.003	8.1	204
		452645		88.063		.5.1	0.022		3286

Courtesy of the Perkin Elmer Corporation

Figure 6 Analytical Report from Computer Data Processing

An example of the print-out from a Perkin Elmer, LC140 liquid chromatograph giving the operating conditions and analytical results is shown in figure 6.

Very sophisticated data processing programs are continually being developed providing, apparently, more information or more precise quantitative data from the sample.

Nevertheless, it must again be stressed that the quality of the information provided by the data processing system can only be as good as that which the chromatographic system provides and, specifically, the accuracy and precision of the measurements made by the detector.

Reference

1. *Data Acquisition and Conversion* (Ed. E. L. Zuch), Mansfield, MA (1979).

Part 2

Gas Chromatography Detectors

CHAPTER 4

GAS CHROMATOGRAPHY DETECTORS: THEIR EVOLUTION AND GENERAL PROPERTIES

The Evolution of GC Detectors

The first GC detector was invented in 1952 by the originators of the technique, James and Martin [1], and took the form of a titration apparatus situated at the end of the column. One of the original applications of GC was the separation of a mixture of fatty acids and consequently, the eluent gas was bubbled through a suitable aqueous liquid to absorb the solutes. The solution contained an indicator, the color of which changed as each solute was eluted, and the solution was then manually titrated. Later the titration process was automated by the inventors (probably the first automatic titration apparatus to be made and certainly the only one available at that time) and an integral chromatogram was formed by plotting the volume of base solution added against time. The chromatogram consisted of a series of steps one for each solute. This rather primitive arrangement clearly and plainly demonstrated that gas chromatography would work but, at the same time, it also indicated that a detector with greater sensitivity and a more catholic response was necessary for the effective use of the technique.

The next detector, the first practical detector to be developed, was also invented by James and Martin but, for some reason, was never formally reported in the literature. Its description, however, did appear in a review by A. T. James [2].

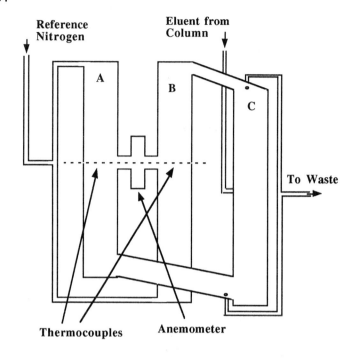
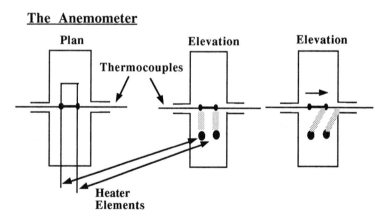

Figure 1 The Martin Gas Density Bridge

A detailed explanation of the function of the detector was given by Munday and Primavesi [3] in the 1956 symposium entitled "Vapor

Phase Chromatography" (the name originally given to the technique by Martin and later changed to Gas Chromatography). The gas density balance was an extremely complicated and ingenious device and, incidentally, the modern so-called gas density bridge bears little or no resemblance to the original design. In view of its technical ingenuity and because it was the first effective GC detector to be developed it will be described in some detail. A diagram of the gas density balance is shown in figure 1.

The detector consisted of a Wheatstone network of capillary tubes that were drilled out of a high conductivity copper block and was fairly compact. The reference flow of mobile phase and the eluent from the column entered at two opposing junctions of the bridge arms (the center of tube (C)) such that the eluent was contained in one vertical arm (C) and the pure mobile phase in a parallel vertical arms (A) and (B). The increase in pressure at the base of tube (C) due to the presence of solute in (C) applied a pressure to the bottom of tube (A). This caused a flow of gas through the anemometer from tube (A) to tube (B) providing an output that was fed to a recording milliammeter. Subsequently all flows exited from the top and bottom of tube (C).

The anemometer was particularly special. It consisted of a cylindrical chamber about 1.5 cm in diameter and about 4 mm wide. A length of 0.001 in O.D. copper wire, containing 2 mm of 0.001 in Constantan wire arc welded to the copper wire in the center, passed through the conduit connecting the chamber to tubes (A) and (B). (To make one of these dual thermocouples with the equipment available in 1952 was a feat in itself. Beneath the copper Constantan junctions was situated a heater loop that raised the temperature of both junctions by convection currents circulating round the cylindrical chamber as shown in the center of the anemometer diagram. When a flow of gas passed through the anemometer as a result of solute vapor being present in tube (C), the convection currents above the heater loop were displaced so that one junction was cooled and the other heated as shown in the right-hand side of the anemometer diagram. The differential output from the two thermocouples was charted on a recording milliammeter.

The detector was robust, but a little difficult to set up initially. It was linear ($0.98<r<1.02$) over about 3 orders of magnitude of concentration range and had a sensitivity (minimum detectable concentration) of about 5×10^{-7} g/ml (n-heptane). It helped to produce the data for many of the fundamental studies in GC. Unfortunately, the detector proved to be very difficult to make for the general chromatographer and, even after many attempts, was never produced commercially as an effective GC detector. The lack of an alternative, simple detector provoked a number of chromatographers to develop alternatives. The development of GC detectors was far easier than the development of LC detectors as organic vapors change the physical characteristics of a gas to a far greater extent than they do a liquid. As a consequence, a number of very effective highly sensitive GC detectors were developed over a relatively short period of time and many of them were manufactured commercially.

The first alternative device to be used as a GC detector was the katherometer introduced by Ray [4] (now more prosaically know as the *hot wire detector* (HWD)). This detector will be described later as it is still a popular device for the detection of permanent gases and so will only be briefly mentioned here. It consists of two heated filaments, one suspended in the eluent gas from the column and the other in a pure reference stream of gas. The filaments are situated in the arms of a Wheatstone bridge. When a solute is eluted, both the thermal conductivity and the heat capacity of the gas change, which alters the heat loss and consequently the temperature of the filament and its resistance This unbalances the bridge and the out-of-balance signal is passed to a suitable monitoring device. This detector is relatively insensitive but responds to all solutes that differ in heat capacity and thermal conductivity from those of the carrier gas. This detector was used extensively in the early days of GC for the analysis of hydrocarbon gases. In the early days of detector development there was much discourse and dissent with regards to the exact mechanism of detection involved in the katherometer [5,6] and even today it is considered to respond to a number of different physical properties of the eluent gas with no one playing a major role.

In the same symposium the "flame thermocouple detector" was first described by Scott [7] and it was, in fact, the forerunner to the flame ionization detector FID. Either hydrogen or a mixture of hydrogen and nitrogen was used as the carrier gas or the eluent from the column was mixed with hydrogen so that the elution products could be burnt at a small jet. A thermocouple was situated above the jet and was heated by the flame. When a solute was present in the eluent, the heat of combustion of the gas increased, raising the flame temperature and the output from the thermocouple. The electronic circuit was simple, consisting of a backing off circuit to offset the output from the hydrogen flame alone and an attenuating circuit, the output from which passed to a potentiometric recorder. The detector had a linear response over about three orders of magnitude of concentration and a sensitivity of about 1×10^{-6} g/ml (n-heptane). Its response was proportional to the heat of combustion of the solute. This detector was also made commercially but enjoyed a very short life as it was quickly supplanted by the FID.

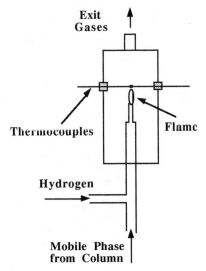

Figure 2   The Flame Thermocouple Detector

The β–ray ionization detector was also introduced in the 1956 symposium by Boer [8,9] and as this was the first ionization detector that utilized a radioactive source it will also be described in some detail here. The design of the detector is shown diagramatically in figure 3. It consisted of two cells, a reference cell, through which pure carrier gas passed, and a sensor cell, which carried the eluent from the column. In each cell was placed a ^{90}strontium β emitting source. The decay of ^{90}strontium is in two stages, each stage emitting a β particle producing the stable atom of ^{90}zirconium.

$$^{90}Sr \xrightarrow[0.6 \text{ MeV}]{\varepsilon} {}^{90}Y \xrightarrow[2.5 \text{ MeV}]{\varepsilon} {}^{90}Zr$$

half life ca 25 y — half life ca 60 h — Stable

The ionization currents formed (collected by appropriate electrode potentials) are arranged to oppose one another and consequently any variation in pressure or temperature of the two cells will be balanced out.

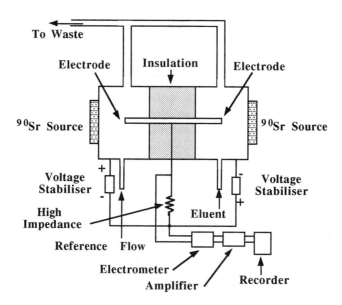

Figure 3 The β-Ray Ionization Detector

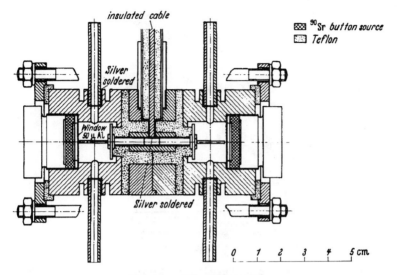

Reproduced from reference 8

Figure 4 Reproduction of Boer's Original Ionization Cell

The differential signal resulting from the presence of a solute in the column eluent sensor cell is amplified and recorded. The ionization detector, as originally described by Boer and taken directly from his original publication, is shown in figure 4.

The ionization current increases as the voltage increases until a plateau is reached where the ionization current is constant from about 50 to 200 volts. The ionization current produced depends on the nature of the gas and the ionization cross sectional area of the gas molecule. Curves relating ionization to applied voltage are shown in figure 5. It is seen that the ionization current for nitrogen is much greater than that for hydrogen. It is also seen that in the presence of organic gases such as 10% v/v butane in hydrogen doubles the ionization current relative to that of pure hydrogen. Similarly, 10% v/v of butane in nitrogen increases the ionization current by about 50%. Providing the ionization voltage is such that the ion current of the eluted material lies on the plateau portion of the response curve, the relationship between ion current and solute concentration is linear.

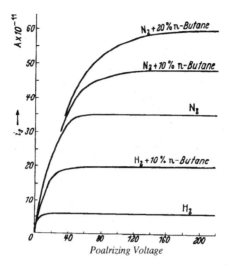

Reproduced from reference 8

Figure 5 Response Curves for the Boer Ionization Detector

As the ionization currents are additive the signal(s) from the compensated cell can be described by the following equation.

$$s = m i_S^c + (1 - m) i_S^g - i_S^g = m (i_S^c - i_S^g)$$

where (m) is the molar fraction of component (c) in carrier gas (g),

(i_S^c) is the saturation current of the pure component,

and (i_S^g) is the saturation current of the carrier gas.

It is seen that it can be shown theoretically that the response of the detector should be linear with concentration.

The sensitivity of the detector was similar to that of the katherometer *i.e.* about 1×10^{-6} g/ml. Unfortunately, the practical lifetime of this detector was also relatively short as it was eclipsed by both the FID and the argon ionization family of detectors introduced by Lovelock. Nevertheless, the development of these early detectors not only helped establish GC as a viable analytical technique but also stimulated the development of other types of vapor sensing devices. The future detectors would prove to have extremely high sensitivities and wide linear dynamic ranges and to function on widely different principles. However, with the exception of the katherometer, they would also render these early detectors virtually obsolete. Nevertheless, they played an important role in the early days of GC and in their time were exciting devices to operate.

The second generation of GC detectors was introduced at the 1958 Symposium on Gas Chromatography, probably the most important and technically exciting of all chromatography symposia. At that symposium, many original GC concepts was confirmed but, more importantly, further developments in detector technology were described including two new detectors, the FID and the emissivity detector. In addition, novel column systems were introduced, including the open tubular column by Golay [10] (which was to become the column of the future) together with the first high efficiency packed columns [11]. The contents of this symposium probably represented the climax of GC development and, although a plethora of symposia would appear over the following half century, none would have the excitement and novel technical content of this one.

Detector technology and new detectors were particularly hot topics at the 1958 symposium. McWilliams and Dewer [12] described the flame ionization detector which was to be the 'work horse' of all future gas chromatographs. Further developments of the flame thermocouple detector were described by Primavesi *et al.* [13], the design of the katherometer was simplified by Stuve [14] and Grant [15] described the first thermal emissivity detector.

The emissivity detector developed by Grant was an interesting and innovative extension of the flame thermocouple detector. It did not prove particularly popular at the time, but in recent years the concept has been revived and commercial detectors based on the emissivity concept of Grant are now available.

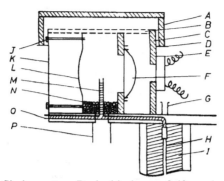

A, draught proof Sindanyo top. B, double layer of 40 mesh copper gauze. C, Sindanyo front for photo-cell mounting. D, metal stops for draught top,. E, selenium photo-cell. F, glass condensing lens. G, column inlet. H, column. I, column heating jacket. J, supports for reflector. L, metal reflector. M, stainless steel jet. N, 2 cm deep layer of porcelain beads. O, coal gas inlet. P, air inlet.

Figure 6 The Emissivity Detector

The original detector was very simple in design and the original diagram of the device is shown in figure 6. In principle the column eluent was mixed with a combustible gas and burnt at a small jet in the same manner as the flame thermocouple detector. In fact Grant used coal gas (largely a mixture of hydrogen, methane and carbon monoxide) with which the column eluent was mixed prior to combustion. Opposite the flame was situated a lens that focused the light emitted onto a photo cell. The output from the coal gas flame light was balanced out by a simple potentiometer network and the signal passed to a potentiometric recorder. The detector gave a partially selective response to aromatic hydrocarbons or any solute that either increased the luminosity or imparted color to the flame. The sensitivity varied widely with the solute being detected. Aromatics that impart strong luminosity to the flame giving a strong response and consequently a high sensitivity ($ca\ 1 \times 10^{-6}$ g/ml). Conversely, saturated hydrocarbons impart little luminosity to the flame and thus have a weak response and a very low sensitivity–the very features for which the detector was designed. The linear range is difficult to determine from the original publication but appears to be in excess of two orders of magnitude. The performance of the detector is illustrated in figure 7.

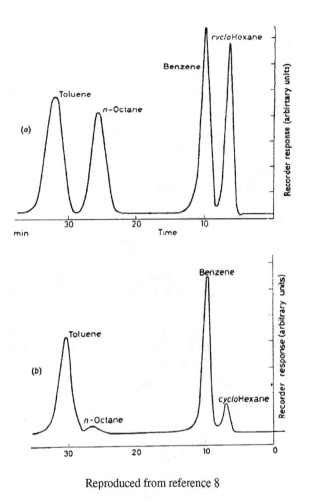

Reproduced from reference 8

Figure 7 Chromatograms Demonstrating the Selectivity of the Emissivity Detector

The separation shown in figure 7 of a mixture of aliphatic and aromatic hydrocarbons was monitored using both a katherometer detector that responded to all solutes and the emissivity detector that selectively responded to the aromatics. It is seen that the emissivity detector clearly discriminates the solutes benzene and toluene from the aliphatic hydrocarbons, which give very little response.

In retrospect the triad of symposia held in 1956, 1958 and 1960 not only established GC as a unique and reliable analytical technique but also contained the first disclosure of 90% of the devices used in modern GC. Between the 1958 and 1960 symposia, Lovelock developed a family of ionization detectors starting with the Argon Detector [16, 17] in 1958 of which three different models were created and demonstrated to function well, and ending in 1959 with the electron capture detector [18], which, second to the FID, is probably the most commonly used detector today.

At the 1960 symposium, research into the mechanism of detection of a number of the more important GC detectors was reported. The FID was carefully examined by Ongkiehong [19] and Desty *et al.* [20], the different argon detectors were described in detail by Lovelock [21] and the performance of the FID and Argon Detectors were compared by Condon *et al.* [22]. The integral and differential ionization detectors were also discussed by Matousek [23]. At the end of the 1960 symposium, the majority of the GC detectors that are used today had been described and their function explained and understood. Since that time few, if any, new GC detectors have been introduced and innovations that have been described have been largely extensions and improvements of older concepts.

Perhaps, before discussing the general properties of detectors one other GC detector might be mentioned as it possibly represents the only GC detector that functions by absolute measurement. This detector, developed by Bevan and Thorburn [24] in the early 1960s, was the mass detector. Unfortunately, this detector never progressed beyond the research prototype. It consisted of an efficient charcoal vapor adsorbent suspended from the arm of a recording microbalance. The column eluent was arranged to pass through the adsorbent without disturbing the balance. As solutes were eluted from the column they were quantitatively adsorbed on the charcoal and the weight recorded by the balance. As a consequence, the balance record produced an integral chromatogram, each step being an absolute measure of the mass of solute eluted. Although it had an absolute response, this detector was never developed seriously as its sensitivity was relatively poor, which placed severe restrictions on the type of column with which it could be used.

The General Properties of GC Detectors

The basic function of the GC detector is to respond to the presence of very small quantities of vapor in a permanent gas. This is tantamount to the detection of relatively high boiling compounds contained at very small concentrations in very low boiling substances. Because the physical and chemical properties of permanent gases differ widely from those of a vapor, a very wide range of detection methods can be employed. Such methods range from the measurement of standard physical properties such as thermal conductivity and light adsorption to ionization potentials and heats of combustion.

The response of a GC detector can be general or specific. A detector with a catholic response such as the FID is used widely in routine analysis. The specific detector, such as the nitrogen-phosphorus detector (NPD), is extremely useful for measuring particular types of compounds such as herbicides and pesticides, where the compounds of interest are not eluted discretely but mixed with a number of other contaminating compounds. Examples of this type of application will be given when the NPD is discussed in detail.

In general GC detectors should be insensitive to changes in flow rate but, as already discussed, few detectors have this attribute although some, for example the FID, are virtually insensitive to changes in *column* flow. This is advantageous as it permits the use of flow programming development. Flow programming attempts to achieve the same result as temperature programming, which is to accelerate the strongly retained peaks through the column. It is less effective than temperature programming as the effect of temperature change on solute retention is *exponential* whereas the effect of flow programming on solute retention is *linear* with change in flow rate. Nevertheless, it is useful where, due to the solutes being thermally labile, temperature programming is not possible. Consequently, detectors that are insensitive to column flow rate change can be advantageous.

Some detectors require no other gas than that used as the carrier gas, other require specific gases to be added to the columns eluent for them to function. In some cases the detector prescribes a certain gas to be used as the carrier gas. It will be seen later that the sensitivity of the katherometer is greater when helium is used as the carrier gas. In addition, if the gas chromatograph is being used for permanent-gas analysis, then helium must be used to differentiate the carrier gas from the other gases being analyzed. The use of extra gases increases the operating costs, particularly if they are relatively expensive such as helium or argon. In Europe helium can be very costly relative to other gases as it is mostly obtained from the United States.

All gas chromatographs are designed to operate over a wide range of temperatures, the extremes being from -20°C to 400°C. Consequently, in order to avoid solute condensation in the detector or detector-connecting tubes the detector should be capable of operating at least 20°C higher than the maximum column temperature. It is clear that only a limited number of detectors will function at 400°C and thus many chromatographs do not have such a wide operating range, nor is it often needed. When choosing equipment, select an instrument that will accommodate the analyses that are envisioned and not those that *might* be needed in the vague future. It is worth remembering that "*But what if...*" rarely ever happens.

Temperature programming is used in almost all analyses and thus to assure temperature stability, the detector is usually thermostatted in a separate oven set at a temperature at least 20°C above the maximum in the column program. Although some GC analyses are carried out at sub–ambient temperatures, they are few and far between and care must be taken to avoid condensation, particularly in areas where there are electrical connections to the detector. All analytical instruments should be operated in an air conditioned temperature controlled environment.

The detector consists primarily of two parts, the sensor and the associated signal conditioning electronics. They can be integral or separate. Although the sensor should be situated as close to the column as possible to minimize extra-column dispersion, the electronic system can be situated some distance from the sensor if this is more convenient. However, the signal must be transmitted in an appropriate manner that ensures no electrical interference that might provide extra noise or signal distortion. This means the route chosen for the connection must be carefully selected to avoid stray electrical and magnetic fields and all connections and cables should be electrically, and if necessary, magnetically screened.

The associated electronics may contain an A/D converter to provide a binary output that can be addressed and acquired by a computer or the analog signal may be passed to a computer that has its own A/D converter. In general the sooner the signal is digitized the better, as digital data is far more immune to external interference than analog signals. Although the original designs of many of the detectors involved amplifiers and attenuators that operated with thermionic valves, all modern detector electronics are exclusively solid state devices. From the author's experience, however, this means that they are smaller, cooler and less expensive but they do not seem to work any better.

Each GC detector will now be described in detail including their design, function and operation; additionally their most dominant areas of application will be discussed.

References

1. A. T. James and A. J. P. Martin, *Biochem. J.* **50**(1952)679.
2. A. T. James, The Times Science Review, **Summer** (1955)8.
3. C. W. Munday and G. R. Primavesi, "*Vapor Phase Chromatography*" (Ed. D. H.Desty and C.L. A. Harbourn), Butterworths Scientific Publications,(1957)146.
4. N. H. Ray, *J. Appl. Chem.*, **4**(1954)21.

5. N. MNellor, "*Vapor Phase Chromatography*" (Ed. D.H.Desty and C.L. A. Harbourn), Butterworths Scientific Publications (1957)63.
6. D.Harvey and G.O. Morgan, "*Vapor Phase Chromatography*" (Ed. D. H.Desty and C.L. A. Harbourn), Butterworths Scientific Publications, (1957)74.
7. R. P. W. Scott, "*Vapor Phase Chromatography*" (Ed. D.H.Desty and C.L. A. Harbourn), Butterworths Scientific Publications, (1957)131.
8. H. Boer, "*Vapor Phase Chromatography*" (Ed. D.H.Desty and C.L. A. Harbourn), Butterworths Scientific Publications (1957)169.
9. C. H. Deal , J. W. Otvos, V. N. Smith and P. S. Zucco (presented at the "*Symposium on Vapor PhaseChromatography*" of the A.C.S. (1956).
10. M. J. E. Golay, "*Gas Chromatography 1958*" (Ed. D. H. Desty), Butterworths Scientific Publications, (1957)36.
11. R. P. W. Scott, "*Gas Chromatography 1958*" (Ed. D. H. Desty), Butterworths Scientific Publications, (1957)189.
12. I. G. McWilliams and R. A. Dewer, "*Gas Chromatography 1958*" (Ed. D. H. Desty), Butterworths Scientific Publications, (1957)142.
13. G. R. Primavesi, G. F. Oldham and R. J. Thompson, "*Gas Chromatography 1958*", (Ed. D. H. Desty), Butterworths Scientific Publications, (1957)165.
14. W. Stuve, "*Gas Chromatography 1958*" (Ed. D. H. Desty), Butterworths Scientific Publications, (1957)178.
15. D. W. Grant, "*Gas Chromatography 1958*" (Ed. D. H. Desty), Butterworths Scientific Publications, (1957)153.
16. J. E. Lovelock, *J. Chromatogr.*, **1**(1958)35
17. J.E.Lovelock, A.T.James and E.A.Piper, *Ann. N. Y. Acad. Sci.*, **72**(1959)720.
18. J. E. Lovelock and S. R. Lipsky, *J. Amer. Chem. Soc.*, **82**(1960)431.
19. L. Ongkiehong, "*Gas Chromatography 1960*" (Ed. R. P. W. Scott) Butterworths, London (1958)9.
20. D. H. Desty, A. Goldup and C. J. Geach, "*Gas Chromatography 1960*" (Ed. R. P. W. Scott) Butterworths, London (1958)156.
21. J.E.Lovelock,"*Gas Chromatography 1960*" (Ed. R. P. W. Scott) Butterworths, London (1958)9.
22. R.D.Condon,P.R.Scholly and W.Averill , "*Gas Chromatography 1960*", (Ed. R. P. W. Scott) Butterworths, London (1958)134.
23. S. Matousek, "*Gas Chromatography 1960*" (Ed. R. P. W. Scott) Butterworths, London (1958)219.
24. S. C. Bevan and S. Thorburn, *J. Chromatogr.*, **11**(1963)136.

CHAPTER 5

THE FLAME IONIZATION DETECTOR AND ITS EXTENSIONS
The Phosphorous Nitrogen Detector and the Emissivity Detector

The flame ionization detector has a very wide dynamic range, has a high sensitivity and, with the exception of about half a dozen small molecular weight compounds, will detect all substances that contain carbon. It is, without doubt, the most useful GC detector available and by far that most commonly used in GC analyses.

At one time there was contention as to who was the first to invent the FID; some gave the credit to Harley and Pretorious [1], others to McWilliams and Dewer [2]. In any event it would appear that both contenders developed the device at about the same time and independent of one another; the controversy had more patent significance than historical interest. The FID is an extension of the flame thermocouple detector and is physically very similar, the fundamentally important difference being that the ions produced in the flame are measured as opposed to the heat generated. Hydrogen is mixed with the column eluent and burned at a small jet. Surrounding the flame is a cylindrical electrode and a relatively high voltage is applied between the jet and the electrode to collect the ions that are formed in the flame. The resulting current is amplified by a high impedance amplifier and the output fed to a data acquisition system or a potentiometric recorder. The disadvantage of the detector is that it normally requires three separate gas supplies together with their precision flow regulators. The gases normally used are hydrogen for combustion, helium or nitrogen for the carrier gas and oxygen or air as the combustion agent. The need for three gas supplies is a decided inconvenience but is readily tolerated in order to take advantage of the many attributes of the FID. The detector is normally thermostatted in a separate oven; this is not because the response of the FID is particularly temperature sensitive but to ensure that no solutes condense in the connecting tubes.

The Design of the FID

As already stated, in this book a detector is considered to be composed of a sensor and associated electronics. In this context the sensor unit is commonly referred to as the FID itself. A diagram of the FID sensor is shown in figure 1. The body and the cylindrical electrode is usually made of stainless steel and stainless steel fittings connect the detector to the appropriate gas supplies. The jet and the electrodes are insulated from the main body of the sensor with appropriate high temperature insulators.

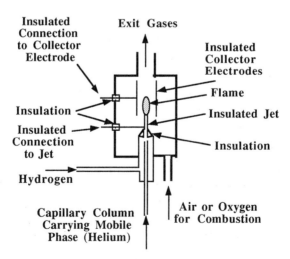

Figure 1   The FID Sensor

101

Care must be taken in selecting these insulators as many glasses (with the exception of fused quartz) and some ceramic materials become conducting at high temperatures (200-300°C) [3]. As a result of the relatively high voltages used in conjunction with the very small ionic currents being measured, all connections to the jet or electrode must be well insulated and electrically screened. In addition, the screening and insulating materials must be stable at the elevated temperature of the detector oven. In order to accommodate the high temperatures that exist at the jet-tip, the jet is usually constructed of a metal that is not easily oxidized such as stainless steel, platinum or platinum/rhodium.

Electrode Configurations

The associated electronics of the FID sensor consist of a high voltage power supply and a high impedance amplifier. The jet and electrode can be connected to the power supply and amplifier in basically two different configurations; the two alternatives are shown in figure 2. The floating jet configuration is the most commonly used and is shown on the left in figure 2.

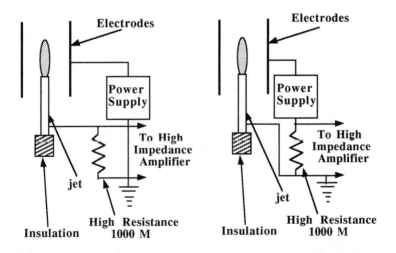

Floating Jet Configuration   Earthed Jet Configuration

Figure 2   Electrode Configurations

One side of the high voltage power supply is earthed (grounded) and the other is taken to the cylindrical electrode. A high resistance is situated between the jet and earth. This resistance is changed to adjust the sensitivity of the detector but at the maximum sensitivity may be as great as 1000 M. The ion current flows through the resistance to complete the circuit and the voltage developed across the resistance is fed to the high impedance amplifier. This is the more stable of the configurations providing the jet insulator has a high enough resistance at the temperature of operation. The other alternative is to earth the jet (which can eliminate the need for a jet insulator) and insert the resistance between the power supply and ground. This means that the power supply circuit must be isolated from earth and thus will be more susceptible to electrical interference. It does, however, reduce the high impedance sensor connections to a single cable. The voltage applied across the jet and electrode ranges from 50 to 250 volts depending on the geometry of the sensor and, in particular, the proximity of the electrode to the jet.

The High Impedance Amplifier

Originally, before the advent of integrated circuits containing field effect transistors (FET), high impedance amplifiers were constructed using specially selected thermionic valves with very low grid currents (for those familiar with thermionic valves, the EF87 was a specially selected 6J5G triode with a grid current of about 10^{-12} amp). Modern integrated circuits (*e.g.* an FET operational amplifier functioning in its noninverting mode) can have an input impedance of 10^{14} ohm. Consequently, there is no significant drain on the current source (such as the FID) and the voltage developed across the high resistance in figure 2 will be truly related to the ion current. The FID produces an ion current when only hydrogen is being burned at the jet, and so in the early stages of the amplifier circuit an adjustable offset voltage is often provided to produce zero output when no solutes are being eluted. If the output is acquired directly by a computer, the offset can also be achieved by suitable software (the base signal from the hydrogen flame is subtracted from all acquired data) providing data for eluted solutes only. The disadvantage of this type of zero offset is that unless the computer employs a D/A converter and an analog output, then a potentiometric recorder can not be used to display the separation. The FID can operate over a concentration range of more than six orders of magnitude and so the amplifier must also provide range switching. This can be manual or automatically controlled by the computer. However, due to the wide output range, adjustment must usually be achieved in two ways. Firstly over a range of about 128, attenuation is executed by changing the amplifier output or alternatively by computer software. Range changes in excess of this, however, require the sensor resistors to be changed, usually in factors of 10. Further, due to the wide dynamic range of the FID, this coarse sensitivity adjustment by the use of different input resistors can not be easily achieved by software (although it can be automatic) and often entails resistor change by mechanical switching.

The Response Mechanism of the FID

The response mechanism of the FID was carefully examined by Ongkiehong [4)] and Desty [5] in 1960 and in the author's opinion no such detailed evaluation of the detector has been carried out since. It was originally thought that the ionization mechanism in the FID flame is similar to the ionization process in a hydrocarbon flame which was studied intensively by Calcote and King [6] and Schuler and Weber [7] in the mid–1950s. However, it became quickly apparent that the ionization in the hydrogen flame is many times higher than could be accounted for by thermal ionization alone. It would appear that the ionization potentials of organic materials become much lower when they enter the flame. The generally accepted explanation of this effect is that the ions are not formed by thermal ionization but by thermal emission from small carbon particles that are formed during the combustion process. Consequently the dominating factor in the ionization of organic material is not their ionization potential but the work function of the carbon that is transiently formed during their combustion. The flame plasma contains both positive ions and electrons which are collected on either the jet or the plate depending on the polarity of the applied voltage.

Initially, the current increases with applied voltage, the magnitude of which depend on the electrode spacing. The current continues to increase with the applied voltage and eventually reaches a plateau at which the current remains sensibly constant. The voltage at which this plateau is reached also depends on the electrode distances. Curves relating ionization current to applied voltage obtained by Ongkiehong are shown in figure 3.

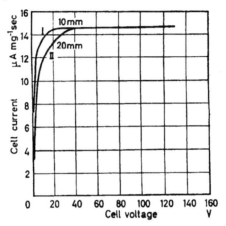

Distance between the electrodes I, 10 mm and II, 20 mm.

Figure 3 Current Voltage Curves for an FID with Different Electrode Geometries and Operating with 5 ppm v/v of Butane in Pure Hydrogen

As soon as the electron/ion pair is produced recombination starts to take place. The longer the ions take to reach the electrode and be collected the more the recombination takes place. Thus, the greater the distance between the electrodes and the lower the voltage the greater the recombination. It is seen that this is substantiated by the curves obtained by Ongkiehong: the plateau is reached at a lower voltage when the electrodes are closer together. It is seen that the plateau level is the same for both electrode conditions and it is assumed that on the plateau, all ions and electrons being produced in the flame are collected. In practice the applied voltage would be adjusted to suit the electrode distance to ensure that the detector operates under conditions where all electrons and ions are collected *viz.* on the plateau shown in figure 3.

Desty *et al.* [5] showed that the performance of the detector was relatively insensitive to electrode distance and hydrogen flow providing the voltage was adjusted to ensure that the detector worked on the current plateau.

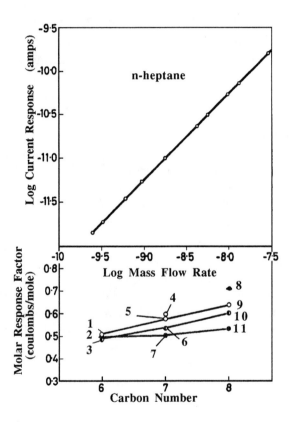

Solutes: 1, n-hexane; 2, benzene; 3, cyclohexane; 4, 2,2,3-trimethylbutane; 5, n-heptane; 6, methylcyclohexane; 7, toluene; 8, 2,2,4-trimethylpentane; 9, n-octane; 10, ethylcyclohexane; 11, ethylbenzene.

Figure 4 The Response of the FID to Different Hydrocarbons

They also showed that the air flow should be at least 6 times that of the hydrogen flow for stable conditions and complete combustion. The base current from the hydrogen flow depends strongly on the purity of the hydrogen. Traces of hydrocarbons significantly increase the base current as would be expected. Consequently, very pure hydrogen should be employed with the FID if maximum sensitivity is required. Employing purified hydrogen Desty *et al.* reported a base current of 1.45×10^{-12} amp for a hydrogen flow of 20 ml/min. This was equivalent to 1×10^{-7} coulomb per mole. The sensitivity reported, for n-heptane, assuming a noise level equivalent to the base current from hydrogen of *ca* 2×10^{-14} amp (a fairly generous assumption), was 5×10^{-12} g/ml at a flow rate of 20 ml/min. It follows that although the sensitivity is amazing high, the ionization efficiency is still very small *ca.* 0.0015%. The general response of the FID to substances of different type varies very significantly from one to another. For a given homologous series the response appears to increase linearly with carbon number but there is a large difference in response between a homologous series of hydrocarbons and a series of alcohols. An example of the response of the FID to a number of different hydrocarbons is shown in the lower curves in figure 4.

Desty suggested that to a first approximation, the response is proportional to the carbon number of the solute which is taken as the number of carbon atoms remaining after as many CO_2 groups as possible have been split off using the internal oxygen contained by the molecule. However, this approximation completely breaks down with halogenated compounds (which incidentally do not extinguish the flame). This means that although the response of the FID to different substances does not differ greatly (except for halogenated compounds) they can not be accurately predicted and for quantitative analysis calibration with standards is essential.

The linear dynamic range of the FID covers at east four to five orders of magnitude for $0.98<r<1.02$. This is a remarkably wide range that also accounts for the popularity of the detector. Examination of the different commercially available detectors shows considerable difference in electrode geometry and operating electrode voltages, yet they all have very similar performance specifications. This supports the claim of Desty *et al.* that the FID is surprisingly forgiving with respect to specific detector geometry.

The Operation of the FID

The FID is one of the simplest, easiest and most reliable detectors to operate. Generally the appropriate flow rates for the different gases are given in the detector manual. Hydrogen flows usually range between 20 and 30 ml per min, air flows about 6 times the hydrogen flow *e.g.* 120 to 200 ml per min. The column flow that can be tolerated is usually about 20-25 ml per min depending on the chosen hydrogen flow. However, if a capillary column is used, the flow rate may be less than 1 ml per min for very small diameter columns. The mobile phase can be any inert gas–helium, nitrogen, argon etc. To some extent the detector is self-cleaning and rarely becomes fouled. However, this depends a little on the substances being analyzed. If silane derivatives are continuously injected on the column, then silica is deposited both on the jet and on the electrodes and may need to be regularly cleaned. In a similar way the regular analysis of phosphate containing compounds may eventually contaminate the electrode system. Electrode cleaning is best carried out by the instrument service engineer.

Typical Applications of the FID

The FID in conjunction with high efficiency capillary columns has found wide application in the analysis of hydrocarbons. It is also employed extensively for pharmaceutical analysis, pesticide analysis, forensic chemistry and essential oil analysis, but its major area of application is in the analytical laboratories of the hydrocarbon industry.

A typical example of a ubiquitous hydrocarbon analysis is the analysis of gasoline shown in figure 5.

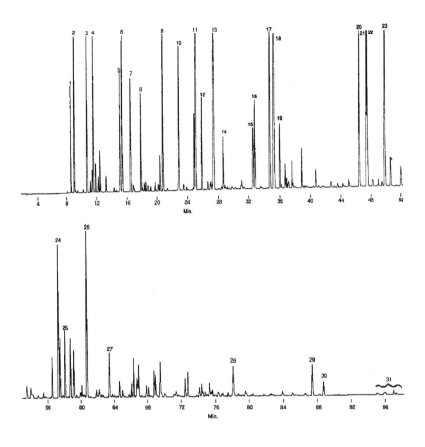

Courtesy of Supelco Inc.

1/ Isobutane
4/ n-Pentane
7/ 3-Methylpentane
10/ Benzene
13/ 2,2,4Trimethylpentane
16/ 2,4-Dimethylhexane
19/ 2,3-Dimethylhexane
22/ p-Xylene
25/ 1,3,5-TriMe-benzene
28/ Naphthalene
31/ Dimethylnaphthalene

2/ n-Butane
5/ 2,3-Dimethylbutane
8/ n-Hexane
11/ 2-Methylhexane
14/ n-Heptane
17/ 2,3,4Trimethylpentane
20/ ethylbenzene
23/ o-Xylene
26/ 1,2,4TriMe-benzene
29/ 2-Methylnaphthalene

3/ Isopentane
6/ 2-Methylpentane
9/ 2,4-Dimethylpentane
12/ 3-Methylhexane
15/ 2,5-Dimethylhexane
18/ 2,3-Dimethylhexane
21/ m-Xylene
24/ -Me-3-Ethylbenzene
27/ 1,2,3TriMe-benzene
30/ 1-Methylnaphthalene

Figure 5 The Analysis of Gasoline

The column used was 100 m long, 250 μm I.D. and carried a film of stationary phase 0.5 μm thick. The stationary phase was Petrocol poly(dimethylsiloxane) that was intra-column polymerized and thus bonded to the surface. The conditions used for the analysis are typical for a wide boiling range hydrocarbon mixture. The column was held at 35°C after injection for 15 min and then programmed to 200°C at 2°C/min and finally held at 200°C for 5 min. The sensor was held at 250°C (50°C) above the maximum column temperature to avoid condensation in the jet and conduits. The sample size was 0.1 μl, which was split 100-1, onto the column and so the total charge on the column was about 1 μg. Helium was used as the carrier gas at a linear velocity of 20 cm/sec. It is seen that the baseline is extremely stable despite the temperature program, and the overall response time of the detecting system easily copes with the rapidly eluting peaks. This separation clearly demonstrates the high sensitivity and wide dynamic range that is available from the FID which make its so useful for quantitative analysis.

Bearing in mind that there are hundreds of different components present in the mixture and the total charge was only 1 microgram the advantages of the high sensitivity are obvious. It should be noted that the columns demands the use of very small charges to function properly and a very poor separation would be obtained if large charges were used. Small charges, in turn, demand special injection techniques which will be briefly mentioned in the chapter on quantitative analysis.

Another type of analysis that is frequently carried out in the hydrocarbon industry is the *paraffin, isoparaffin, aromatic, naphthalene and olefin* estimation (the PIANO analysis), an example of which is shown in figure 6. The sample is a PIANO standard mixture. The column used was fused silica, 50 m long and 0.5 mm I.D., and the stationary phase was also Petrocol DH 50.2. The column temperature was held at 35°C for 5 minutes and then programmed up to 200°C at 2°/min. The carrier gas was helium and its velocity through the column 20 cm/sec.

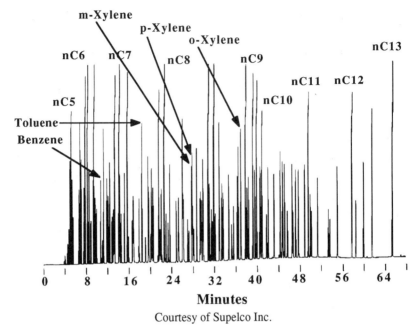

Courtesy of Supelco Inc.

Figure 6 The Separation of a PIANO Standard Mixture

The Nitrogen Phosphorus Detector (NPD)

The nitrogen phosphorus detector (NPD) (sometimes called the *thermionic detector*) is another sensitive, but, in this case, a specific detector, that is based on the FID. Physically the sensor appears to be similar to the FID but, in fact, operates on an entirely different principle. A diagram of an NPD detector is shown in figure 9.

The essential change that differentiates the NPD sensor from that of the FID is a rubidium or cesium bead contained inside a heater coil close to the hydrogen flame. The bead, heated by a current through the coil, is situated above a jet, through which the nitrogen carrier gas mixed with hydrogen passes. If the detector is to respond to both nitrogen and phosphorus, then the hydrogen flow should be minimal so that the gas does not ignite at the jet. If the detector is to respond to phosphorus only, a large flow of hydrogen can be used and the mixture burnt at the jet.

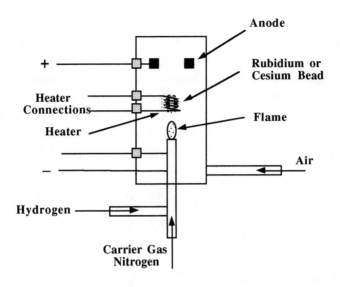

Figure 9. The Nitrogen Phosphorus Detector

The heated alkali bead emits electrons by thermionic emission which are collected at the anode and provides background current through the electrode system. When a solute that contains nitrogen or phosphorus is eluted, the partially combusted nitrogen and phosphorus materials are adsorbed on the surface of the bead. This adsorbed material reduces the work function of the surface and, as a consequence, the emission of electrons is increased, which raises the current collected at the anode. The sensitivity of the NPD is very high and about an order of magnitude less than that of the electron capture detector ($ca.10^{-12}$ g/ml for phosphorus and 10^{-11} g/ml for nitrogen).

The main disadvantage of this detector is that its performance deteriorates with time. Reese [8] examined the performance of the NPD in great detail. The alkali salt employed as the bead is usually a silicate and Reese demonstrated that the loss in response was due to water vapor from the burning hydrogen converting the alkali silicate to the hydroxide and free silica. Unfortunately, at the normal operating temperature of the bead, the alkali hydroxide has a significant vapor pressure and consequently, the rubidium or cesium is continually lost during the operation of the detector. Eventually all the alkali is evaporated, leaving a bead of inactive silica. This is an inherent problem with all NP detectors and as a result the bead needs to be replaced regularly if the detector is in continuous use. The detector can be made "linear" over three orders of magnitude although no values for the response index appear to have been reported. Like the FID it is relatively insensitive to pressure, flow rate and temperature changes but is usually thermostatted at 260°C or above.

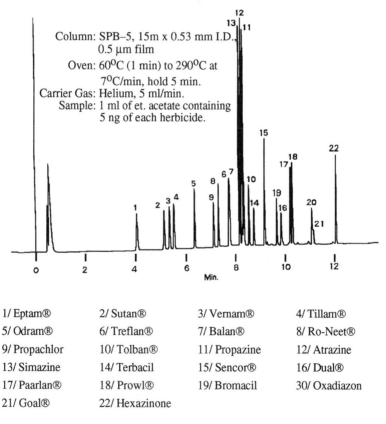

1/ Eptam®	2/ Sutan®	3/ Vernam®	4/ Tillam®
5/ Odram®	6/ Treflan®	7/ Balan®	8/ Ro-Neet®
9/ Propachlor	10/ Tolban®	11/ Propazine	12/ Atrazine
13/ Simazine	14/ Terbacil	15/ Sencor®	16/ Dual®
17/ Paarlan®	18/ Prowl®	19/ Bromacil	30/ Oxadiazon
21/ Goal®	22/ Hexazinone		

Courtesy of Supelco Inc.

Figure 10 The Separation and Specific Detection of Some Herbicides Using the Nitrogen Phosphorus Detector

The specific response of the NPD to nitrogen and phosphorus, coupled with its relatively high sensitivity, makes it especially useful for the analysis of many pharmaceuticals and in particular in environmental analyses involving herbicides. Employing appropriate column systems traces of herbicides at the 500 pg level can easily be determined. An example of the separation of a series of herbicides monitored by the NPD is shown in figure 10.

Another interesting application of the NPD is in the GC analysis of the basic drugs.

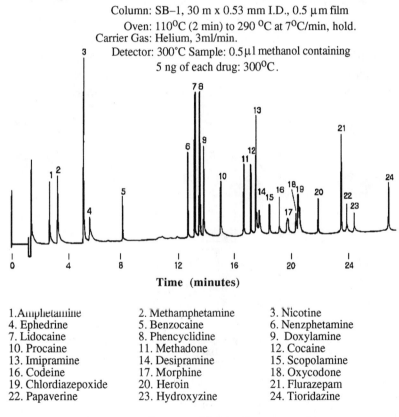

1. Amphetamine
4. Ephedrine
7. Lidocaine
10. Procaine
13. Imipramine
16. Codeine
19. Chlordiazepoxide
22. Papaverine

2. Methamphetamine
5. Benzocaine
8. Phencyclidine
11. Methadone
14. Desipramine
17. Morphine
20. Heroin
23. Hydroxyzine

3. Nicotine
6. Nenzphetamine
9. Doxylamine
12. Cocaine
15. Scopolamine
18. Oxycodone
21. Flurazepam
24. Tioridazine

Courtesy of Supelco Inc.

Figure 11 The Separation of Some Basic Drugs Using the NPD to Selectively Detect Compounds Containing Nitrogen

A chromatogram of a synthetic mixture of basic drugs using the NPD is shown in figure 11. Virtually all the basic drugs contain nitrogen and thus can be specifically detected among a large number of other unresolved compounds not containing nitrogen.

The Emissivity or Photometric Detector

The emissivity detector or what is now known as the Flame Photometric detector (FPD) was originally developed by Grant (9) in 1958 but was not produced commercially at the time as it could not compete in sensitivity with the ionization detectors. The emissivity detector, however, has some unique properties that can make its response quite specific, thus giving it certain unique areas of application. Grant originally employed it to differentiate aromatic from paraffinic hydrocarbons in coal tar products by measuring the luminosity that the aromatic nucleus imparted to the flame.

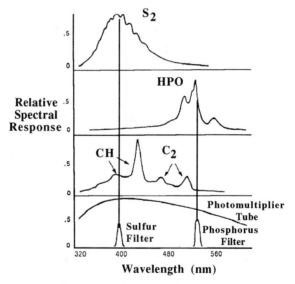

Courtesy of the Hewlett–Packard Corporation

Figure 12 Light Emission Wavelengths of Carbon and Hydrocarbons Containing Sulfur and Phosphorus

Modern photometric detectors do not usually monitor the total light emitted but only that emitted at selected wavelengths. Phosphorus and sulfur containing hydrocarbons produce chemi-luminescence at specific wavelengths when burnt in the hydrogen flame. The wavelengths of the light emitted by carbon and hydrocarbons containing sulfur and phosphorus are shown in figure 12. Included in the diagram is the transmission band for the filters used in the Hewlett-Packard FPD together with the response of the photo-multiplier sensor with which it is used. It is seen that the light emitted by sulfur compounds covers the range from about 320 nm to 480 nm and that of hydrocarbons alone between 390 nm and 520 nm. By choosing a filter having a transmission window between wavelengths of about 390 nm and 410 nm, the maximum emission from sulfur compounds is sensed by the photo-multiplier. Conversely, the emission from the background hydrocarbons is very small (albeit around a minor transmission maximum). In a similar way the emission of phosphorus compounds is seen to lie between 480 nm and 580 nm. Thus, by choosing a filter having a transmission window between wavelengths of about 520 nm and 560 nm, the maximum emission from phosphorus compounds can be sensed, and the emission from the background hydrocarbons will be almost negligible. It follows that by the use of appropriate filters sulfur and phosphorus compounds can be selectively detected.

A diagram of the Hewlett-Packard flame photometric detector sensor is shown in figure 13. In principle, the system is identical to that originally devised by Grant. The end of the capillary column is led into the flame jet where the column eluent mixes with the hydrogen flow and is burnt. The jet and the actual flame is shielded to prevent light from the flame itself falling directly on to the photo-multiplier. The base of the jet is heated to prevent vapor condensation. The light emitted above the flame, first passes through two heat filters and then through the wavelength selector filter and finally on to the photo-multiplier. The response of the detector to sulfur is fairly insensitive to changes in hydrogen flow rate but its response to phosphorus compounds shows a maximum at an optimum hydrogen flow rate, the magnitude of which varies with the air flow through the sensor.

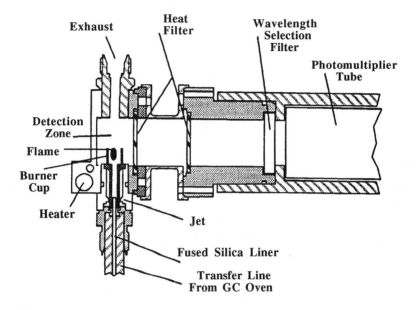

Courtesy of the Hewlett–Packard Corporation

Figure 13 The Hewlett–Packard Flame Photometric Detector

One serious problem that can occur in the FPD is the quenching or re-absorption of the light emitted by the selected species. Hydrocarbon quenching can occur when the peak of interest containing sulfur is co-eluted with another hydrocarbon in relatively high concentration. The high concentration of carbon dioxide appears to suppress the characteristic light emission from a sulfur compound. Quenching can also occur from excess of the selected species itself. It has been suggested that this can be due to collisional energy absorption, competing chemical reactions or the re-absorption of the photon by inactivated species. The latter, in the author's opinion, is the most likely.

The flame conditions can be critical and as the detection zone is above the flame, the gas flows and jet diameter must be optimized to ensure emission in the detection zone. Hydrogen flow rates, air flow rates and jet temperature must also be optimized to ensure maximum sensitivity and selectivity and to avoid sample condensation in the burner conduits. The whole of the burner system must be maintained at an elevated temperature to prevent water condensation that could fog the windows transmitting light to the photo-multiplier. An example of the use of the FPD in the analysis of thiophene and substituted thiophenes in a hydrocarbon mixture is shown in figure 14.

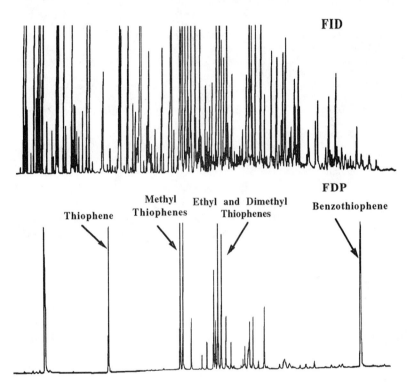

Courtesy of the Hewlett–Packard Corporation

Column: Ultra-1, Cross-linked methylsilicone, 50m long, 0.5 μm I.D.; Carrier gas linear velocity 26 cm/sec; Sample 3 ml split 400:1; Detectors, FPD, 393 nm filter an FID 250°C.

Figure 14 The Selective Detection of Some Sulfur Compounds in a Hydrocarbon Mixture Employing Flame Photometric Detection

It is seen that excellent selectivity is afforded by the FPD; the sulfur compounds are clearly and unambiguously singled out from the multitude of hydrocarbons present in the mixture thus simplifying the analysis and improving its accuracy.

References

1. J. Harley, W. Nel and V. Pretorious, *Nature, London,* **181**(1958)177.
2. I. G. McWilliams and R. A. Dewer, *"Gas Chromatography 1958"*, (Ed. D. H. Desty), Butterworths Scientific Publications (1957)142.
3. S. A. Beres, C. D. Halfmann, E. D. Katz and R. P. W. Scott, Analyst, **112**(1987)91.
4. L. Ongkiehong, *"Gas Chromatography 1960"* (Ed. R. P. W. Scott) Butterworths, London (1958)9.
5. D. H. Desty, A. Goldup and C. J. Geach, *"Gas Chromatography 1960"* (Ed. R. P. W. Scott) Butterworths, London (1958)156.
6. H, F. Calcote and I. R. King, "The Fifth Symposium on Combustion", New York (1955).
7. K. E. Schuler and J. J. Weber, J. Chem. Phys., 22(1994)491.
8. C. H. Reese, *Ph. D. Thesis,* University of London (Birkbeck College) (1992)
9. D. W. Grant, *"Gas Chromatography 1958"* (Ed. D. H. Desty), Butterworths Scientific Publications (1957)153.

CHAPTER 6

THE ARGON IONIZATION FAMILY OF DETECTORS
Argon Detectors, Helium Detectors and the Electron Capture Detector

Introduction

Around the late 1950s and early 1960s, Lovelock [1] developed a family of four detectors, starting with the argon ionization detector and ending with the electron capture detector [2]. The first of the family was called the *macro* argon detector (for reasons that will shortly become apparent) and although it is rarely used in GC today, it will be briefly described, as it is an extremely sensitive detector and its function has a bearing on that of the electron capture detector, which is a very important contemporary detector.

In the noble gases, the outer octet of electrons is complete and as a consequence, collisions between argon atoms and electrons are perfectly elastic. Consequently, if a high potential is set up between two electrodes in argon, and ionization is initiated by a suitable radioactive source, electrons will be accelerated towards the anode and will not be impeded by energy absorbed from collisions with argon atoms. However, if the potential of the anode is high enough, the electrons will develop sufficient kinetic energy that on collision with an argon atom energy can be absorbed and a *metastable* atom can be produced. A metastable atom carries *no* charge but adsorbs its energy from the collision with a high energy electron by the displacement of an electron to an outer orbit. This gives the metastable atom an energy of about 11.6 electron volts. Now 11.6 electron volts is sufficient to ionize most organic molecules. Hence, collision between a metastable argon atom and an organic molecule will result in the outer electron of the metastable atom collapsing back to its original orbit, followed by the expulsion of an electron from the organic molecule. The electrons produced by this process are collected at the anode generating a large increase in anode current.

However, when an ion is produced by collision between a metastable atom and an organic molecule, the electron, simultaneously produced, is immediately accelerated toward the anode. This results in a further increase in metastable atoms and a consequent increase in the ionization of the organic molecules. This cascade effect, unless controlled, results in an exponential increase in ion current.

The relationship between the ionization current and the concentration of vapor was deduced by Lovelock [3, 4] to be

$$I = \frac{CA(x+y) + Bx}{CA\{1 - a\exp[b(V-1)]\} + B}$$

where, A), (B), (a) and (b) are constants,
 (V) is the applied potential,
 (x) is the primary electron concentration,
and (y) is the initial concentration of metastable atoms.

The equation indicates (and experiments confirm) that the rapid increase in current flow with increasing vapor concentration will, in theory, lead to an infinite current at some finite vapor concentration. This would appear to render the argon ionization system impractical as a GC detector. It will be seen however that this cascade effect can be controlled by suitable electronic circuitry.

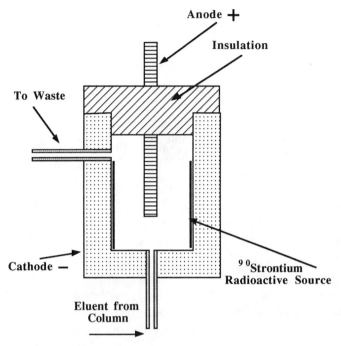

Figure 1 The Macro Argon Detector

The Simple or Macro Argon Detector Sensor

A diagram of the macro argon detector sensor is shown in figure 1. The cylindrical body is usually made of stainless steel and the insulator made of silica filled PTFE or, for high temperature operation, a suitable ceramic. The very first argon detector sensors used a tractor sparking plug as the electrode, the ceramic seal being a very efficient insulator at very high temperatures.

Inside the main cavity of the sensor is a ^{90}strontium source contained in silver foil. The surface layer on the foil that contained the radioactive material must be very thin or the β particles will not be able to leave the surface. This tenuous layer protecting the radioactive material is rather vulnerable to mechanical abrasion, which could result in radioactive contamination. The radioactive strength of the source is about 10 millicuries which for ^{90}strontium could be considered a *hot* source. The source had to be inserted under properly protected conditions. As already stated, the decay of ^{90}strontium is in two stages, each stage emitting a β particle producing the stable atom of ^{90}zirconium.

$$^{90}Sr \xrightarrow[0.6 \text{ MeV}]{\varepsilon} {}^{90}Y \xrightarrow[2.5 \text{ MeV}]{\varepsilon} {}^{90}Zr$$

half life ca 25 y — half life ca 60 h — Stable

The electrons produced by the radioactive source were accelerated under a potential that ranged from 800 to 2000 volts depending on the size of the sensor and the position of the electrodes. From the authors experience, if the main body had an I.D. of about 2 cm, a length of about 5 cm and a sparking plug was used as the electrode, the sensor would function well with an electrode potential of about 1200 volts. The power supply circuit that was used is shown in figure 2. The signal is taken across the 1×10^8 ohm resistor and as the standing current from the ionization of the argon is about 2×10^{-8} amp there is a standing voltage of 2 volts across it that requires 'backing off'.

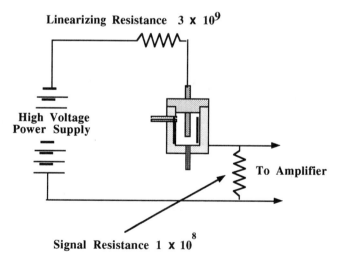

Figure 2 Power Supply Circuit to the Macro Argon Detector

As the current increases due to the presence of organic vapor, the voltage drop across the linearizing resistance also increases and reduces the voltage across the electrode. For example if 1300 volts is applied to the detector and when a solute is eluted, the current increases to 10^{-7} amp, this will cause a 300 volt drop across the linearizing resistance of 3×10^9 and consequently reduce the voltage across the electrodes to 1000 volts. In this way the natural exponential response of the detector can be made sensibly linear.

In a typical detector, the primary current consists of about 10^{11} electrons per second. Taking the charge on the electron as 1.6×10^{-19} coulombs this gives a current of 1.6×10^{-8} amp. According to Lovelock [1], if each of these electrons can generate 10,000 metastables on the way to the electrode, the steady state concentration of metastables will be about 10^{10} per ml (this assumes a life span for the metastables of about 10^{-5} seconds at NTP). From the kinetic theory of gases it can be calculated that the probability of collision between a metastable atom and an organic molecule will be about 1.6 : 1. This would lead to a very high ionization efficiency and Lovelock claims that in the more advanced sensors described below, ionization efficiencies of 10% have been achieved. The author can confirm that the macro argon detector can achieve ionization efficiencies of at least 0.5 %, which, compared with that of the FID, is very large indeed.

However the implied higher sensitivity is not realized. The sensitivity of the macro argon detector is about 10 times less than the FID (the minimum detectable concentration is an order of magnitude higher). The large increase that would appear to be expected by the much greater ionization efficiency is not realized. This is because the large primary current carries a high noise level compared with the FID (more than two orders of magnitude greater) and thus the signal to noise (which determines the sensitivity) is ten times less.

Although the argon detector is a very sensitive detector and can achieve ionization efficiencies of greater than 0.5%, the detector was not popular, largely because it was not linear over more than two orders of magnitude of concentration (0.98 < r > 1.02) and its response was not predictable. Furthermore, the early detectors employed a "hot radioactive source" to provide the ionization (^{90}strontium) and this was also viewed with considerable disfavor. Today, less active sources are available that will work perfectly well with the argon detector and are completely acceptable in the laboratory from the point of view of safety. In addition, modern solid state electronic linearizing circuits might well give it a much wider linear dynamic range and make it worthwhile to reconsider the argon detector for use in GC. In practice nearly all organic vapors and most inorganic vapors have ionization potentials of less than 11.6 electron volts and thus are detected. The short list of substances that are not detected include H_2, N_2, O_2, CO_2, $(CN)_2$, H_2O and fluorocarbons. The compounds methane, ethane, acetonitrile and propionitrile have ionization potentials well above 11.6 electron volts; nevertheless, they do provide a slight response between 1 and 10% of that for other compounds. The poor response to acetonitrile makes this substance a convenient solvent in which to dissolve the sample before injection on the column. It is also interesting to note that the inorganic gases H_2S, NO, NO_2, NH_3, PH_3, BF_3 and many others respond normally in the argon detector. As these are the type of substances that are important in environmental contamination, it is surprising the argon detector, with its very high sensitivity for these substances, has not been re-examined for use in environmental analysis. The sensitivity of the macro argon detector is 4×10^{-11} g/ml about an order of magnitude less than the FID and, as already discussed, this decrease, despite the much greater ionization efficiency, largely results from its extremely high noise level. The main technical disadvantage of the argon detector was its large sensor volume and this led Lovelock to design the micro argon detector.

The Micro Argon Detector

A diagram of the micro argon detector sensor is shown in figure 3. This sensor was designed to have a very small *"effective"* volume so that it could be used with capillary columns where the flow rate may be as low as 0.1 ml/min or less. In the micro argon detector sensor, the anode is withdrawn into a small cavity about 2.5 mm in diameter.

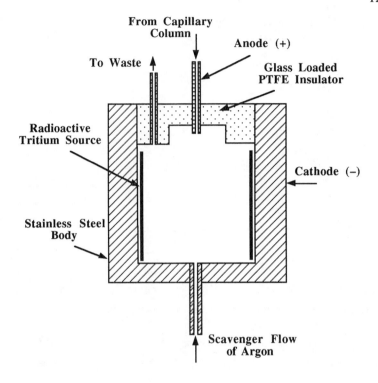

Figure 3 The Micro Argon Detector

This ensures that the electrons can only reach the anode along a restricted path and also changes the electric field around the electrode so that a greater part of it resides within a few diameters of the anode tip. The anode is tubular in form and the capillary column can slide up inside the anode until it is within a millimeter or so of the electric field. A cloud of metastable argon atoms is formed around the anode tip and thus any solute molecules eluted from the column immediately pass through this cloud and are ionized. At the other end of the sensor is another inlet that provides a scavenger flow of argon that rapidly removes the solute from the cell through two holes at the bottom of the anode cavity. This procedure reduces the effective sensor volume to less than a microliter and thus allows the efficient use of a capillary column. The radioactive source originally used was about 25 microcuries of radium (an α particle source).

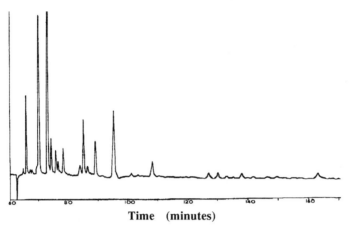

Time (minutes)

Column: 1000 ft of Nylon tubing 0.020 in I.D. Temperature 23°C. Column efficiency for butane eluted at 65 minutes 0.75 x 10^6 theoretical plates

Figure 4 The Separation of a Hydrocarbon Mixture on a Nylon Capillary Column Using the Micro Argon Detector

Although the radioactive source was very small (defined in those days as about a "wrist watch", as equivalent quantities of radium were used to produce the luminous dials of many watches at that time) it was subsequently recognized that exposure to α particles could cause a health problem. Eventually radium was replaced by tritium (a very weak β ray emitter) which, although a fairly strong source (sometimes as much as one curie was used), it was relatively harmless from the point of view of radiation energy. However, it is also somewhat unstable at high temperatures causing loss of tritium to the air and consequent atmospheric contamination. In due course the tritium was replaced by Ni^{63} (another, more energetic β–ray source) but a fairly safe source that can be operated at relatively high temperatures without fear of contamination. Employing radium as the active source, the micro argon detector was used by Scott [5], to examine the possibilities of long Nylon capillary columns for the separation of hydrocarbon mixtures. The capillary column used was 1000 ft long giving 0.75×10^6 theoretical plates. A chromatogram of the separation is shown in figure 4. It is seen that the sensor volume has no effect on the column performance and the detector is now very suitable for use with capillary columns. The modifications carried out to reduce the effective sensor volume, unfortunately, did not improve its linearity nor increase its linear range. However, the noise level was reduced by about two orders of magnitude and thus the sensitivity was commensurably increased by the same amount making it 10 times more sensitive than the FID. A distinct improvement over the macro argon detector.

The Triode Detector

The triode detector is a modification of the micro argon detector and a diagram of the triode detector sensor is shown in figure 5.

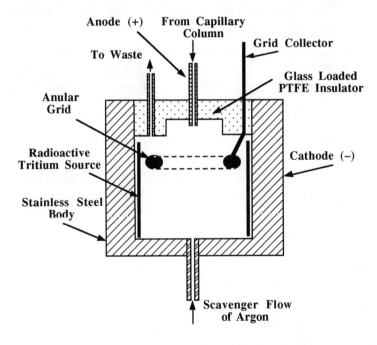

Figure 5 The Triode Detector

The primary electron stream approaching the anode in the micro argon detector sensor is confined to a narrow tubular channel less than 2 mm in diameter. In the triode detector sensor, a ring electrode is introduced co-axially with the anode but about 4 mm inside the chamber. This electrode collects very few of the primary electrons but nearly all the positive ions liberated in the reaction between the organic molecules and the metastable atoms. The performance of the triode is identical to that of the micro argon detector except that the background current of the primary electron stream is separated from the signal current. As the noise level is related to the magnitude of the background current, it is significantly reduced in the triode detector and the sensitivity is increased by another order of magnitude relative to the micro argon detector. This means that the triode detector is nearly two orders of magnitude more sensitive than the FID.

Table 1 The Relative Characteristics of the Three Different Argon Detectors

Measurement Conditions and Characteristics	Detector		
	Macro	**Micro**	**Triode**
Gas Flow Rate ml/min.	60	2.2	2.2
Applied Potential	1,200	1,200	1,200
Frequency Range (Hz)	0 – 0.3	0 – 1	0 – 1
Radiation Source	Radium R 0.07 mcurie	Tritium ca 50 mcurie	Tritium ca 50 mcurie
Background Current (A)	7×10^{-9}	1.2×10^{-8}	3×10^{-10}
Noise Level (A)	10^{-11}	10^{-12}	$<10^{-13}$
Ionization Efficiency(%)	0.05	0.5	0.5
Min. Detectable (g/s)	4×10^{-11}	4×10^{-13}	2×10^{-14}
Min. Detectable (g/ml)	4×10^{-11}	1.2×10^{-12}	6×10^{-14}
Max. Current Consistent with a Linear Response	7×10^{-8}	3×10^{-7}	4×10^{-7}
Linearity Deviation for 1000-fold concentration change (%)	3.1	1.2	1.2

In operation the cathode is connected to earth, the collector gives a positive signal to the amplifier and the anode is connected to the positive power supply. The specifications of all three argon detectors are given in Table 1.

The Thermal Argon Detector

Beres *et al* (6) showed that the argon detector could be made to function without a radioactive source or other electron producing device providing the argon and sensor system was operated at temperatures above 150°C. Glass becomes conducting at temperatures of 150°C and above, and so glass could be employed as one of the electrodes. A diagram of one form of the sensor is shown in figure 6.

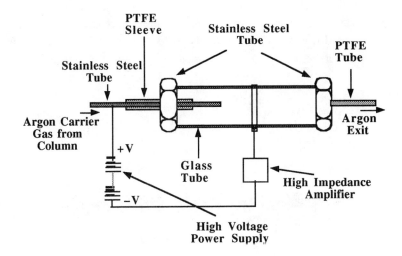

Figure 6 The Thermal Argon Detector Sensor

The argon from the column passes through a stainless steel tube, which acts as the anode, into a cylindrical glass tube held at 150°C or above. The tube is insulated from the glass tube by a PTFE sleeve. The argon exits from the sensor by a length of PTFE tube. A metal band round the glass acts as an electrical connection to the amplifier, the other input of the amplifier being connected to the −ve side of the power supply. The +ve side of the power supply is connected to the metal tubular anode. Electrons, thermally emitted from the glass surface, are accelerated under the high potential and on collision with argon atoms produce metastable atoms in the usual manner which collect round the anode. Organic vapors are sensed in exactly the same way as the normal argon detector, ionization being produced by collision between the organic molecules and the metastable argon atoms. The electrons and organic ions produced are collected and the resulting current is monitored by the high impedance amplifier. The performance of the detector using potentials ranging from about 600 V to 1500 V and sensor temperatures of 150°C, 200°C and 250°C were reported. The authors claimed sensitivities at least as good as those of the FID and a linear range of three orders of magnitude or more (0.98 < r < 1.02). Employing the propensity for glass to conduct at elevated temperatures they also designed a sensor completely free of metal and demonstrated it worked as effectively as the sensor shown in figure 6. A diagram of the metal-free detector sensor is shown in figure 7.

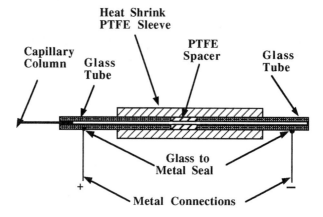

Figure 7 The Sensor of the Metal–Free Argon Detector

It is seen that a very inert type of sensor can be constructed having no radioactive source or other separate electron generator and yet provide a high sensitivity to all compounds that have ionization potentials of less than 11.5 electron volts. The system was found to work equally effectively with helium as the carrier and sensor gases showing that metastable helium atoms could also be produced in the same manner.

A chromatogram demonstrating the separation of some hydrocarbons using the thermal-helium detector and the metal-free sensor displayed in figure 7, is shown in figure 8. Included in figure 8 is the same separation monitored simultaneously with an FID.

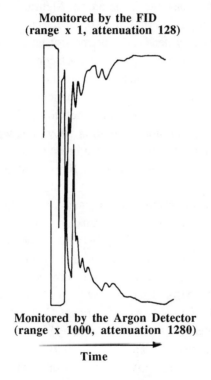

Monitored by the FID
(range x 1, attenuation 128)

Monitored by the Argon Detector
(range x 1000, attenuation 1280)

Time

Figure 8 The Separation of a Mixture of Hydrocarbons Monitored by the Thermal Argon Detector and the FID

It is seen that the performance of the two detectors is very similar. The thermal argon detector is not, presently, commercially available. However, its high sensitivity, freedom from radioactivity and electron producing ancillaries make it a very simple detector to fabricate and operate. The fact that the sensor can be constructed from very inert material and thus accommodate very corrosive gases could also be an advantage for certain applications.

The Helium Detector

The helium detector works on exactly the same principle as the argon detector, but metastable helium atoms are produced by the accelerated electrons instead of metastable argon atoms. Metastable helium atoms, however, have an energy of 19.8 and 20.6 electron volts and thus can ionize, and consequently detect, the permanent gases and, in fact, the molecules of all other volatile substances. As a consequence, the helium must be extremely pure or the metastable helium atom production will be quenched by traces of any other permanent gases that may be present. Originally a very complicated helium purifying chain was necessary to ensure its optimum operation. However, with high purity helium becoming generally available, the detector can now be used to detect concentrations of organic vapors at 10^{-13} g/ml or less.

The metastable atoms that must be produced in the argon and helium detectors need not necessarily be generated from electrons induced by radioactive decay. Electrons can be generated by electric discharge or photometrically and these can be accelerated in an inert gas atmosphere under an appropriate electrical potential to produce metastable atoms. This procedure is the basis of a highly sensitive helium detector that is manufactured by the GOW-MAC Instrument Company. The detector does not depend solely on metastable helium atoms for ionization and for this reason is called the Helium Discharge Ionization Detector (HDID). A diagram of the GOW-MAC sensor is shown in figure 9. The sensor consists essentially of two cavities, one carrying electrodes with a potential difference of about 550 volts which initiates a gas discharge when a flow of helium gas passes through the chamber. The discharge gas passes into a second chamber that acts as the ionization chamber and any ions formed are collected by two plate-electrodes having a potential difference of about 160 v. The column eluent enters the top of the ionization chamber and mixes with the helium from the discharge chamber and exits at the base of the ionization chamber.

Ionization probably occurs as a result of a number of ionization processes.

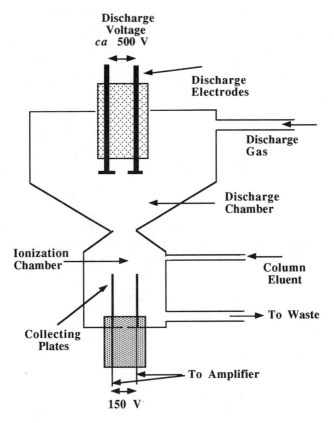

Courtesy of GOW-MAC Instruments

Figure 9 The Discharge Ionization Detector

The electric discharge produces both electrons and photons. The electrons can be accelerated to produce metastable helium atoms, which in turn can ionize the components in the column eluent. However, the photons generated in the discharge have, themselves, sufficient energy to ionize many eluent components and so ions will probably be produced by both mechanisms. It is possible that other ionization processes may also be involved, but the two mentioned are likely to account for the majority of ions produced. The response of the detector depends on the collecting voltage and is very sensitive to traces of inert gases in the carrier gas. Peak reversal is often experienced at high collecting voltages, which may also indicate some form of electron capturing may take place between the collecting electrodes. This peak reversal appears to be controllable by the introduction of traces of neon in the helium carrier gas. The helium discharge ionization detector is a relatively new detector but has already demonstrated high sensitivity to the permanent gases and has been very successfully used for the analysis of trace components in ultra-pure gases. Nevertheless, the conditions necessary for the efficient ionization of eluted solutes in the sensor will, without doubt, be further optimized in due course, which will make the detector easier to operate. Linearity data is a little scarce as yet, but it would appear that the detector response is linear over at least two and possible three orders of magnitude with a response index probably lying between 0.97 and 1.03.

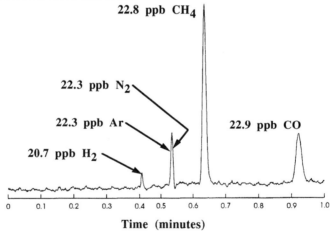

Courtesy of GOW-MAC Instruments

System: Capillary Chromatograph Series 590; Column: GS MoleSeve®, 30m x 0.55 mm; Carrier gas: helium, ionizing gas 78.6 ml/min, ionizing flow, 21.1 ml/min. Ionization voltage 524 V, sample volume 0.25 ml

Figure 10 The Analysis of a Sample of Helium

135

Any inherent non–linearity of the sensor can be corrected by an appropriate signal modifying amplifier. An example of the use of the detector to analyze a sample of helium is shown in figure 10. The high sensitivity of the detector to traces of the permanent gases is clearly demonstrated.

The Pulsed Helium Discharge Detector

The pulsed helium discharge detector [7,8] is an extension of the helium detector, a diagram of which is shown in figure 11.

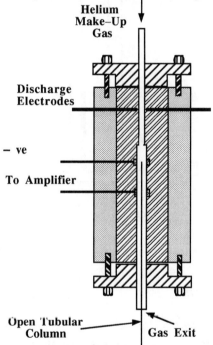

Courtesy of Valco Instruments Company Inc.

Figure 11 The Pulsed Helium Discharge Detector

The detector has two sections: the upper section consisting of a tube 1.6 mm I.D. (where the discharge takes place) and the lower section, 3 mm I.D. (where reaction with metastable helium atoms and photons takes place). Helium make–up gas enters the top of the sensor and passes into the discharge section. The potential (about 20 V) applied across the discharge electrodes is pulsed at about 3 kHz with a discharge pulse-width of about 45 μs for optimum performance. The discharge produces electrons and high energy photons (that can also produce electrons), and probably some metastable helium atoms.

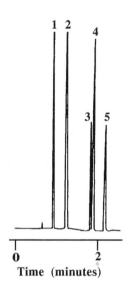

Column; J & W DB–1701, 10 m x 0.05 mm, film thickness 0.05 μm; Flow rate 20 ml/min. Sample split 1:150;1, benzene; 2, toluene; 3, ethylbenzene; 4, *m* and *p* xylene; 5, *o*-xylene

Figure 12 The Separation of Some Aromatic Hydrocarbons Monitored by the Pulsed Helium Discharge Detector

The photons and metastable helium atoms enter the reaction zone where they meet the eluent from the capillary column. The solute molecules are ionized and the electrons produced are collected at the lower electrode and measured by an appropriate high impedance amplifier. The distance between the collecting electrodes is about 1.5 mm. The helium must be 99.9995 pure. The base current ranges from

1 x 10^{-9} to 5 x 10^{-9} amp, the noise level is about 1.2 x 10^{-13} amp and the ionization efficiency is about 0.07%. It is claimed to be about 10 times more sensitive than the flame ionization detector and to have a linear dynamic range of 10^5. An example of the use of a pulsed helium discharge detector for monitoring the separation of some aromatics on a capillary column is shown in figure 12. The pulsed helium discharge detector appears to be an attractive alternative to the flame ionization detector and would eliminate the need for three different gas supplies. It does, however, require equipment to provide specially purified helium, which diminishes the advantage of using a single gas.

The Electron Capture Detector

The development of the ionization detectors by Lovelock that evolved from the original argon detector culminated in the invention of the *electron capture detector* [2]. However, the electron capture detector operates on a different principle from that of the argon detector. A low energy β-ray source is used in the sensor to produce electrons and ions. The first source to be used was tritium absorbed into a silver foil but, due to its relative instability at high temperatures, this was quickly replaced by the far more thermally stable ^{63}Ni source.

The detector can be made to function in two ways: either a constant potential is applied across the sensor electrodes (the DC mode) or a pulsed potential is used (the pulsed mode). In the DC mode, a constant electrode potential of a few volts is employed that is just sufficient to collect all the electrons that are produced and provide a small standing current. If an electron capturing molecule (for example a molecule containing a halogen atom which has only seven electrons in its outer shell) enters the sensor, the electrons are captured by the molecules and the molecules become charged. The mobility of the captured electrons are much reduced compared with the free electrons and, furthermore, are more likely to be neutralized by collision with any positive ions that are also generated. As a consequence, the electrode current falls dramatically.

In the pulsed mode of operation, which is usually the preferred mode, a mixture of methane in argon is usually employed as the carrier gas. Pure argon can not be used very effectively as the carrier gas as the diffusion rate of electrons in argon is ten times less than that in a 10% methane-90% argon mixture. The period of the pulsed potential is adjusted such that relatively few of the slow negatively charged molecules reach the anode, but the faster moving electrons are all collected.

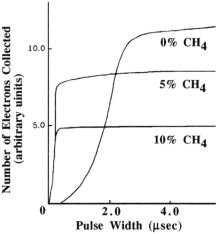

Figure 13 Curves Relating Electron Collection to Pulse Width for Carrier Gases Containing Different Amounts of Methane

During the "off period" the electrons re-establish equilibrium with the gas. The three operating variables are the pulse duration, pulse frequency and pulse amplitude. The relationship between the number of electrons collected and the collecting time (the pulse width) is shown in figure 13. It is seen that with no methane present electron collection takes nearly 3 μsec to complete. However, with 5% or 10% of methane present in the argon all the electrons are collected in less than 1 μsec. This reflects the increased diffusion rates of the electrons in argon-methane mixtures. By appropriate adjustment of the pulse characteristics, the current can be made to reflect the relative mobilities of the different charged species in the cell and thus exercise some discrimination between different electron capturing materials. In general use, however, the pulse width is usually set at about 1 μsec and the frequency of the pulses at about 1 kHz. This allows about 1 millisecond for the sensor to re-establish equilibrium in the cell before the next electron collection occurs. The sampling and equilibrium periods are depicted in figure 14.

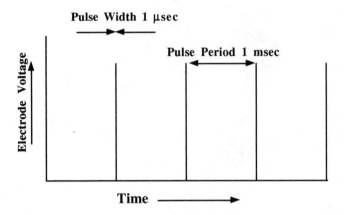

Figure 14 Common Pulse Pattern for the Electron Capture Detector Operated in the Pulsed Mode

The collecting potential applied across the electrodes of the sensor at each pulse is usually about 30 V but depends on the geometry of the sensor and the strength of the radioactive source. The average current resulting from the electrons collected at each pulse gives a standing current of about 1×10^{-8} amp with an associated noise level of about 5×10^{-12} amp both of which will also vary with the strength of the radioactive source that is used. A diagram of the electron capture detector as developed by Lovelock *et al* is shown in figure 15. Their sensor consisted of a small chamber, one or two milliliters in volume, with metal ends separated by a suitable insulator. The metal ends acted both as electrodes and as fluid conduits for the carrier gas to enter and leave the cell. The cell contained the radioactive source, usually electrically connected to the conduit through which the carrier gas enters and to the negative side of the power supply. A gauze "diffuser" was connected to the exit of the cell and to the positive side of the power supply. The electrode current was measured by a high impedance amplifier using thermionic valves similar to those originally used in the original FID and the argon detectors.

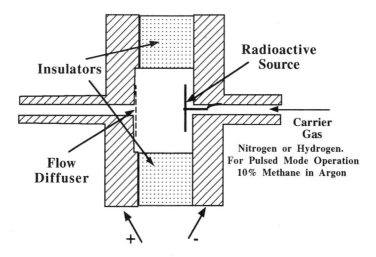

Figure 15   The Electron Capture Detector

An example of a more modern sensor design is that used in the Perkin Elmer electron capture detector and is shown in figure 16. The sensor is designed to operate with oxygen–free nitrogen or argon/methane mixtures. The active source is ^{63}Ni that has a long life and is stable up to 450ºC. The detector can be used for the analysis of aqueous samples as steam has no effect on the source in the sensor. The sensor in the Perkin Elmer system is thermostatted in a separate oven which can be operated at temperatures ranging from 100ºC to 450ºC. The column is connected to the sensor at the base and make–up gas can be introduced into the base of the detector if open tubular columns are employed as these columns are usually operated with hydrogen or helium as the carrier gas. The electron capture detector is extremely sensitive, probably one of the most sensitive GC detectors available (minimum detectable concentration $ca.\, 10^{-13}$ g/ml) and is widely used in analysis of halogenated compounds, in particular, pesticides. Unfortunately, its sensitivity is often given in terms of the minimum mass of solute eluted, which can be a little misleading. The detector is concentration sensitive and thus the concentration of the solute for a given mass will vary with the position it is eluted in the chromatogram (for a *given mass of solute*, an early peak would be narrow and have a small volume and a high concentration at the peak maximum: if eluted as a late peak it would be broad, have a relatively large volume and a lower concentration at the peak maximum). Consequently, a mass of solute just identifiable (signal to noise = 2) when eluted as an early peak would not be detected or discerned when eluted as a late peak

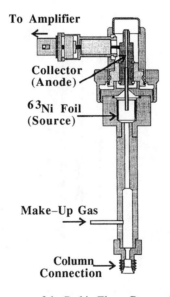

Courtesy of the Perkin Elmer Corporation

Figure 16 The Perkin Elmer Electron Capture Detector

This again emphasizes the need for an improved procedure for defining detector specifications. The linear dynamic range of the electron capture detector is again ill-defined by many manufacturers. In the DC mode the linear dynamic range is usually relatively small, perhaps two orders of magnitude, with the response index lying between 0.97 and 1.03. The pulsed mode has a much wider linear dynamic range and values up to 5 orders of magnitude have been reported. The linear dynamic range will change with the strength of the radioactive source and the detector geometry. The values reported will also depend on how the linearity is measured and defined. If a response index lying between 0.98 and 1.02 is assumed, then a linear dynamic range of at least three orders of magnitude should be obtainable from most electron capture detectors.

An example of a pesticide analysis employing an electron capture detector to monitor the separation is shown in figure 17.

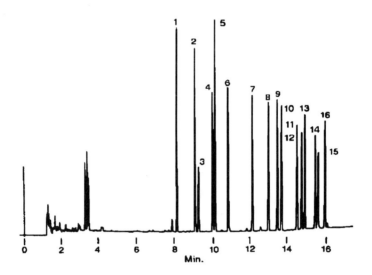

1 α–BHC	2 γ-BHC (Lindane)	3 β-BHC	4 Heptachlor
5 δ-BHC	6 Aldrin	7 Heptachlor Epox.	8 Endosulphan
9 p,p'-DDE	10 Dieldrin	11 Endrin	12 p,p'-DDD
13 Endosulphan 11	14 p,p'-DDt	15 Endrin Aldehyde	16 Endosulp. Sulf.

Courtesy of Supelco Inc.

Figure 17 Analysis of Priority Pollutant Pesticides by Method 608

The column used was a SPB-608 fused silica capillary column, 30 m x 0.53 mm I.D. with a 0.5 µm film of stationary phase. It was programmed from 50°C at 1°/min to 150°C and then at 8°/min to 260°C. Helium was used as the carrier gas at a flow rate of 5 ml/min and the sample consisted of 0.6 µl of a solution of the pollutants in n-decane. The mass of each pollutant present was about 120 pg. The electron capture detector is also used extensively for monitoring the separation of polychlorinated hydrocarbons and in particular the herbicides.

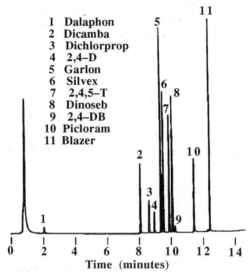

Column: SPB–608, 15 m x 0.53 mm I.D., 0.5 mm film, Oven: 60°C (1 min.) to 289°C at 16°C/min, hold 5 min. Carrier gas: helium, 5 ml/min. Detector temperature 310°C. Sample: chlorophenoxymethylesters

Figure 18 The Separation of Some Herbicides Using Electron Capture Detection

The separation of some common herbicides monitored by the electron capture detector is shown in figure 18.

The Pulsed Discharge Electron Capture Detector

The pulsed discharge electron capture detector is an extension of the previously discussed pulsed discharge helium ionization detector, a diagram of which is shown in figure 19. The detector functions in exactly the same as that of the traditional electron capture detector but differs in the method of electron production.

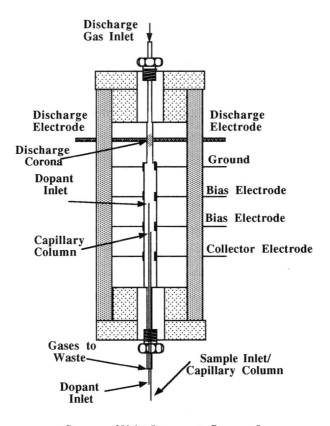

Courtesy of Valco Instruments Company Inc.

Figure 19 The Pulsed Discharge Electron Capture Detector

The sensor consists of two sections: the upper section where the discharge takes place has a small diameter and the lower section where the column eluent is sensed and the electron capturing occurs has a wider diameter. As with the pulsed discharge helium ionization detector, the potential across the discharge electrodes is pulsed at about 3 kHz with a discharge pulse width of about 45 μsec for optimum performance. The discharge produces electrons and high energy photons (which can also produce electrons), and some metastable helium atoms. The helium doped with propane enters just below the second electrode, metastable atoms are removed and electrons are generated both by the decay of the metastable atoms and by the photons. The electrons are attracted by appropriate potentials applied to each electrode into the section between the third and forth electrode and finally collected on the forth electrode. The collector electrode potential (the potential between the third and forth electrodes) is pulsed at about 3 kHz with a pulse width of about 23 μsec and has a pulse height of about 30 V. The device functions in the same way as the conventional electron capture detector with a radioactive source. The column eluent enters just below the third electrode, any electron capturing substance present removes some of the free electrons and the current collected by the fourth electrode falls. A photograph of the Valco pulsed discharge electron capture detector is shown in figure 20.

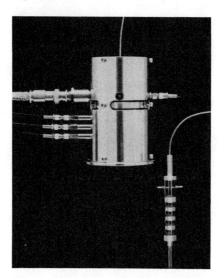

Courtesy of Valco Instruments Company Inc.

Figure 20   The Valco Pulsed Discharge Electron Capture Detector

The sensitivity claimed for the detector is 0.2 to 1.0 ng [9] but (as already discussed) this is not very informative as its significance depends on the characteristics of the column used and on the k' of the solute peak on which the measurements were made.

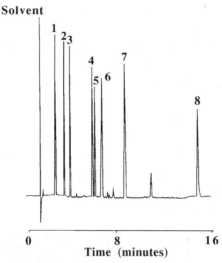

Column DB-6, 30m x 0.25 mm; df 0.25 μm

1	Lindane	81 pg	2	Heptachlor	162 pg
3	Aldrin	162 pg	4	p,p'-DDE	162 pg
5	Dieldrin	162 pg	6	Endrin	324 pg
7	o,p'-DDT	324 pg	8	Mirex	486 pg

Courtesy of Valco Instruments Company Inc.

Figure 21 The Separation of Some Pesticides Monitored by the Valco Pulsed Discharge Electron Capture Detector

The sensitivity should be given as that solute *concentration* that produces a signal equivalent to twice the noise. Such data allows a rational comparison between detectors. The linear dynamic range is also not precisely clear from the original publication but appears to be at least three orders of magnitude for a response index of (r) where $0.97 < r < 1.03$, but this is an estimate from the data published. The modified form of the electron capture detector, devoid of a radioactive source, is obviously an attractive alternative to the conventional device and appears to have similar, if not better, performance characteristics. Its real value will emerge after it has been used for some time in routine analysis. An example of the use of the pulsed discharge electron capture detector to monitor the separation of a mixture of pesticides is shown in figure 21.

In general, the electron capture detector is used extensively in forensic analyses and in environmental chemistry. It is very simple to use and is one of the least expensive, selective detectors available.

References

1. J. E. Lovelock, "Gas Chromatography 1960", Butterworths Scientific (Ed. R. P. W. Scott) Publications Ltd., London, (1960)9.
2. J.E.Lovelock and S.R.Lipsky, *J. Amer. Chem. Soc.* **82**(1960)431.
3. J. E. Lovelock, *J. Chromatogr.* **1**(1958)35.
4. J. E. Lovelock, *Nature, Lond.* *181(1958)1460.*
5. R. P. W. Scott, *Nature, Lond*, **183**(1959)1753.
6. S. A. Beres, C. D. Halfman, E. D. Katz and R. P. W. Scott, **112**(1987)91.
7. W. E. Wentworth, S. V. Vasnin, S. D. Stearns and C. J. Meyer, *Chromatographia,* **34** (1992)219.
8. W. E. Wentworth, Huamin Cai and S. D. Stearns, *J. Chromatogr.*, **688**(1994)135.
9. W. E. Wentworth, Ela Desal D'Sa and Huamin Cai, *J. Chromatogr. Sci.,* **30**(1992)478

CHAPTER 7

THE KATHEROMETER AND SOME OF THE LESS WELL KNOWN DETECTORS

The katherometer detector [sometimes spelt catherometer and often referred to as the *thermal conductivity detector* (TCD) or the *hot wire detector* (HWD)] is the oldest GC detector still in use and commercially available. Unfortunately, it is a relatively insensitive detector and has survived largely as a result of its catholic response. In particular, it is sensitive to the permanent gases that are not easily detected by other means. The frequent need for permanent gas analysis in industry probably accounts for it becoming the fourth most commonly used GC detector despite its relatively low sensitivity. Besides its use in gas analysis, the katherometer is at times also used for general GC analysis. It is simple in design and requires minimal electronic support and, as a consequence, is also relatively inexpensive compared with other detectors, which may also account for its somewhat surprising popularity. Universities providing programs on separation science commonly use the katherometer in their practical GC courses.

The katherometer in the late 1940s and early 1950s was originally used for measuring the amount of carbon dioxide in flue gases. However, with the advent of GC, its possible use as a detector was investigated by Ray [1]. He found it to be a very effective GC detector and, at the time, simpler to make than the gas density bridge and with about the same sensitivity and linearity. For a while it was the only detector that was commercially available. Initially, the manner in which it functioned was the subject of some controversy and it was not clear whether it responded to changes in the *thermal conductivity* or the *specific heat* of the column eluent. The response of the detector was examined in detail by Mellor [2] and Harvey and Morgan [3] in 1956 and it would appear that no such detailed studies has been carried out since. The net conclusion appears to have been that the katherometer responds to both changes in he thermal conductivity *and* the specific heat of its surroundings. It is possible that one or the other property may dominate in any particular system depending on the operating conditions employed. However, the relationship is not simple and it is not possible to accurately predict the response of the detector from a knowledge of the specific heat and thermal conductivity of the gases or vapors involved.

The katherometer functions in the following manner. A filament carrying a current is situated in the column eluent and, under equilibrium conditions, the heat generated in the filament is equal to the heat lost and consequently the filament assumes a constant temperature. The filament is constructed from a metal that has a high temperature coefficient of resistance and at the equilibrium temperature, the resistance of the filament and thus the potential across it will be constant. As already stated, the heat lost from the filament will depend on both the thermal conductivity of the gas and its specific heat. Both these parameters will change in the presence of a different gas or solute vapor and as a result the temperature of the filament changes, causing a change in potential across the filament. This potential change is amplified and either fed to a suitable recorder or passed to an appropriate data acquisition system.

As the detector filament is in thermal equilibrium with its surroundings and the device actually responds to the heat lost from the filament, the katherometer detector is extremely *flow* and *pressure* sensitive. Consequently, all katherometer detectors must be carefully thermostatted and must be fitted with reference cells to help compensate for changes in pressure or flow rate.

There are two types of katherometer, the "in-line" cell where the column eluent actually passes directly over the filament and the "off-line" cell where the filaments are situated away from the main carrier gas stream and the gases or vapors only reach the sensing element by diffusion. Due to the high diffusivity of vapors in gases, the diffusion process can be considered as almost instantaneous. A diagram of an in-line katherometer is shown in figure 1.

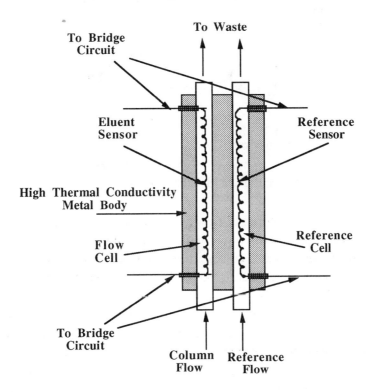

Figure 1 The Katherometer Sensor Cells (In-Line Cell)

The filament wire is usually tungsten or platinum as both metals have reasonably high temperature coefficients of resistance and at the same time are very inert and unlikely to interact chemically with the column eluent. Both filaments are situated in the arms of a Wheatstone Bridge as shown in figure 2 and a suitable current is passed through the filaments to heat them significantly above ambient temperature. The sensors and their conduits are installed in a high thermal conductivity metal block which is thermostatted by means of a separate oven. This is necessary to ensure temperature stability.

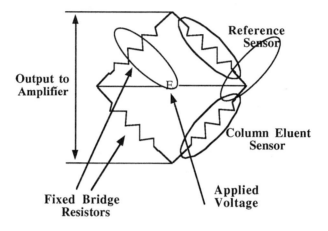

Figure 2   Katherometer Bridge Circuit

The out-of-balance signal caused by the presence of sample vapor in contact with the sensor is amplified and fed to a recorder or computer data acquisition system. For maximum sensitivity hydrogen should be used as the carrier gas, but to reduce fire hazards, helium can be used with very little compromise in sensitivity. The sensitivity of the katherometer is only about 10^{-6} g/ml (probably the least sensitive of all GC detectors) and has a linear dynamic range of about 500 (the response index being between 0.98 and 1.02). Although the least glamorous, this detector can be used in most GC analyses that utilize packed columns and where there is no limitation in sample availability. The device is simple, reliable, and rugged and, as already stated, relatively inexpensive.

The off–line sensor behaves in an exactly similar manner to the in–line device but is less flow sensitive as the sensors are situated out of the main flow of carrier gas. The off–line arrangement, however, does not change the sensitivity of the device to changes in pressure in the sensor. An example of the off-line katherometer cell is shown in figure 12. It consists of two conduits formed by two cylindrical holes drilled in a high conductivity metal block that carry the reference and column flow. The sensor elements are situated in small chambers adjacent to, and connected with, the main conduits of gas through the block.

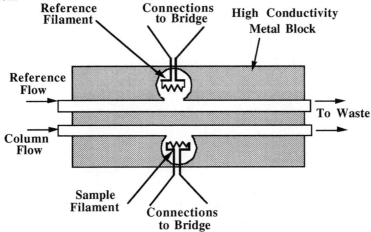

Figure 3   The Katherometer Sensor Cells (Off-Line Cell)

The filaments are situated in the arms of an electrical bridge in a similar manner to the in–line sensor, and the out-of-balance signal is amplified and fed to a recorder and/or an A/D converter and thence to a computer. Both systems work well and there seems to be no consensus favoring one more than the other.

An example of the use of a katherometer in the analysis of a mixture of the various compounds of hydrogen, deuterium and tritium, employing gas solid chromatography is shown in figure 4. The stationary phase was activated alumina [treated with $Fe(OH)_2$], and the column was 3 m long and 4 mm I.D. The carrier gas was neon, the flow rate 200 ml/min (at atmospheric pressure) and the column temperature was -196°C.

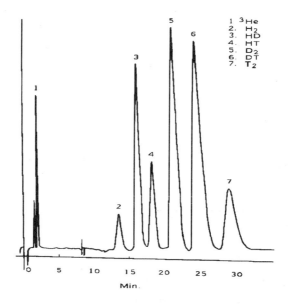

Courtesy of Supelco Inc.

Figure 4 The Separation of the Compounds of Hydrogen, Deuterium and Tritium

Note this high flow rate is necessary to ensure the correct flow rate at the column temperature of -196°C which will be approximately 40 ml/min. It is seen that an excellent separation is obtained and that the katherometer responds well to the various compounds of hydrogen and deuterium.

Another typical example of the use of the katherometer in the analysis of gas mixtures is afforded by the separation of the components of the Scott gas mixture 237. This is a standard mixture which consists of a mixture of oxygen, carbon monoxide, methane, and carbon dioxide in an excess of nitrogen. The sample is a typical mixture of gases that are liable to be found in automobile exhaust fumes and is used to test emission analyzing equipment and gas analyses apparatus. An example of such a separation carried out on a proprietary packing at 25°C is shown in figure 5.

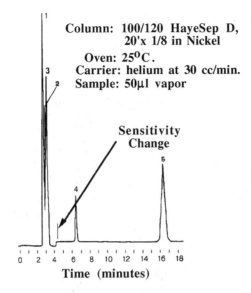

Courtesy of Hayes Separations Inc.

1 Nitrogen (balance)
2 Oxygen (7%)
3 Carbon Monoxide (7%)
4 Methane (4.5%)
5 Carbon Dioxide (15%)

Figure 5 The Separation of Scott Gas Mixture 237

It is seen that a sensitivity change is made on the katherometer detector after the carbon monoxide had been eluted. The separation is adequate and is efficiently monitored by the katherometer.

The Simple Gas Density Balance

The original gas density balance was invented by James and Martin [4] and, as already stated, was complicated and difficult to fabricate and its manufacture was not a commercial success. In the early days of chromatography GOW-MAC developed some elegantly designed filaments for use in the construction of katherometers, which, in due course, were used in many other manufacturer's katherometer products. These sensing filaments were rugged and highly reliable and were used by GOW–MAC to emulate Martin's density balance in a simple form. A diagram of the GOW-MAC gas density balance is shown in figure 6.

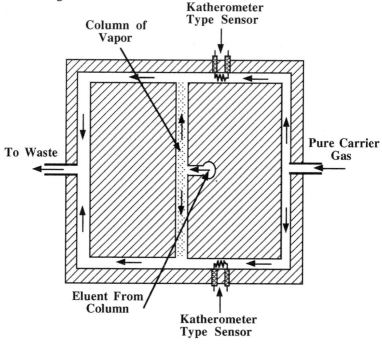

Courtesy of GOW-MAC

Figure 6 The GOW-MAC Gas Density Balance

The sensor consists of a bridge of tubes with three vertical tubes all connected by horizontal tubes at the top and the bottom. Pure carrier gas enters at the center of the right hand vertical tube and splits into two streams one passing along the lower horizontal tube and the other along the upper horizontal tube. The eluent from the column enters the center of the middle tube and the flow also splits into two streams and each meets the respective flow from the right-hand tube. The flows in the two horizontal tubes finally pass up and down the left-hand vertical tube to meet at the center and then exit to waste. Flow sensors are situated in the horizontal tubes between the right-hand vertical tube and the center vertical tube. When only carrier gas is present in the system, the horizontal flows are equal and the temperature and thus the potential across the filaments of the two sensors are the same. When a solute is eluted from the column, vapor will be present in the center vertical tube and the pressure at the top and bottom of the tube will differ. This will result in a differential flow through the horizontal tubes with a consequent change in the output from the sensors. As the differential flow will be proportional to the pressure difference between the right-hand column of pure carrier gas and the center column full of vapor, the output from the sensor filaments will be proportional to the vapor density of the solute and consequently be related to the molecular weight. In fact with a second detector that measured the concentration of the solute, the gas density balance can be used to determine molecular weight of an eluted solute.

This device works well, has about the same sensitivity and linearity as the katherometer but, unfortunately, is no longer manufactured. It was one of the very few simple and inexpensive methods available for measuring the molecular weight of an eluted solute.

The Radioactivity Detector

The development of the technique of chromatography by James and Martin was originally evoked by their work on the synthesis of fatty acids in plants. To aid in their research, a method was needed to separate the fatty acids extracted from plant tissue and to quantitatively estimate the different fatty acids present. As a consequence, the technique suggested by Martin and Synge in 1941 [5] (GC) was developed into a practical separation procedure. Subsequently, the synthetic pathways for the different fatty acids were examined using ^{13}C and ^{3}H markers. Thus, having established a technique to separate the fatty acids, those that were radioactive needed to be identified and the relative activity of each peak compared. To do this successfully a proportional radioactive detector was required. James and Piper developed such a detector in 1961-3 [6,7] for this purpose and their basic system has been used to this day, although the detector has been fabricated by a number of different manufacturers. A diagram of the radioactivity detector based on the device of James and Piper is shown in figure 7.

There are two basic forms of the radioactivity detector, one that measures ^{13}C only and the other that measures both ^{13}C and ^3H. In both systems the carrier gas used must be helium or argon and the column eluent is fed through a furnace packed with copper oxide to oxidize all the solutes to carbon dioxide and water. Then, if only ^{13}C is being counted, the combustion products are passed through a drying tube and then mixed with 10% of propane and passed into the counting tube. In the counting tube the radioactive particles cause ionization and the electrons produced are accelerated towards the anode and, in doing so, produce further ionization of the carrier gas which enhances the signal. Normally this would result in a stable discharge being formed but the presence of the propane prevents this happening and for this reason the propane is sometimes called the *quench* gas. The counting tube is a metal cylinder carrying and insulated central electrode in the form of a rod. The outer case is usually grounded and a high potential is applied between the central electrode and the case. The signals received from the counter are integrated with respect to time and thus the output current from the integrator is proportional to the total number of disintegrations occurring per second. Hence the integration of the signal over the duration of the peak will a give a value that is proportional to the total activity of the peak. The ^{13}C counting apparatus is shown in the upper part of figure 7.

If both ^{13}C and ^3H are to be counted the apparatus shown in the lower part of figure 7 is used. After oxidizing all the solute to carbon dioxide and water some hydrogen is fed into the gas stream, the mixture is passed over heated iron powder in another furnace and the water is reduced to hydrogen and tritium. The added hydrogen saturates any adsorptive sites in the system and reduces the adsorption of the tritium to a satisfactory minimum. 10% of propane is then added to the exit gas from the reducing furnace and passed into the counter which operates in the same way but now counts tritium as well as well as ^{13}C. Unfortunately, the counting efficiency for ^3H usually differs from that for ^{13}C, therefore appropriate corrections may need to be made to the final count. The device has been used with considerable success in identifying synthetic pathways in biological systems using radioactive tracer techniques.

Radioactive Counting for Carbon Only

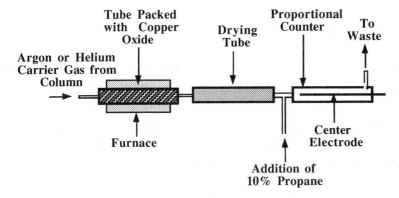

Radioactive Counting for Carbon and Tritium

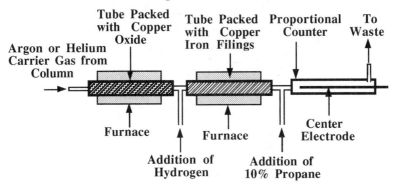

Figure 7 The Radioactivity Detector

Some Less Common GC Detectors

Between the years 1956 and 1980, a considerable number of novel GC detecting systems were developed and, at one time or another, each was strongly acclaimed for some specific application. Very few of these detectors have survived and even fewer are still being manufactured and commercially available. However, one or two have recently been rediscovered and found suitable for new areas of application. A selected number of this fairly large group of "lost detectors" will be briefly described to illustrate the large variety of sensing techniques that have been applied to GC detection.

The Thermionic Ionization Detector

Electrons produced by a heated filament can be accelerated under an appropriate potential to attain sufficient energy to ionize any gas or vapor molecules in their path. In 1957, the early days of gas chromatography, Ryce and Bryce [8,9] modified a standard vacuum ionization gauge to examine its possibilities as a GC detector. A diagram of the device is shown in figure 8.

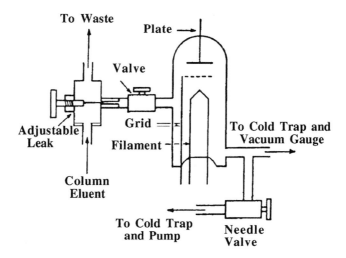

Figure 8 The Ionization Gauge Detector

The sensor consisted of a vacuum tube containing a filament, grid and anode, very similar in form to the thermionic triode valve. An adjustable leak was arranged to feed a portion of the column eluent into the gauge which was operated under reduced pressure. The sensor was fitted with its own pumping system and vacuum gauge and the usual necessary cold traps. Helium was used as a carrier gas and the grid collector–electrode was set at +18 V with respect to the cathode and the plate at -20 V to collect any positive ions that are formed. As the ionization potential of helium is 24.5 volts, the electrons would not have sufficient energy to ionize the helium gas. However, most organic compounds have ionization voltages lying between 9.5 and 11.5 V and consequently would be ionized by the 18 V electrons and provide a plate current. The plate current was measured by an impedance converter in much the same way as the FID ionization current. The detection limit was reported to be 5×10^{-11} moles, but unfortunately the actual sensitivity in terms of g/ml is not known and is difficult to estimate. The sensitivity is likely to be fairly high, probably approaching that of the FID. The response of the detector is proportional to the pressure of the gas in the sensor from about 0.02 mm to 1.5 mm of mercury. In this region of pressure it was claimed that the response of the detector was linear [10]. Hinkle *et al.* [11], who also examined the performance of the detector, suggested the sensor must be operated under conditions of molecular flow *i.e.* where the mean free path of the molecules is about the same as the electrode separation. It was found that the use of very pure helium was necessary to ensure a low noise and base signal. The detector had a "fast" response but its main disadvantage was the need to operate at very low pressures so that it required a vacuum pump; furthermore, for stability, the sensor pressure needed to be very precisely controlled.

The Discharge Detector

About the same time that Ryce and Bryce were developing the thermionic ionization detector, Harley and Pretorious [12] and (independently) Pitkethly and his co-workers [13,14] were developing the discharge detector. Under a suitable potential, a discharge can be maintained between two electrodes if the gas pressure is maintained between 0.1–10 mm of mercury. Once the discharge has been initiated the voltage across the electrodes can be reduced and the discharge continues to be maintained. Under these circumstances, the potential across the electrodes then remains constant and independent of the gas pressure and the electrode current. The potential across the electrodes, however, depends strongly on the composition of the gas. It follows that the system has the propensity for being used as a GC detector.

Pitkethly modified a small domestic neon lamp for this purpose and a diagram of his sensor is shown in figure 9.

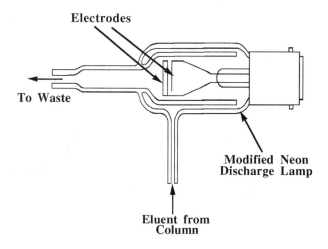

Figure 9 The Discharge Detector

The lamp was operated at about 3 mm of mercury pressure with a current of 1.5 mA. under which conditions the potential across the electrodes was 220 V. Pitkethly reported that a concentration of 10^{-6} g/l gave an electrode voltage change of 0.3 V. Now the noise level was reported as about 10 mV thus at a signal–to–noise level of 2 the minimum detectable concentration would be about 3×10^{-11} g/ml. This sensitivity is comparable to that of the FID and the argon detector.

The detector was claimed to be moderately linear over a dynamic range of three orders of magnitude but values for the response index are not known. It is also not clear whether the associated electronics contained signal modifying circuitry or not. The disadvantages of this detector included erosion of the electrodes due to "spluttering", contamination of the electrodes from sample decomposition and the need for a well–controlled vacuum system.

The Spark Discharge Detector

The voltage at which a spark will occur between two electrodes situated in a gas will depend on the composition of the gas between the electrode tips. Lovelock [15] suggested that this could form the basis for a GC detector. The system suggested by Lovelock is shown in figure 10.

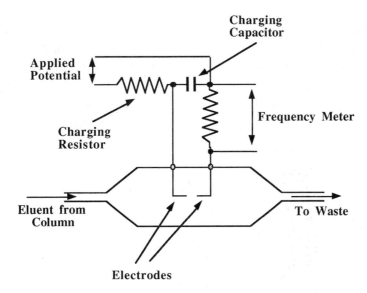

Figure 10 The Spark Discharge Detector

The sensor consists of a glass tube in which two electrodes are sealed. The electrodes are connected to the circuit as shown in figure 10. The voltage across the electrodes is adjusted to a value that is just less than that required to produce a spark. When a solvent vapor enters the sensor, the sparking voltage is reduced and a spark discharge occurs. This discharges the capacitor until its voltage falls below that which will maintain the spark discharge. The capacitor is then charged up through the charging resistor until the breakdown voltage is again reached and another spark is initiated. Thus the spark frequency will be proportional to (or at least be a monotonic function of) the vapor concentration. The total counts in a peak will be proportional to the peak area and, if a digital–to–analog converter is also employed, the output will be proportional to the concentration in the detector and thus, plotted against time, will provide the normal chromatogram. This detector does not appear to have been developed further but is an interesting example of a sensor that, in effect, produces a digital output.

The Radio Frequency Discharge Detector

When an RF discharge occurs across two electrodes between which the field is diverging (i.e. within a coaxial electrode orientation) a DC potential appears across the electrodes, the magnitude of which depends on the composition of the gas through which the discharge is passing. Karman and Bowman (16) developed a detector based on this principle. A diagram of their detector is shown in figure 11.

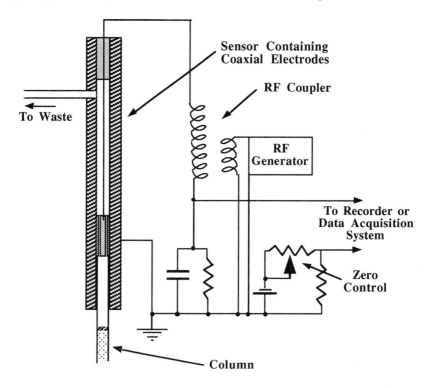

Figure 11 The Radio Frequency Discharge Detector

The sensor consisted of a metal cylinder that formed one electrode with a coaxial wire passing down the center that formed the other. A 40 MHz radio frequency was applied across the electrodes and the DC potential that developed across them was fed *via* simple electronic circuit to a potentiometric recorder. It would be necessary to decouple the radio frequency to prevent it interfering with the function of the potentiometric recorder (or signal processing amplifier). The resistance capacity decoupling shown in their circuit appears hardly sufficient to achieve this in a satisfactory manner and consequently, the circuit shown in figure 11 may be only schematic. The column was connected directly to the sensor and the eluent passed through the annular channel between the central electrode and the sensor wall.

The response of the radio frequency discharge detector was reported as 10^6 mV for a concentration change of 10^{-3} g/ml of methyl laureate. The noise level was reported to be 0.05 mV, which would give the minimum detectable concentration for a signal–to–noise ratio of 2 as about 6×10^{-10} g/ml. This detector operated at atmospheric pressure and so no vacuum system was required. The effect of temperature on the detector performance was not reported, nor was its linearity over a significant concentration range. This detector was not made available commercially.

The Ultrasound Whistle Detector

The presence of a solute vapor in a gas changes its density and consequently the velocity of the propagation of sound through it. This property can be utilized as a basis for detection in GC. The frequency of a whistle, consisting of an orifice which directs a stream of gas against a jet edge proximate to a resonant cavity, is related to the velocity of sound in the gas passing through it.

A diagram of such a whistle is shown in figure 12. Nyborg *et al.* [17] demonstrated that the frequency (f_n) of the whistle can be described by the following equation.

$$f_n = \frac{\left(n - \frac{1}{2}\right)c}{2(L+e)}$$

where (n) is an integer,
(c) is the velocity of sound in the gas,
(L) is the cavity length,
and (e) is the end effect.

Testerman and McLeod [18] engineered and built a detector based on the whistle principle. In their detector sensor design, typical values taken for the dimensions in the diagram, and variables in the equation, were (t), 0.064 mm, (d), 0.74 mm, (h), 1.676 mm and (L) 3.81 mm.

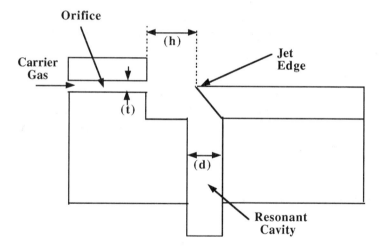

Figure 12 The Supersonic Whistle Detector Sensor

Operating at gas flow rates appropriate for GC separations, frequencies ranging from 30-50 kHz (supersonic frequencies) were observed. The actual sensor contained two sound generators, one operating with pure carrier gas and the other with the eluent from the column. The two frequencies were allowed to *beat* together, the *beat frequency* being directly related to the frequency difference between the two whistles and consequently the density difference between the contents of the two sensors. An example of the use of the whistle detector to monitor the separation of a mixture of hydrocarbons is shown in figure 13. The sample size was 7.5 µl of gas mixture and the carrier gas flow rate was 180 ml /min. This chromatogram illustrates the effective use of the detector and the operating conditions shows its limitations. The sensitivity appears somewhat less than that of the katherometer but the very high flow rates necessary to activate the whistle restrict the use of this type of detector very severely. In the original report the linearity was stated to cover 2 orders of magnitude of concentration but with modern electronics it is likely that this linear range could be extended by at least another order of magnitude.

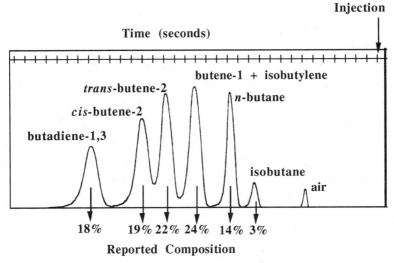

Figure 13 The Separation of a Mixture of Hydrocarbons Monitored by the Supersonic Whistle Detector

The Dielectric Constant Detector

In 1961 Winefordner *et al.* [19] described a GC detector that functioned on the change in dielectric constant of the carrier gas when a vapor or another gas was present. The detector responded to most gases and vapors providing a suitable carrier gas was chosen. The sensor consisted of a variable capacitor mounted in a special cell which allowed the column eluent to pass between its plates. The capacitor constituted part of the "tank" circuit of a Clapp oscillator. The output of the oscillator was allowed to beat against a reference oscillator and the beat frequency was taken as the sensor signal. The detector was reported to respond linearly to changes in vapor concentration. However, the device was found to be no more sensitive than the katherometer and considerably more complex.

In 1965 Winefordner *et al.* [20] developed the detector further. They employed basically the same system except that they used a miniature coaxial type sensor with very small spacing between the internal and external cylindrical conductors (0.005 in.). Sensitivities of 10^{-10} g/ml were reported for the detection of oxygen, nitrogen, hydrogen, carbon dioxide, carbon monoxide, methane, nitrogen dioxide, nitrous oxide and sulfur dioxide. In addition, it was claimed that this sensitivity could be increased by one or two orders of magnitude by employing solid state electronics in place of the thermionic tubes to reduce the electronic noise. Subsequently Williams and Winefordner [21] claimed they had demonstrated a "nearly" linear response from the detector over a concentration range in excess of 4 orders of magnitude for ethylene, ethane, propane and ammonia. This detector was not made available commercially but has interesting possibilities for use in gas analyzers.

The Piezoelectric Adsorption Detector

The response of this detector is based on the fact that the frequency output from piezoelectric material is influenced by the weight of the coatings or layers on its surface. This effect has been used for many years to measure trace concentrations of water vapor in a gas and xylene vapor in air has been detected by this means at concentrations as low as 10^{-6} g/ml. It was first introduced as a GC detecting system by King [22]. The detector consists of a quartz crystal (coated with a high boiling liquid) that is appropriately situated in an electronic circuit that causes it to oscillate at its natural frequency. The oscillation frequency is continuously monitored by a separate circuit.

As material is absorbed by the coating, the weight of the crystal plus coating changes and the natural frequency falls. The relationship between oscillation frequency (f) and weight (w) absorbed is given by the following equation:

$$f = f_0 - 2.3 \times 10^{-6} \times f_0^2 \frac{w}{A}$$

where (f_0) is the natural frequency of the coated crystal,
and (A) is the total area of the coated crystal surface.

and thus the change in frequency will be

$$\Delta f = 2.3 \times 10^{-6} \times f_0^2 \frac{w}{A}$$

It follows that a standard crystal with a natural frequency of 9 MHz and a surface area of about one square centimeter will manifest a change in frequency of about 200 Hz for each microgram of adsorbed solute. Now frequency changes can be measured to within 0.1 Hz with normal equipment; consequently a change in mass adsorbed of 0.2 ng (10^{-9}g) should be detectable. It follows that this type of device should be very sensitive but it appears that, so far, it has not been made available commercially, at least, not as a GC detector. One attractive feature of this detecting system is that it basically measures *mass* and therefore could be considered to be an *absolute* detector.

The Absolute Mass Detector

The absolute mass detector was devised by Bevan and Thorburn [23,24], who adsorbed the eluent from a GC column on to the coated walls of a vessel supported on a recording balance. A diagram of their apparatus is shown in figure 14. The adsorption vessel was 1.4 cm I.D. and about 5 cm high. The walls of the vessel were coated with a high boiling absorbent such as polyethylene glycol or an appropriate normal hydrocarbon depending on the samples being trapped. The solutes separated had to be relatively low boiling otherwise they would condense in the capillary connecting tube to the adsorption vessel.

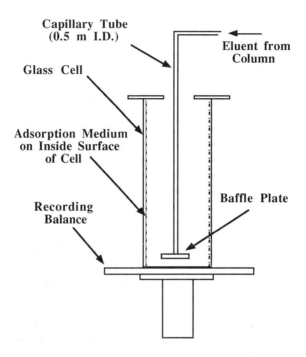

Figure 14 The Absolute Mass Detector

The tube dipped to the base of the absorber where a baffle was situated to direct the eluent to the walls of the adsorption vessel. The balance record represented an integral chromatogram, the step height giving directly the mass of solute eluted. It would appear that the adsorption was quite efficient and, with 10 mg charges on the column, an accuracy of 1% could be easily achieved. Later Bevan *et al* [25,26], reduced the size of the absorber and employed charcoal as the adsorbing material. Although this improved the performance of the detector and reduced the necessary sample size, the detecting system was never made commercially. Even after modification, its sensitivity was relatively poor and despite it being an absolute detecting system, it placed too many restrictions on the operation of the chromatograph to be generally useful.

The Surface Potential Detector

The surface potential detector was developed by Griffiths and Phillips [27,28] in the early 1950s and consisted of a cell containing two parallel metal plates between which flowed the column eluent. One plate was mechanically attached to an oscillator that vibrated the plate at about 10 kHz. If the plates are identical, the surface charge on each plate is the same and so no potential is induced into the second plate by the vibrating plate. If however the surfaces are dissimilar, then the surface charge on each plate will differ and the vibrating plate will induce a potential on the other plate. A diagram of the detector is shown in figure 15.

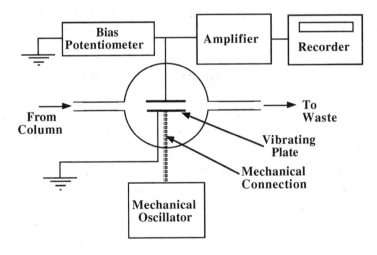

Figure 15 The Surface Potential Detector

Both plates were constructed of the same metal but one plate was coated with a monolayer of a suitable substance that would absorb any vapors present in the column eluent. The absorbing layer caused the charge on the two plates to be dissimilar and thus a potential appeared across the two plates which was balanced out by the bias potentiometer.

When a solute vapor passes through the detector, some is distributed into the absorbent layer, changing the surface charge and thus inducing a change in potential between the electrodes. This voltage is then amplified and the output passed to a recorder (or to a data acquisition system). The signals provided by the detector sensor could be as great as several hundred millivolts.

The sensitivity of the detector was similar to that of the katherometer, *i.e.* about 10^{-6} g/ml. Its response was partly determined by the distribution coefficient of the solute vapor between the carrier gas and the absorbing layer as well as the chemical nature of the solute vapor. As a consequence, the response varied considerably between different solutes. Within a given homologous series, the response increased with the molecular weight of the solute but this was merely a reflection of the increase in the value of the distribution coefficient with molecular weight. Although an interesting alternative method of detection, this detector has been little used in GC and is not commercially available.

Gas chromatography is an analytical technique that has reached what might be termed a "steady state" and this is also true for the development of GC detectors. The four detectors FID, ECD, NPD and the katherometer are now well established, are the most popular and are employed in over 95% of all GC applications. Of the four, the FID is the most versatile, sensitive and linear, and should be the first to be considered when facing the challenge of a new, and hitherto unknown sample containing unfamiliar substances. Nevertheless, specific and unusual samples can demand special detecting conditions particularly for certain environmental and forensic analyses. In such cases, one of the many other detectors described may well be found more appropriate. Fortunately, there is a fairly wide range of practical GC detectors from which to choose that are commercially available.

It will be interesting to see how the various detecting systems based on the different helium ionization processes that were discussed in the previous chapter fare commercially over the next few years.

References

1. N. H. Ray, *J. Appl. Chem.*, **4**(1954)21.
2. N. Mellor, *"Vapor Phase Chromatography"* (Ed. D.H.Desty and C.L. A. Harbourn), Butterworths Scientific Publications (1957)63.
3. D.Harvey and G.O. Morgan, *"Vapor Phase Chromatography"*, (Ed. D. H.Desty and C.L. A. Harbourn), Butterworths Scientific Publications (1957)74.
4. A. T. James and A. J. P. Martin, *Biochem. J.* **50**(1952)679.
5. A.J.P. Martin and R.L.M. Synge, *Biochem. J.* , **35**(1941)1358.
6. A. T. James and E. A. Piper, *J. Chromatogr.* **5**(1961)265.
7. A. T. James and E. A. Piper, *Anal. Chem.* **35**(1963)515.
8. S. A. Ryce and W. A. Bryce, *Nature* **179**(1957)541.
9. S. A. Ryce and W. A. Bryce, *Can. J. Chem.*, **35**(1957)1293.
10. L. V. Guild, M. I. Lloyd and F. Aul, *"Gas Chromatograph, 2 nd International I. S. A. Symposium, June, 1969"* (Ed., H. J. Noebels, R. F. Wall and N. Brenner) Academic Press New York (1961)91.
11. E. A. Hinkle, H. C. Tucker, R. F. Wall and J. F. Combs, *"Gas Chromatograph, 2 nd International I. S. A. Symposium, June, 1969"*, (Ed., H. J. Noebels, R. F. Wall and N. Brenner) Academic Press New York, (1961)55.
12. J. Harley and V. Pretorious, *Nature,* **178**((1957)1244.
13. R. C. Pitkethly, *132 nd Am. Chem. Soc. Meeting, New York, September, 1957)*
14. R. C. Pitkethly, *Anal. Chem.* **30**(1958)1460.
15. J. E. Lovelock, *Nature*, **181**(1958)1460.
16. A. Karman and R. L. Bowman, *Ann. N. Y. Acad. Sci.*, **72**(1959)714.
17. W. L. Nyborg, C. L. Woodbridge and A. K. Schilling, *Acoust. Soc. Am.,* **25**(1953)138.
18. M. K. Testerman and P. C. McLeod, "Gas Chromatography" (Eds. N. Brenner, J. E. Callen and M. D. Weiss, Academic Press, New York (1962)183.
19. J.D.Winefordner, D. Steinbrecher and W. E. Lear, *Anal. Chem.,* **33**(1961)515.
20. J.D.Winefordner, H. P. Williams and C. D. Miller, *Anal. Chem.* **37**(1965)161.
21. W. H. King, *Anal Chem.*, **36**(1964)1735.
23. S. C. Bevan and S. Thorburn, *J. Chromatogr.*, **111**(1963)301
24. S. C. Bevan and S. Thorburn, *Chem. in Britain*, 1(1965)206.
25. S. C. Bevan, T. A. Gough and S. Thorburn, *J. Chromatogr.*, **43**(1969)192.
26. S. C. Bevan, T. A. Gough and S. Thorburn, *J. Chromatogr.*, **42**(1969)336.
27. J. H. Griffiths and C. S. G. Phillips, *J. Chem. Soc.*, (1954)3446.
28. J. H. Griffiths, D. J. James and C. S. G. Phillips, *Analyst*, **77**(1952).

Part 3

Liquid Chromatography Detectors

CHAPTER 8

INTRODUCTION TO LC DETECTORS
The UV Detectors

The first chromatography detector ever to be used was an LC detector and it was the human eye. Tswett in his pioneering chromatographic separation of plant pigments used the eye to determine the nature of the separation he achieved and even today, as one of the more widespread separation techniques is thin-layer chromatography, the human eye is still one of the more frequently used detectors. The human eye as an LC detector has, however, some severe limitations. The majority of substances that are chromatographed are colorless and thus must be chemically changed to render them visible; furthermore, the retinic response of the eye is not linear, which makes the eye a poor detecting system for quantitative estimation. The human eye, in fact, can only be used for quantitative work as a null sensing device where closely similar light intensities are being matched as in the use of a comparator.

Detector development has gone hand in hand with column development in all forms of chromatography. The rapid development of GC arose solely from the swift availability of sensitive detecting systems and it was not until sensitive LC detectors were introduced that could the rapid advances in LC could take place. Highly sensitive detectors have provided accurate concentration profiles of eluted solutes and allowed the extent to which solute dispersion occurs in the column to be evaluated. The information so obtained has allowed the development and confirmation of theories that describe the various processes that lead to band dispersion, which, in turn, has lead to improved column performance. In fact, the impressive separations that are achieved in both GC and LC today can be directly attributed to the use of sensitive in-line detecting systems, *i.e.* detecting systems that can be connected directly to the outlet.

The first attempt to develop an alternative detecting system to that of the human eye was to collect the column eluent in a large number of fractions and to subsequently analyze each fraction by appropriate techniques such as colorimetry or titration. The concentration of each solute in each fraction was then plotted against fraction number and a form of chromatographic histogram was produced. This procedure was very time consuming and tedious and only effective for well–resolved mixtures. Furthermore, a chromatographic histogram did little to help improve column technology. The concept of on-line detection, where an appropriate sensing device is connected directly to the column outlet, was established in the late 1930s and early 1940s. Such a device could provide an output that was directly related to the concentration of the solute in the eluent. Examples of two early on-line detectors are afforded by the electrical conductivity detector described by Martin and Randall [1], and the refractive index detector described by Tiselius and Claesson [2]. However, as a result of the invention of GC and its sensational growth, interest in LC generally waned during the 1950s. As a consequence, the major progress in LC detector development did not commence until the 1960s and early 1970s and, even then, was probably the direct result of the impetus provided by the dramatic development of GC detectors. In some cases the principles of GC detection were applied directly to LC detectors. In due course, many of the scientists involved in the development of GC detectors in the late 1950s, having exhausted the possibilities of GC, turned their attention to the development of LC detectors. Between 1956 and 1960, at least six GC detectors were invented and developed to their full potential; in contrast the development of LC detectors has been slow and arduous and, to this day, none have sensitivities that compare with those of their GC counterparts. As already stated, detection in LC is far more difficult than in GC as the presence of a solute in a liquid does not modify the overall properties of the liquid to anything like the same extent as the same concentration of solute would do to the properties of a gas. It follows that a detector that functions on the measurement of some physical property of the liquid will provide a response from a very low concentration of a solute that is commensurate with that which would arise from small changes in ambient conditions or fluctuations in column flow rate. The detector signal will thus be commensurate with the detector noise and exhibit very limited sensitivity.

There has been a continuous interaction between improved detector performance and improved column performance, each advance being mutually dependent on the other. Initially, separations monitored by high sensitivity detectors permitted a precise column theory to be developed and experimentally substantiated. This allowed new columns to be designed having reduced dispersion and consequently higher efficiencies. The improved column efficiency, however, furnished peaks having a greatly reduced volume, small that is compared with the volume of the detector sensor and the dispersion that took place in connecting tubing and other conduits in the detector system. At this point, the ultimate efficiency obtainable from the column was determined by the geometry of the fluid conduits of the detector and not its sensitivity. As a result, the detectors were redesigned with smaller sensor volumes, different geometry and shorter connecting tubes between the column and sensor. In turn, these detector modifications allowed the use of much smaller particles in the column with even lower dispersion and higher efficiencies. Such columns produced very rapid separations (five components resolved and eluted in 3.5 seconds [3]) that demanded detector amplifiers with a very fast response. This situation provoked a complete redesign in detector electronics. In this way detector design and column design have interacted over the years to a point where the resolutions obtained from LC columns are commensurate with those obtainable from GC columns albeit the very high sensitivities readily obtainable from GC detectors are still not obtainable from LC detectors.

Although over the years a large number of LC detectors have been developed and described, over 95% of all contemporary LC analyses are carried out using one of four detectors, the UV detector in one of its forms, the electrical conductivity detector, the fluorescence detector and the refractive index detector. In addition, some form of the UV detector probably accounts for 80% of those analyses.

The UV Absorption Detectors

UV absorption detectors respond to those substances that absorb light in the range 180 to 350 nm. A great number of substances absorb light in this wavelength range, including those substances having one or more double bonds (π electrons) and substances having unshared (unbonded) electrons, *e.g.* all olefins, all aromatics and compounds, for example, containing $>C=O$, $>C=S$, $-N=N-$ groups. The detector normally consists of a short cylindrical cell having a capacity of between 2 µl and 5 µl through which passes the column eluent. The UV light also passes through the cell and falls on a photo–electric cell (or array) the output of which is conveyed to an appropriate signal modifying amplifier and thence to a recorder or data acquisition system.

The relationship between the intensity of UV light transmitted through a cell (I_T) and the concentration of solute in it (c) is given by Beer's Law.

$$I_T = I_0 e^{-klc} \tag{1}$$

or $\quad \ln(I_T) = \ln(I_0) - kcl$

where (I_0) is the intensity of the light entering the cell,
(l) is the path length of the cell,
and (k) is the molar extinction coefficient of the solute for the specific wavelength of the UV light.

If equation (1) is put in the form,

$$I_T = I_0 10^{-k'lc} \qquad (2)$$

then (k') is termed the *molar extinction coefficient.*

Differentiating equation (2),

$$\frac{d\left(\log \frac{I_T}{I_0}\right)}{dc} = -kl$$

It is seen that the sensitivity of the detector as measured by the transmitted light will be directly proportional to the value of the extinction coefficient (k) and the path length of the cell (l). It follows, that to increase the sensitivity of the system, (l) should be increased however, there is a limit to which the path length can be increased as the total volume of the cell and, in particular, the length of the cell must be restricted. This is necessary to ensure minimum peak dispersion in the detector and to avoid more than a small fraction of a peak existing in the cell at any one time. To restrict peak dispersion in the cell to a reasonable level and maintain a small sensor volume, the radius of the cell must also be reduced as (l) is increased. This results in less light falling on the photo–cell which in turn will reduce the signal–to–noise ratio and thus the detector sensitivity or minimum detectable concentration. Thus, increasing the detector sensitivity by increasing the path length has limitations and a well–designed cell involves a careful compromise between cell radius and length to provide the maximum sensitivity. Most modern UV detector sensors have path lengths that range between 1 and 10 mm and internal radii that range from about 0.5 to 2 mm

From equation (2),

$$\log \frac{I_T}{I_0} = k'lc = A$$

where (A) is termed the absorbance

Now (ΔA) is commonly employed to define the detector sensitivity where the value of (ΔA) is the change in absorbance that provides a signal-to-noise ratio of two.

$$\text{Thus} \quad \Delta A = k' l \Delta c$$

where (Δc) is the detector concentration sensitivity or minimum detectable concentration.

$$\text{Thus} \quad \Delta c = \frac{\Delta A}{k' c}$$

It is clear that two detectors, having the same sensitivity defined as the minimum detectable change in absorbance, will not necessarily have the same sensitivity with respect to solute concentration. Only if the path lengths of the two sensors are identical will they also exhibit the same concentration sensitivity. This can cause some confusion to the chromatographer who would expect that two instruments defined as having the same spectroscopic sensitivity would also have the same chromatographic sensitivity. To compare the sensitivity of two detectors given in units of absorbance the path lengths of the cells in each instrument must also be taken into account.

UV detectors can be used with gradient elution providing the solvents do not absorb significantly over the wavelength range that is being used for detection. In reversed phase chromatography, the solvents usually employed are water, methanol, acetonitrile and tetrahydrofuran (THF), all of which are transparent to UV light over the total wavelength range normally used in UV detectors. In normal phase operation, however, more care must be taken in solvent selection as many solvents that would be appropriate as far as the chromatographic phase system is concerned are likely to absorb UV light very strongly. The *n*-paraffins, methylene dichloride, small quantities of the aliphatic alcohols (large quantities are likely to deactivate the stationary phase and produce very rapid elution of the components) and THF are useful solvents that are transparent in the UV and can be used with normal distribution systems (*e.g.* a polar stationary phase such as silica gel).

The Fixed Wavelength UV Detector

The fixed wavelength detector, as its name implies, operates with light of a single wavelength (or nearly so) which is generated by a specific type of discharge lamp. The most popular lamp for this purpose is the low pressure mercury vapor lamp, which generates the majority of its light at 254 nm. There are a number of other lamps that could be used in fixed wavelength UV detectors, the low-pressure cadmium lamp which generates the majority of its light at 225 nm and the low pressure zinc lamp that emits largely at 214 nm. None of the lamps emits strictly monochromatic light and light of other wavelengths is always present but usually at a significantly lower intensity. The emission spectra of the mercury, cadmium and zinc lamps are shown in figure 1. It is seen that if a completely monochromatic source of light was required, then an appropriate filter would be needed. The low pressure mercury light source (wavelength 253.7 nm) is the lamp normally used in the fixed wavelength detector and provides the closest to true monochromatic light of all three lamps. However, it does provide light of significant intensity below 200 nm, but light of such wavelengths is often absorbed and eliminated by the mobile phase.

The zinc lamp has a major emission line at 213.9 but the emission line at 307.6 is also of comparable intensity and would probably need to be removed by a suitable filter if detection was required to be exclusively at the lower wavelength. The cadmium lamp has a major emission line at 228.8 but light is emitted at both lower wavelengths and at substantially higher wavelengths and so an appropriate filter would again be desirable. Suitable interference filters can be quite expensive to construct, which may account for the unpopularity of these two lamps. They do, however, emit light at wavelengths which would give an increased sensitivity to substances such as proteins and peptides, which might make their use worthwhile in the biotechnology field.

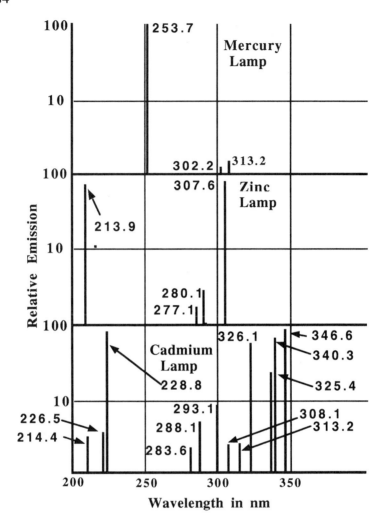

Figure 1 Emission Spectra for Three Discharge Lamps

A diagram of the typical optical system of a fixed wavelength UV detector is shown in figure 2. Light from the UV source is collimated by a suitable lens through both the sample cell and the reference cell and then on to two photo cells The cells are cylindrical with quartz windows at either end. The reference cell compensates for any absorption that mobile phase might have at the sensing wavelength. The output from the two photo cells are passed to a signal modifying amplifier so that output is linearly related to the concentration of solute being detected. It is seen from equation (1) that a simple linear amplifier would not be appropriate.

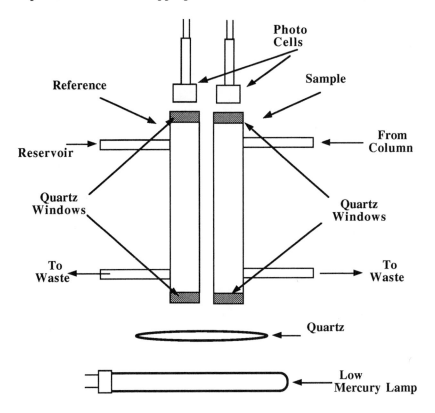

Figure 2 The Fixed Wavelength UV Detector

One of the first UV detectors suitable for use with packed LC columns was described in 1966 by Horvath and Lipsky [4] and about a year afterward Kirkland [5] described a miniaturized version of the detector; in fact, sensor design has changed little since that time. The cell was 10 mm long and 1 mm I.D. having a total volume of about 7.5 µl. Kirkland claimed a noise level of 0.0002 absorbance units, and an upper limit of linear dynamic range of 1.2 absorbance units, which was equivalent to a concentration of about 10^{-4} g/ml. This gives a sensitivity, or minimum detectable concentration, of about 3×10^{-7} g/ml at a signal–to–noise ratio of 2. A sensor volume of 7.5 µl would be considered excessive for use with modern columns, particularly microbore columns, and modern sensor volumes are about 2 or 3 µl or even less. For reasons already discussed, modern sensor cells have angular conduits that form a 'Z' shape to reduce dispersion. The UV sensor can be fairly sensitive to both flow rate and pressure changes but this instability can be greatly reduced if the sensor is well thermostatted [6]. The fixed wavelength UV detector is probably one of the most commonly used LC detectors; it is sensitive, linear and relatively inexpensive. Sensitivity (minimum detectable concentration) can be expected to be about 5×10^{-8} g/ml with a linear dynamic range of about three orders of magnitude for $0.98 < r < 1.02$.

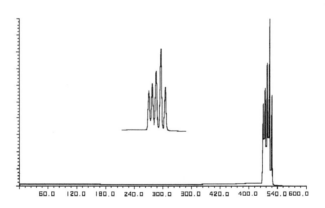

Column, length, 10m, I.D., 1 mm, mobile phase tetrahydrofuran, flow-rate 30 ml/min, adsorbent Partisil 10. Column efficiency *ca.* 250,000 theoretical plates. Solutes benzene, ethyl benzene, butyl benzene, hexyl benzene, octyl benzene and decyl benzene. Such columns must be packed in short lengths (about 1 m) and subsequently joined and are thus somewhat tedious to construct.

Figure 3 The Separation of Some Alkyl Benzenes by Exclusion on a High Efficiency Column

The separation of some aromatic hydrocarbons by exclusion chromatography on a very high efficiency column (efficiency 250,000 theoretical plates) monitored by a fixed wavelength detector is shown in figure 3. It is seen that all the solutes are distinctly resolved despite their having molecular weight differences equivalent to only two methylene groups. However, the peaks from such columns are only a few microliters in width and so a *specially reduced volume sensor cell* was necessary to truly represent the resolution that was obtained. The molecular weight of decyl benzene is 218 and, thus, one methylene group would represent a differential of only 6.4% of the molecular weight.

An interesting example of a very simple fixed wavelength detector suitable for use in preparative chromatography is shown in Figure 4. This detector was invented by Miller and Strusz [7] and originally manufactured by GOW-MAC Instruments.

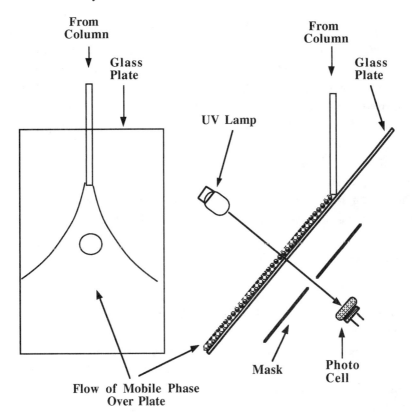

Figure 4 Fixed Wavelength Detector for Preparative Work

In complete contrast to detectors used for analytical purposes, detectors for preparative work must have a very low sensitivity as the sample size and the eluent solute concentrations are very large. Analytical detectors can be used for preparative purposes but a portion is usually split from the column eluent, diluted with more mobile phase and then passed through the detector. In practice this becomes a rather clumsy procedure.

In the device shown in figure 4 the column eluent passes through a delivery tube and over a supporting plate that is either sufficiently thin or made of fused quartz, so that adequate UV light can penetrate and reach the photo cell situated on the other side of the plate. The liquid flows over the plate and the effective path length is the film thickness peculiar to the flowing solvent layer. The UV lamp is situated on one side of the plate and the photo cell on the other side, facing normal to the surface. A reference photo cell (not shown) is also placed close to the lamp to compensate for changes in light intensity due to variations in lamp emission. The very short path length produces a very low sensitivity and the detector can operate satisfactorily at concentrations as great as 10^{-2} g/ml (1% w/w), which is ideal for preparative chromatography. Another advantage of the device is that it has a very low flow impedance and thus can easily cope with the high flow rates used in preparative LC. The film thickness does depend, among other things, on the column flow rate and thus precise flow control is necessary for the satisfactory performance of this detector.

The Multi–Wavelength UV Detector

The multi–wavelength detector employs a source that provides light over a wide range of wavelengths and consequently, with the aid of an appropriate optical system, light of a specific wavelength can be selected for detection purposes. For example, light of a wavelength where the solute has the maximum absorption can be chosen and this would provide the maximum sensitivity. Alternatively, the absorption spectra of eluted substances can be obtained for identification purposes. The latter procedure, however, differs with the type of multi-wavelength detector being used.

There are two basic types of multi-wavelength detector, the *dispersion* detector and the *diode array detector*, the former being the more popular. In fact, today very few dispersion instruments are sold but there are many still used in the field and so their characteristics will be discussed. Both types require a broad emission light source such as deuterium or the xenon lamp, the use of the deuterium lamp being the most widespread.

The two types of multi-wavelength detectors have important differences. In the dispersive instrument the light is dispersed before it enters the sensor cell and thus virtually monochromatic light passes through the cell. However, if the incident light can excite the solute and cause fluorescence at another wavelength, then the light falling on the photo cell will contain that incident light that has been transmitted through the cell together with any fluorescent light that may have been generated in the cell. Consequently the light monitored by the photo cell will not be solely monochromatic and light of another wavelength, if present, could impair the linear nature of the response. In most cases this effect would be negligible but with some substances the effect could be quite significant. The diode array detector operates in quite a different manner. Light of all wavelengths generated by the deuterium lamp is transmitted through the cell and the transmitted light is then dispersed over an array of diodes. In this way, the absorption at discrete groups of wavelengths is continuously monitored at each diode. However, the light falling on a discrete diode may not be solely that transmitted through the cell from the source but may contain light that was the result of fluorescence excited by light of a shorter wavelength. In this case the situation is exacerbated by the fact that the cell contents are exposed to all the light emitted from the lamp and so fluorescence is more likely. In general this means that under some circumstances measurement of transmitted light may also contain fluorescent light and as a consequence the absorption spectrum obtained for a substance may be degraded from the true absorption curve.

The ideal multi-wavelength detector would be a combination of both the dispersion instrument and the diode array detector, which, employing the vernacular of the day, would probably be dubbed LC/UV/UV. This system would allow a monochromatic light beam to pass through the detector and then the transmitted beam would itself be dispersed again onto a diode array. Only that diode corresponding to the wavelength of the incident light would be used for monitoring the transmission. In this way any fluorescent light would strike other diodes, the true absorption would be measured and accurate absorption spectra could be obtained.

The Multi–Wavelength Dispersive UV Detector

A diagram of the multi–wavelength dispersive UV detector is shown in figure 5.

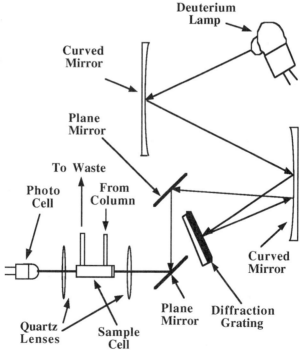

Courtesy of the Perkin Elmer Corporation

Figure 5 The Multi–Wavelength Dispersive UV Detector

Light from the source is collimated by two curved mirrors onto a holographic diffraction grating. The dispersed light is then focused by means of a curved mirror, onto a plane mirror and light of a specific wavelength is selected by appropriately positioning the angle of the plane mirror. Light of the selected wavelength is then focused by means of a lens through the flow cell and, consequently, through the column eluent. The exit beam from the cell is then focused by another lens onto a photo cell which gives a response that is some function of the intensity of the transmitted light. The detector is usually fitted with a scanning facility that, by arresting the flow of mobile phase, allows the spectrum of the solute contained in the cell to be obtained.

Due to the limited information given by most UV spectra and the inherent similarity between UV spectra of widely different types of compounds, this procedure is not very reliable for the identification of most solutes. The technique is useful, however, for determining the homogeneity of a peak by obtaining spectra from a sample on both sides of the peak. The technique is to normalize both spectra, then either subtract one from the other and show that the difference is close to zero, or take the ratio and show it is constant throughout the peak.

A more common use of the multi-wavelength detector is to select a wavelength that is characteristically absorbed by a particular component or components of a mixture in order to enhance the sensitivity of the detector to those particular solutes. Alternatively, by choice of characteristic wavelength, the detector can be rendered more specific and thus not respond significantly to substances of little interest in the mixture.

An example of the use of the variable wavelength UV detector to select a specific wavelength to give a high sensitivity for a certain group of compounds is afforded by the separation of some carboxylic acids that is monitored by UV absorption at 210 nm. The separation is shown in figure 6. The separation of a series of common fatty acids was carried out on a reversed phase column using water buffered with phosphoric acid as the mobile phase.

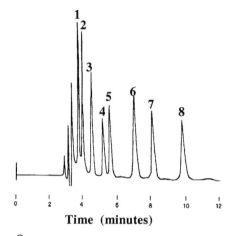

Column: Spherisorb® Octyl, 25 cm x 4.6 mm I.D., 5 μm particles. Mobile Phase: 0.2 M phosphoric acid. Flow rate 0.8 ml/min. monitored at 210 nm. 1. tartaric acid, 2. lactic acid, 3. malic acid, 4. formic acid, 5. acetic acid, 6. citric acid, 7. succinic acid, 8. fumaric acid.

Courtesy of Supelco Inc.

Figure 6 The Separation of Some Carboxylic Acids Monitored by UV Absorption at 210 nm

In the past the multi-wavelength dispersive detector has proved extremely useful, providing adequate sensitivity, versatility and a linear response. It is, however, somewhat bulky (due to the need for a relatively large internal "optical bench"), has *mechanically operated* wavelength selection and requires a stop/flow procedure to obtain spectra "on-the-fly". The diode array detector has the same advantages but none of the disadvantages, although, as one might expect, it is somewhat more expensive.

The Diode Array Detector

The diode array detector also utilizes a deuterium or xenon lamp that emits light over the UV spectrum range. Light from the lamp is focused by means of an achromatic lens through the sample cell and onto a holographic grating. The dispersed light from the grating is arranged to fall on a linear diode array.

The resolution of the detector ($\Delta\lambda$) will depend on the number of diodes (n) in the array, and also on the range of wavelengths covered ($\lambda_2 - \lambda_1$).

Thus
$$\Delta\lambda = \frac{\lambda_2 - \lambda_1}{n}$$

It is seen that the ultimate resolving power of the diode array detector will depend on the semi-conductor manufacturer and on how narrow the individual photo cells can be commercially fabricated.

A diagram of a diode array detector is shown in figure 7. Light from the broad emission source is collimated by an achromatic lens system so that the total light passes through the detector cell onto a holographic grating. In this way the sample is subjected to light of all wavelengths generated by the lamp.

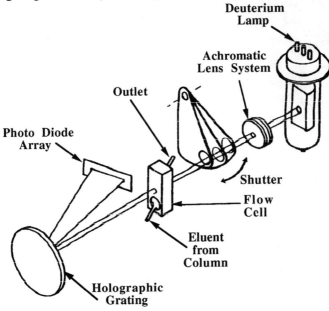

Figure 7 The Diode Array Detector

The dispersed light from the grating is then allowed to fall onto a diode array. The array may contain many hundreds of diodes and the output from each diode is regularly sampled by a computer and stored on a hard disc. At the end of the run, the output from any diode can be selected and a chromatogram produced using the UV wavelength that was falling on that particular diode.

Most instruments will permit the monitoring of at least one diode in real time so that the chromatogram can be followed as the separation develops. This system is ideal, in that by noting the time of a particular peak, a spectrum of the solute can be obtained by recalling from memory the output of all the diodes at that particular time. This gives directly the spectrum of the solute *i.e.* a curve relating adsorption and wavelength. The diode array detector can be used in a number of unique ways and an example of the use of a diode array detector to verify the purity of a given solute is shown in figure 8.

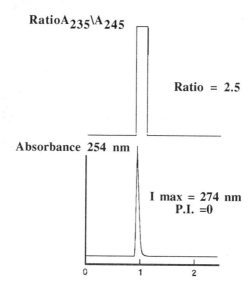

The chlorthalidone was isolated from a sample of tablets and separated by a reverse phase (C18) on a column 4.6 mm I.D., 3.3 cm long, using a solvent mixture consisting of 35% methanol, 65% aqueous acetic acid solution (water containing 1% of acetic acid). The flow rate was 2 ml/min.

Courtesy of the Perkin Elmer Corporation

Figure 8 Dual Channel Plot from a Diode Array Detector Confirming Peak Purity

The chromatogram monitored at 274 nm is shown in the lower part of figure 8. As a diode array detector was employed, it was possible to ratio the output from the detector at different wavelengths and plot the ratio simultaneously with the chromatogram monitored at 274 nm. Now, if the peak was pure and homogeneous, the ratio of the adsorption at the two wavelengths (those selected being 225 and 245 nm) would remain constant throughout the elution of the entire peak. The upper diagram in figure 8 shows this ratio plotted on the same time scale and it is seen that a clean rectangular peak is observed which unambiguously confirms the purity of the peak for chlorthalidone. It should be pointed out that the wavelength chosen to provide the confirming ratio will depend on the UV adsorption characteristics of the substance concerned, relative to those of the most likely impurities to be present.

Another interesting example of the use of the diode array detector to confirm the integrity of an eluted peak is afforded by the separation of the series of hydrocarbons shown in figure 9.

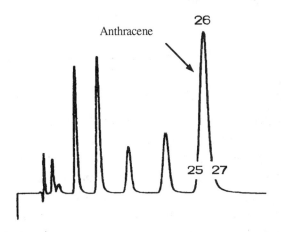

The separation was carried out on a column 3 cm long, 4.6 mm in diameter and packed with a C18 reversed phase on particles 3 μ in diameter.

Courtesy of the Perkin Elmer Corporation

Figure 9 The Separation of Some Aromatic Hydrocarbons

It is seen that the separation appears to be satisfactory and all the peaks represent individual solutes. Furthermore, without further evidence, it would be reasonable to assume that all the peaks were pure. However, by plotting the adsorption ratio, $\frac{250\,\text{nm}}{255\,\text{nm}}$, for the anthracene peak it becomes clear that the peak tail contains an impurity as the clean rectangular shape of the other peaks is not realized. The absorption ratio peaks are shown in figure 10.

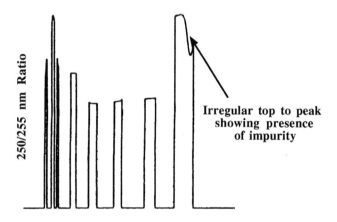

Figure 10 Curves Relating the Adsorption Ratio, $\frac{250\,\text{nm}}{255\,\text{nm}}$, and Time

The ratio peaks in figure 10 clearly show the presence of an impurity by the sloping top of the anthracene peak and again this is confirmed unambiguously by the difference in the spectra for the leading and trailing edges of the peak. Spectra taken at the leading and trailing edge of the anthracene peak are shown superimposed in figure 11. Further work identified the impurity as t-butyl benzene at a level of about 5%. Another example of the use of the diode array detector to demonstrate peak purity is shown in figure 12. A chromatogram is shown containing five peaks and spectra have been taken of peak (a) and peak (b) halfway up the rising side of each peak, at the top of each peak and halfway down the trailing edge of each peak.

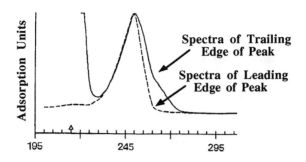

Figure 11 Superimposed Spectra Taken at the Leading and Trailing Edges of the Anthracene Peak

The spectra are also included in the figure. The impure and the pure peak are unambiguously identified demonstrating the power of the detecting system for analytical purposes.

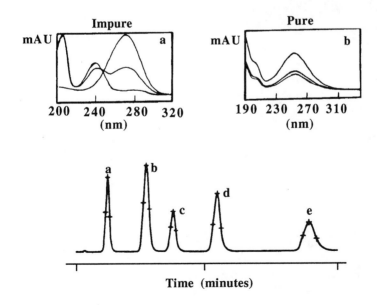

Figure 12 Diode Array Spectra Demonstrating Peak Purity

The versatility and advantages of the diode array detector are obvious. Although originally, due to its expense, it was used mainly as a research instrument and in method development, its cost has been reduced significantly and it is now considered the best UV detector for use in general LC analysis. The performance of both types of multi-wavelength detectors are very similar and typical values for their more important specifications are as follows: they have a sensitivity of about 1×10^{-7} g/ml and a linear dynamic range of about 1×10^{-7} to 5×10^{-4} g/ml with a response index range of 0.97 - 1.03.

References

1. A. J. P. Martin and S. S. Randall, *Biochem. J.* **49**(1951)293.
2. A. Tiselius and D. Claesson, *Arkiv. Kemi Minerl. Geol.,* **15B (No.18)**(1942).
3. E. Katz and R. P. W. Scott, *J. Chromatogr.*, **253**(1982)159.
4. C. G. Horvath and S. R. Lipsky, *Nature*, **211**(1966)748.
5. J. J. Kirkland, *Anal. Chem.*, **43**(1971)1095.
6. G. Brooker, *Anal. Chem.* **43**(1971)1095.
7. J. M. Miller and R. Strusz, *Am. Lab.* **Jan**(1970)29

CHAPTER 9

THE FLUORESCENCE DETECTOR AND OTHER LIGHT PROCESSING DETECTORS

Fluorescence is a specific type of luminescence. When molecules are excited by electromagnetic radiation to produce luminescence, the phenomena is termed photoluminescence. If the release of electromagnetic energy is immediate, or stops on the removal of the excitation radiation, the substance is said to be fluorescent. If, however, the release of energy is delayed, or persists after the removal of the exciting radiation, then the substance is said to be phosphorescent. Due to its persistence (even with a short but significant lifetime), phosphorescence is of little use for LC detection. This is because the persistence of the signal will cause unacceptable apparent peak broadening with the consequent loss of resolution and the loss of quantitative accuracy. In contrast, fluorescence has been shown to be extremely useful and detectors based on fluorescent measurement have provided some of the highest sensitivities available in LC.

When light is absorbed by a molecule, a transition to a higher electronic state takes place and this absorption is highly specific for the molecules concerned; radiation of a particular wavelength or energy is only absorbed by a particular molecular structure. If electrons are raised, due to absorption of light energy, to an upper excited single state, and the excess energy is not dissipated rapidly by collision with other molecules or by other means, the electron will return to the ground state with the emission of light at a lower frequency and the substance is said to fluoresce. Some energy is always lost before emission occurs and thus, in contrast to Raman scattering, the wavelength of the fluorescent light is always greater than the incident light. Excellent discussions on the theoretical basis of fluorescence have been given by Guilbault (1), Udenfriend (2), and Rhys-Williams (3) and those interested in learning more about fluorescence should refer to these books.

In comparison with other detection techniques fluorescence affords greater sensitivity to sample concentration, but less sensitivity to instrument instability, *e.g.* sensor temperature and pressure. In part this is due to the fact that the fluorescent light is measured against a very low light background and thus against a very low noise level. This is in contrast to light absorption measurements where the signal is imposed on a strong background signal carrying a high noise level. The major disadvantage of fluorescence detection is that relatively few compounds fluoresce in a practical range of wavelengths. Nevertheless a number of compounds, including products from foods, drugs, dye intermediates etc., do exhibit fluorescence and their chromatographic separation can be monitored by this means. Furthermore, the scope of fluorescence detection can be extended by forming fluorescent derivatives. For example, the reagents Fluoropa (*o*–phthalaldehyde) and Fluorescamine (4–phenylspiro(furan–2–(3H),1'–phthalan)–3',3'–dione) provide fluorescent derivatives of primary amines. The general techniques of derivatization will be discussed in a later chapter.

Most fluorescent detectors are configured in such a manner that the fluorescent light is viewed at an angle (usually at right angles) to the direction of the exciting incident light beam. This geometric arrangement minimizes the amount of incident light that may provide a background signal to the fluorescent sensor. As a consequence, the fluorescent signal is sensed against a virtually black background and hence provides the maximum signal to noise ratio. The background signal can often be further reduced still further by the use of an appropriate filter to remove any stray scattered incident light. It should be recalled that reducing the noise will proportionally reduce the minimum detectable concentration.

The fluorescence signal (If) is given by

$$I_f = \phi I_o \left(1 - e^{-kcl}\right)$$

where (ϕ) is the quantum yields (the ratio of the number of photons emitted and the number of photons absorbed),
(I_o) is the intensity of the incident light,
(c) is the concentration of the solute,
(k) is the molar absorbance,
(l) is the path length of the cell.

Fluorescence detectors vary widely in complexity, the simplest consisting of a single wavelength excitation source and a sensor that monitors UV light at all wavelengths. For selected applications, this simple form of fluorescence detector can be very sensitive and very inexpensive. However, with only a single excitation wavelength and no means of selecting the emission wavelength, it obviously has limited versatility. At the other extreme is the fluorescence spectrometer that has been fitted with a sensor cell of appropriate dimensions. Such a detector system is highly complex and versatile and allows both the excitation wavelength and the emission wavelength to be chosen. Furthermore, excitation spectra can be obtained at any fixed fluorescent wavelength or an emission spectrum can be obtained for any fixed excitation wavelength.

The Fluorescence Detectors

The Single Wavelength Excitation Fluorescence Detector

The single wavelength excitation fluorescence detector is probably the most sensitive detector generally available to LC, but, as already stated, it is so, at the sacrifice of versatility. A diagram of a simple form of the fluorescence detector that is excited by light from a "single" wavelength UV source is shown in figure 1.

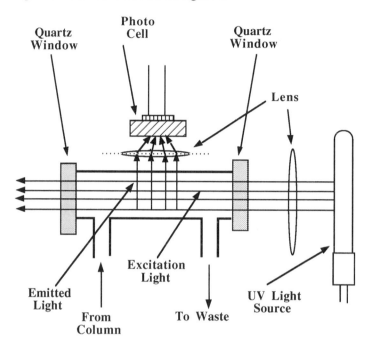

Figure 1 The Single Wavelength Excitation Fluorescent Detector

The UV lamp that provides the excitation radiation is usually a low pressure mercury lamp as it is comparatively inexpensive and provides relatively high intensity UV light at 253.7 nm. Many substances that fluoresce will, to a lesser or greater extent, be excited by light at this wavelength. The light is focused by a quartz lens through the cell and another lens, situated normal to the incident light, focuses the fluorescent light onto a photo cell. Typically, a fixed wavelength fluorescence detector will have a sensitivity (minimum detectable concentration at an excitation wavelength of 254 nm) of 1×10^{-9} g/ml, a linear dynamic range of about 1×10^{-9} to 5×10^{-6} g/ml where the response index is defined as $0.96 < r < 1.04$.

An example of the use of a simple fluorescence detector is afforded by the separation of the mixture of priority pollutants shown in figure 2. The separation was actually monitored by the fluorescence sensor of the TriDet detector the design of which will be discussed later. The excitation light was approximately monochromatic at 254 nm and all the fluorescent light was focused on the photo cell. It is seen that an excellent sensitivity is obtained.

There are detectors that serve as a compromise between the expensive fluorescence spectrometer detector that can select both the excitation and the emission wavelengths and the single wavelength excitation fluorescence detector where the excitation wavelength is fixed and invariant and the total fluorescent light is monitored. A typical example of this compromise type of detector is the Schöffel fluorescence detector that utilizes the monochromater of a dispersive UV spectrometer. The Schöffel detector comprises a UV dispersion spectrometer that is fitted with a special absorption cell having reduced dimensions. This is to ensure that the sensor cell is suitable for sensing the narrow peaks eluted from high efficiency LC columns without undue loss of chromatographic resolution. The wavelength of the excitation light is selected by the monochromator within the normal UV range of the spectrometer. The excitation light passes through the cell and the fluorescent light emitted at right angles to the path of the excitation light is focused on to a photo electric cell. So far, the sensor system is very similar to the fixed wavelength fluorescence sensor. However, in the Schöffel detector, provision is also made for the insertion of appropriate light filters between the sensor cell and the lens that focuses the fluorescent light on to the photo cell. In this way the wavelength of the light monitored by the sensor can be selected by the choice of an appropriate filter. This is rather a crude way of selecting the emission wavelength, but can be quite effective in practice and certainly eliminates the need for a second monochromator with an advantageous reduction in cost. The use of this type of detector in monitoring the separation of amino acids using their fluorescent *o*-phthalaldehyde derivatives is shown in figure 3.

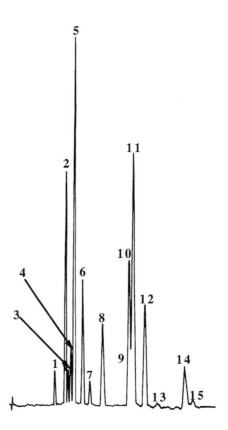

Column: 2 Pecosphere™–5C C18 (150 mm x 4.6 mm) in series. Mobile Phase: 90% acetonitrile/10% water. Flow rate: 2.0 ml/min. Detector Fluorescence (Excitation 254 nm total emission sensed). Sample: 20 µl of NBS Standard.

1. Naphthalene
2. Fluorene
3. Acenaphthene
4. Phenanthrene
5. Anthracene
6. Fluoranthracene
7. Pyrene
8. Benzo(a)anthracene
9. Chrysene
10. Benzo(b)fluoranthene
11. Benzo(k)fluoranthene
12. Benzo(k)fluoranthene
13. Dibenz(a,h)anthracene
14. Indeno(1,2,3,cd)pyrene
15. Benzo(ghi)perylene

Courtesy of the Perkin Elmer Corporation

Figure 2 Separation of the Priority Pollutants Monitored by the Simple Fluorescence Detector

It is seen that a very high sensitivity is realized and the integrity of the chromatographic resolution is well maintained.

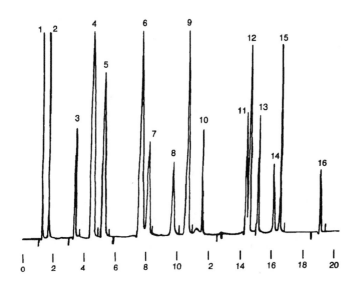

Column.: Supercosil LC-18, 5 cm x 4.6 mm, 5 μm particles. Mobile Phase: methanol : tetrahydrofuran: 0.02 M sodium acetate (pH 5.9 with acetic acid) (A), 22.5:2.5:77.5, B, 80:2.5:17.5. 2 minutes at 100% (A) to 100% (B) in 20 minutes. Flow Rate: 2 ml/min. Sample: 50 to 100 pmol of each derivative in solvent (A).

1. Aspartic acid	2. Glutamic Acid	3. Asparagine	Serine
5. Glutamine	6. Glycine	7. Threonine	8. Arginine
9. Alanine	10. Tyrosine	11. Methionine	12. Valine
13. Phenylalanine	14. Isoleucine	15. Leucine	16. Lysine

Courtesy of Supelco Inc.

Figure 2 The Separation of Some Amino Acids by Monitoring their *o*-Phthalaldehyde Derivatives with a Fluorescence Detector

A final comment: despite its high sensitivity and its relatively low cost, the simple fluorescence detector is not in common use. It would appear that the chemist applies the old adage that 'you get what you pay for' or probably more correctly interpreted, 'unless an instrument is complex and expensive it cannot be accurate and reliable. This popular contemporary concept in many cases is quite incorrect. In fact, the converse is likely to be true, the more simple the device the more reliable it is likely to be and the more precise will be the results it produces. Nevertheless, the complex instrument is usually more versatile and in a relatively small proportion of analyses this versatility may well be essential and the necessary investment in an expensive instrument necessary.

The Multi Wavelength Fluorescence Detector

The multi wavelength fluorescence detector consists of two monochromators, the first that selects the wavelength of the excitation light and the second disperses the fluorescent light and provides a fluorescence spectrum or allows the separation to be monitored at a selected fluorescence wavelength, A diagram of the multi wavelength fluorescence detector is shown in figure 4.

Basically the detector comprises a fluorescent spectrometer fitted with suitable absorption cell that can be used with LC columns without degrading the resolution of the column. The spectrometer incorporates two distinctly different light paths and as a result the optical system appears quite complicated. If the different light paths are considered separately, that is firstly, the path of the excitation light and secondly, the path of the fluorescent light, the diagram is easier to understand.

The excitation source that emits UV light over a wide range of wavelengths (usually a deuterium lamp) is situated at the focal point of an ellipsoidal mirror shown at the top left hand corner of the diagram. The parallel beam of light is arranged to fall on a toroidal mirror that focuses it onto the grating on the left-hand side of the diagram. This grating allows the wavelength of the excitation light to be selected or the whole excitation spectrum scanned providing a complete range of excitation wavelengths. The selected wavelength then passes to a spherical mirror and then to a ellipsoidal mirror at the base of the diagram which then focuses it onto the sample. The excitation light path is largely situated on the left-hand side of the diagram.

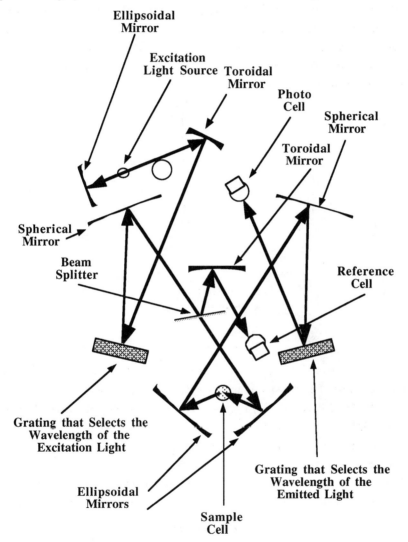

Figure 4 The Fluorescence Spectrometer Detector

Between the spherical mirror and the ellipsoidal mirror, in the center of the diagram, is a beam splitter that reflects a portion of the incident light onto another toroidal mirror. This toroidal mirror focuses the portion of incident light onto the reference photo cell, providing an output that is proportional to the strength of the incident light. The path of the fluorescent light is largely on the right hand side of the diagram. Fluorescent light from the cell is focused by an ellipsoidal mirror on to a spherical mirror at the top right-hand side of the diagram. This mirror focuses the light onto a grating situated at about center right of the figure.

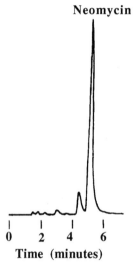

Column: Supercosil LC-8, 15 cm x 4.6 mm, 5 μm particles: Mobile Phase: tetrahydrofuran : 0.0056M sodium sulfate/0.007M acetic acid/0.01M pentane sulfonate, 3:97. Flow rate: 1.75 ml/min. Post Column reagent: 1L 0.4M boric acid/0.38M potassium hydroxide containing 6 ml 40% Brij-35, 4 ml mercaptoethanol, 0.8g o-phthalaldehyde. Flow rate 0.4 ml/min. Mixer 5 cm x 4.6 mm column packed with glass beads. Reactor 10 ft x 0.5 mm knitted Teflon capillary tubing. Reaction Temperature 40°C. Sample 20 ml of a mobile phase extract of a commercial sample. Excitation wavelength 365 nm; emission wavelength 418 nm.

Courtesy of Supelco Inc.

Figure 5 Detection of Neomycin OPA Derivative at an Excitation Wavelength of 365 nm and an Emission Wavelength of 418 nm

This grating can select a specific wavelength of the fluorescent light to monitor, or scan the fluorescent light and provide a fluorescent spectrum. Light from the grating passes to a photoelectric cell which monitors its intensity. The instrument is quite complex and, as a result, rather expensive; however, from the point of view of measuring fluorescence it is extremely versatile. The use of the detector to optimize both the wavelength of the excitation light and that of the fluorescence light to provide high selectivity for the Fluoropa derivative of neomycin is shown in figure 5. It is a very good example of the selection of a specific excitation light wavelength and the complementary emission light wavelength to provide maximum sensitivity.

The principle of optimizing excitation and emission light wavelengths to obtain maximum sensitivity can become quite complex as shown by the separation of some priority pollutants depicted in figure 6. The separation was carried out on a column 25 cm long, 4.6 mm in diameter and packed with a C18 reversed phase. The mobile phase was programmed from a 93% acetonitrile, 7% water to 99% acetonitrile, 1% water over a period of 30 minutes. The gradient was linear and the flow rate was 1.3 ml/min. All the solutes are separated and the compounds, numbering from the left, are given in the table. The separation illustrates the clever use of wavelength programming to obtain the maximum sensitivity. The program that was used is shown below.

Fluorescence Detector Program

Time (seconds)	Wavelength of Excitation Light	Wavelength of Emitted Light
0	280 nm	340 nm
220	290 nm	320 nm
340	250 nm	385 nm
510	260 nm	420 nm
720	265 nm	380 nm
1050	290 nm	430 nm
1620	300 nm	500 nm

During the development of the separation, both the wavelength of the excitation light and that of the emission light was changed to an optimum for the particular solute. This ensured that each solute, as it was eluted, was excited at the most effective wavelength and then monitored at the strongest fluorescent wavelength.

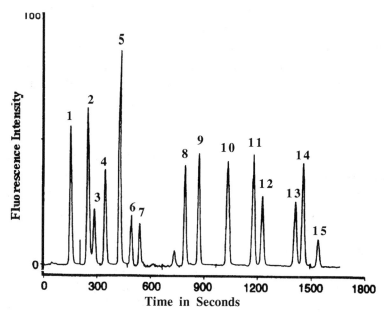

Courtesy of the Perkin Elmer Corporation

Fifteen Priority Pollutants

1 Naphthalene
2 Acenaphthene
3 Fluorene
4 Phenanthrene
5 Anthracene
6 Fluoranthene
7 Pyrene
8 Benz(a)anthracene

9 Chrysene
10 Benzo(b)fluoranthene
11 Benzo(k)fluoranthene
12 Benzo(a)pyrene
13 Dibenz (a,h)anthracene
14 Benzo(ghi)perylene
15 Indeno(123-cd)pyrene

Figure 6 Separation of a Series of Priority Pollutants with Programmed Fluorescence Detection

It is seen that the analysis involves an elaborate procedure that is carried out with a very complex and expensive instrument. Nevertheless, as already implied, if the analysis is sufficiently important it may be necessary to resort to this type of instrumentation. The system can also provide fluorescence or excitation spectra, should they be required, by arresting the flow of mobile phase when the solute resides in the detecting cell and scanning the excitation or fluorescent light. (This is the same technique as that used to provide UV spectra with the variable wavelength UV detector). In this way, it is possible to obtain excitation spectra at any chosen fluorescent wavelength or fluorescent spectra at any chosen excitation wavelength. Consequently, even with relatively poor spectroscopic resolution many hundreds of spectra can be produced, any or all of which (despite many spectra being very similar) can be used to confirm the identify a compound.

The Evaporative Light Scattering Detector

The evaporative light scattering detector incorporates a spray system that continuously atomizes the column eluent into small droplets that are allowed to evaporate, leaving the solutes as fine particulate matter suspended in the atomizing gas. The atomizing gas may be air or an inert gas if so desired. The suspended particulate matter passes through a light beam and the light scattered by the particles viewed at 45° to the light beam using a pair of optical fibers. The scattered light entering the fibers falls on to a photomultiplier, the output of which is electronically processed and passed either to a computer or to a potentiometric recorder. A diagram of the light scattering detector is shown in figure 7. The detector responds to all solutes that are not volatile and as the light dispersion is largely Raleigh scattering, the response is proportional to the mass of solute present; as a consequence, it is sometimes refered to as the *mass detector*. To ensure good linearity the droplet size must be carefully controlled as it, in turn, controls the particle size of the dried solutes. The sensitivity of the detector is claimed to be between 10 and 20 ng of solute. However, in these terms it is difficult to compare it with other detectors.

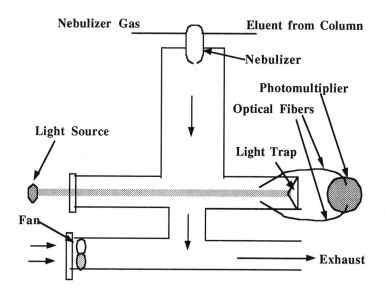

Courtesy of Polymer Laboratories Inc.

Figure 7 A Diagram of the Evaporative Light Scattering Detector

If the solute is assumed to be eluted at a (k') of 1 from a column 15 cm long, 4.6 mm in diameter and packed with 5 m particles, then this would indicate a sensitivity in terms of concentration of approximately 2×10^{-7} g/ml, which is equivalent to the response of a very sensitive refractive index detector. The great advantage of the detector, however, is its catholic response and that its output is linearly related to the mass of solute present.

A diagram of the sensor of the evaporative light scattering detector manufactured by Polymer Laboratories is shown in figure 8. The eluent is atomized in a stream of nitrogen and the finely divided spray passes down a heated chamber to evaporate the solvent. The removal of the solvent converts the stream of droplets into a stream of particles which then pass through a collimated light beam. The light scattered by the particles at an angle to the incident light is focused onto a photomultiplier and the output is processed in an appropriate manner electronically. The device is fairly compact and relatively simple to operate.

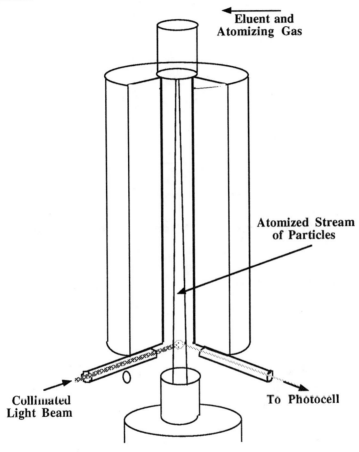

Courtesy of Polymer Laboratories Inc.

Figure 8 The Sensor of a Commercial Evaporative Light Scattering Detector

An example of the results obtained when used for monitoring a general lipid class analysis is shown in figure 9.

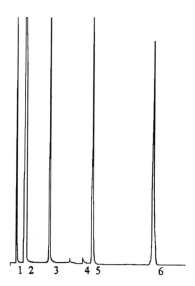

Courtesy of Polymer Laboratories Inc.

Peak	Compound	Mass (µg)	Retention Time (min)
1	cholesterol ester	5	0.717
2	triglyceride	18	1.746
3	cholesterol	10	4.687
4	unknown	10	8.860
5	phosphatidyl choline	10	10.028
6	phosphatidyl ethanolamine		17.390

Figure 9 The Separation of Some Lipid Class Materials Monitored by a Evaporative Light Scattering Detector

Estimation of the minimum detectable mass estimated from this chromatogram was again made to be about 10 ng of solute. To some extent, this detector provides a replacement for the transport detector as it detects all substances irrespective of their optical or electrical properties.

Liquid Light Scattering Detectors

Light scattering detectors respond to the light scattered by a polymer or large molecular weight substances present in the column eluent as it passes through an appropriate sensor cell while illuminated by a high intensity beam of light. The light source is usually a laser (light amplification by the stimulated emission of radiation) that generates light at an appropriate wavelength for measurement. There are two basic forms of the detector: the *low angle laser light scattering* (LALLS) detector and the *multiple angle laser light scattering* (MALLS) detector. Both devices are in common use but the multiple angle laser light scattering detector has greater versatility as it can also provide molecular dimensions as well as molecular weights. As the scattered light is measured at a very small angle to the incident light (virtually 0°), the low angle laser light scattering detector signal can also be affected by scattering from contaminating particulate matter that is always present in the eluent. This can result in considerable noise, which, in turn, will reduce the detector sensitivity. Discussions on the subject have been given by Wyatt [5] and some early experimental results reported by D. T. Phillips [6].

The ratio of the intensity of the light scattered at an angle (ϕ), (I_ϕ) to the intensity of the incident light (I_o), for Rayleigh light scattering is given by

$$\frac{I_\phi}{I_o} = \alpha \omega R_\phi \qquad (1)$$

where, (α) is the attenuation constant,
(ω) is a function of the refractive index,
and (R_ϕ) is Rayleigh's constant

Thus,
$$R\phi = \frac{I_\phi}{\alpha \omega I_o} \qquad (2)$$

Now the molecular weight (M_W) of the solute is related to the Rayleigh factor by the following expression,

$$M_w = \frac{R\phi}{c(K - 2A_2 R\phi)} \qquad (3)$$

where (c) is the concentration of the solute,
(A$_2$) is a function of polymer-polymer interactions,
and (K) is the polymer optical constant.

Substituting for (Rf) in (3) from (2)

$$M_w = \frac{\dfrac{I_\phi}{\alpha \omega I_o}}{c\left(K - 2B_2 \dfrac{I_\phi}{\alpha \omega I_o}\right)} = \frac{I_\phi}{c(\alpha \omega I_o K - 2B_2 I_\phi)} \qquad (4)$$

where

$$K = \frac{2\pi^2 \eta^2}{\lambda^4 N \left(\dfrac{d\eta}{dc}\right)^2} \qquad (5)$$

where (η) is the solvent refractive index,
(l) is the wavelength of the light in vacuum,
and (N) is Avogadro's number.

Equation (4) gives the basic relationship between the molecular weight of the scattering material, the intensity of the scattered light and the physical properties of the materials and equipment being involved. Equation (4), however, contains constants, the magnitude of which are difficult to determine. Consequently, in practice a simple graphical procedure is used to determine the molecular weight of the solute without the need to determine all the pertinent constants. Rearranging equation (3)

$$\frac{1}{M_w} = \frac{c(K - 2A_2 R\phi)}{R\phi} = \frac{cK}{R\phi} - 2cA_2$$

or
$$\frac{cK}{R\phi} = 2cA_2 + \frac{1}{M_w} \tag{6}$$

Now (c), (K), and ($R\phi$) are either known or can all be calculated from known data and light scattering measurements; thus, by plotting $\left(\frac{cK}{R\phi}\right)$ against (c) a straight line will be produced with the intercept being $\left(\frac{1}{M_w}\right)$.

The Low Angle Laser Light Scattering Detector

The optical system of the low angle laser light scattering detector produced by LDC Analytical of the Thermo Instruments Corporation is shown diagramatically in figure 10. In order to conserve space and make the optical system compact, a folding prism is used that allows the device to be contained to a reasonable size and still accommodate the length of the laser generator.

Light from the laser passes through a diverging lens, through a chopper and then through the folding prism. The laser beam then passes out of the folding prism, through some measuring attenuators, then through a calibrating attenuator shutter and then through the cell. Between the cell and the relay lens is an annular mask that only allows light scattered in the cell at a low angle to pass to the relay lens. Between the annular shutter and the relay lens is a safety attenuator that ensures that none of the potentially damaging laser light can reach the photomultiplier. The scattered light is focused through a field stop onto the forward detector lens. Between the field stop and the forward detector lens is a prism that allows the scattered light to be viewed through a microscope. Between the forward detector lens and the rear detector lens is a filter holder and an analyzer/polarizer. Finally the light is focused through a sensor aperture to an opal diffuser that spreads the scattered light through a red filter and onto the photomultiplier.

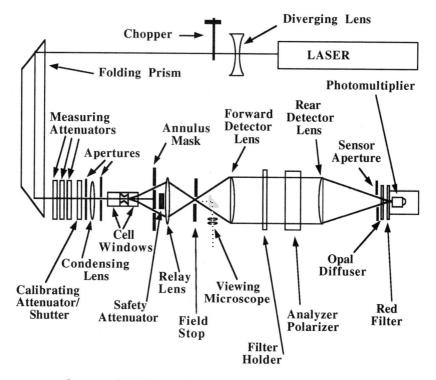

Courtesy of LDC Analytical, Thermo Instruments Corporation.

Figure 10 Optical Diagram of a Low Angle Laser Light Scattering Detector

The device is often operated with a refractive index detector in series in order to coincidentally measure the refractive index of the eluent. This is necessary to calculate (K) from the refractive index as given in equation (5). A common refractive index detector used for this purpose is that manufactured by the Wyatt Technology Corporation and it is described as the interferometer detector in chapter 11. As discussed above the molecular weight of a solute is determined from the intercept of the graph relating ($\frac{cK}{R\phi}$) to the solute concentration (c) as shown in figure 11. The concentration is usually determined by the refractive index detector from prior calibration.

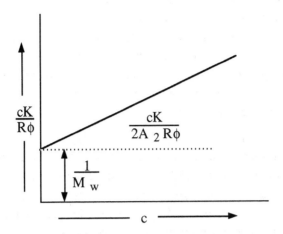

Figure 11 Determination of Molecular Weight from Low Angle Light Scattering Measurements

The overall sensitivity of the detector appears to be very similar to that of the refractive index detector with about the same linearity. However, the most important characteristic of this detector is not its propensity for accurate quantitative analysis but its proficiency in providing molecular weight data for extremely large molecules.

The Multiple Angle Laser Light Scattering (MALLS) Detector

The multiple angle laser light scattering detector differs from the previously described low angle device, in that scattering measurements are made at a number of different angles, none of which are close to the incident light. This reduces the problem associated with scattering from particulate contaminants in the sample. Data taken at a series of different angles to the incident light allows the *root-mean-square* (*rms*) of the molecular radius $\langle r^2 \rangle^{1/2}$ to be calculated in addition to the molecular weight of the substance. The relationship that is used is as follows:

$$\frac{cK}{R\phi} = a\langle r^2 \rangle^{1/2} \sin(\theta)^2 + bM_w \qquad (7)$$

In fact, theory can provide explicit functions for (a) and (b) but values for these constants are usually obtained from calibrating substances of known molecular weights and molecular radii. Furthermore, each photocell will not have precisely the same response to low levels of light intensity and consequently calibration procedures are also necessary to take their different responses into account to provide appropriate correction factors..

The total number of different angles at which the scattered light is measured differs with different instruments, and commercial equipment that measure the intensity of the scattered light at 16 different angles are available. It is clear that the greater the number of data points taken at different angles, the more precise the results will be. A diagram of a (MALLS) detector system which measures the light scattered at *three* different angles is shown in figure 12.

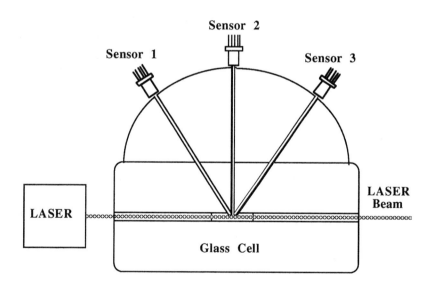

Courtesy of Wyatt Technology Corporation

Figure 12 The Multiple Angle Laser Light Scattering Detector (miniDawn®)

This device, manufactured by Wyatt Technology Corporation, is called the miniDawn®. It contains no mirrors, prisms or moving parts and is designed such that the light paths are direct and not "folded". As seen in figure 12, light passes from the laser (wavelength 690 nm) directly through a sensor cell. Light scattered from the center of the cell passes through three narrow channels to three different photocells, set at 45° and 90° and 135° to the incident light. Thus scattered light is continuously sampled at three different angles during the passage of the solute through the cell.

A continuous analog output is provided from the 90° sensor and all the sensors are sampled every 2 sec. The molecular weight range extends from 10^3 to 10^6 Daltons and the *rms* radii from 10 to 50 nm. The total cell volume appears to be about 3 µl and the scattering volume is 0.02 µλ. The detector has a sensitivity, defined in terms of the minimum detectable excess Rayleigh ratio of 5 x 10^{-8} cm^{-1} which is difficult to translate into normal concentration units but appears to be equivalent to a minimum detectable concentration of about 10^{-6} g/ml.

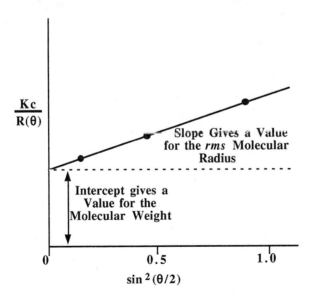

Figure 13 Calibration Curves

The relationship between the intensity of the scattered light, the scattering angle and the molecular properties are as follows:

$$\frac{cK}{R\phi} = 2cA_2 + \frac{1}{M_w P(\phi)}$$

where $P(\theta)$ describes the dependence of the scattered light on the angle of scatter and the other symbols have the meanings previously attributed to them.

In fact, the relationship between the angle of scattering, (θ), the molecular weight and the *rms* molecular radius of the solute is obtained using equation (7). Employing appropriate reference materials, graphs of the form shown in figure 13 can be constructed to evaluate constants (a) and (b) and thus permit the measurement of the molecular weight and molecular radius of unknown substances.

References

1. G. G. Guilbault, *Practical Fluorescence*, Marcel Dekker, Inc., New York (1973).
2. S. Udenfriend, *Fluorescence Assay in Biology and Medicine*, Academic Press, New York and London (1962).
3. A. T. Rhys-Williams, *Fluorescence Detection in Liquid Chromatography*, Perkin Elmer Ltd., Beaconsfield, England (1980).
4. W. W. Christie, *J. Lipid Research*, **26**(1985)507
5. P. J. Wyatt, *Applied Optics* **7**(196801879.
6. D. T. Phillips, *Nature* **221**(1969)1257.
7. P. J. Wyatt, *LC-GC*, **13, No 4**(1995)286.

CHAPTER 10

THE ELECTRICAL CONDUCTIVITY DETECTOR AND THE ELECTROCHEMICAL DETECTOR

The electrical conductivity detector was, at one time, considered to be an electrochemical detector and was included in the group of solute property detectors. The electrical conductivity detector actually measures the conductivity of the mobile phase and gives a small output when only water is present in the sensor. Such background signals from the mobile phase are backed off by suitable electronic adjustments. If the mobile phase contains buffers, the detector can give an output that completely overwhelms that from any solute being eluted, making detection impossible. It follows that the electrical conductivity detector is not a *solute property* detector but a *bulk* property detector. In fact, due to it being a bulk property detector, it senses all ions whether they are from a solute or from the mobile phase and this disadvantage evoked the development of the *ion suppresser*, the details of which will be discussed later.

In contrast, the electrochemical detector responds only to substances that can be oxidized or reduced and thus, providing the mobile phase is free of such materials, it will only detect oxidizable or reducible substances when they are eluted. It follows that this detector is not only a solute property detector but is also a specific detector. The electrical conductivity detector is a non-specific detector and used widely in ion chromatography where it occupies a unique and almost exclusive position. In contrast, the electrochemical detector, in its present form, is usually employed for very special applications but has the advantage of having a very high sensitivity, some claiming it to be more sensitive than the fluorescence detector.

The Electrical Conductivity Detector

An acid base or salt, when dissolved in water, ionizes into charged ions. If an undissociated acid, base and salt are represented by HA, BOH and BA, respectively, the ionization can be described by the following equations,

$$HA \rightarrow H^+ + A^-$$
$$BOH \rightarrow B^+ + OH^-$$
$$BA \rightarrow B^+ + A^-$$

If the acid or base is weak, ionization is incomplete and an equilibrium condition occurs where

$$K_A (HA) \leftrightarrow \left(H^+\right) + \left(A^-\right)$$
$$K_B (BOH) \leftrightarrow \left(B^+\right) + \left(OH^-\right)$$

where (K_A) and (K_B) are the dissociation constants of the acid and base respectively.

In any event, some ions are produced, the ionic concentration depending on both the original concentration of the acid or base, the respective dissociation constant and the physical properties of the solvent, *e.g.* the pH and dielectric constant. Under the influence of a potential gradient, the ions in solution can carry an electric charge; consequently, if a potential is applied across two electrodes situated in the solvent, a current will flow and the solvent is said to be conducting. It is clear that such a system could be used to detect ionic species in LC.

However, early investigations into the electrical properties of solutions were somewhat frustrating as, initially, direct current (DC) measurements were made which resulted in the hydrolysis of the solvent and the evolution of oxygen at the anode and hydrogen at the cathode. The presence of bubbles of gas changed the nature of the electrode surface and hence the electrode resistance and confused the results. It was not until alternating voltages (AC) were applied to the electrodes that the polarization was eliminated and consistent results were obtained. Theoretically, it is the *impedance* not the *resistance* of the electrode system that should be measured. When an AC voltage is applied across a conductor, there can be an inductive current controlled by the inductance of the conductor, a capacity current controlled by its capacity and a resistive current controlled by the conductor's resistance. The situation is further complicated by the fact that there is a phase difference between the three currents, the inductive current leads the resistive current and the capacity current lags behind the resistive current and thus the measurement of the electrode impedance would appear to be rather difficult. In practice, however, due to the geometry of the electrode system, the inductive and capacity components of the current are made extremely small compared with the resistive current and thus the measurement of resistance can be made using the total current that is in phase with the applied voltage.

Employing AC potentials it has been shown that Ohm's law is obeyed by ionic solutions:

$$V = RI$$

where (V) is the potential applied across the electrodes in volts,
(I) is the current flowing between the electrodes in amps,
and (R) is the resistance between the electrodes in ohms.

Generally, in physical chemistry the conductivity of a solution has greater significance than its resistance. However, it is the resistance of the electrode system that determines the current across it and it is therefore of interest to see how the detector output is related to solute concentration.

The resistance (R) of any conductor varies directly as its length (L) and inversely as its cross sectional area (a).

Thus, $$R = \frac{\rho L}{a} \quad (1)$$

where (ρ) *specific resistance* of the conductor.

The *specific resistance* of a conductor is the resistance across two opposite faces of a 1 cm cube of the conductor material.

The *specific conductance* (κ) of a solute is defined as the reciprocal of the specific resistance, thus

$$\kappa = \frac{1}{\rho} \quad (2)$$

Thus the conductance of a given solution (C) is given by,

$$C = \frac{1}{R} = \frac{\kappa a}{L} \text{ ohms}^{-1} \quad (3)$$

Now the conduction of all the ions produced by 1 g equivalent of an electrolyte at any particular concentration can be evaluated by imagining two large parallel electrodes, 1 cm apart, and the whole of the solution placed between them. The conductivity of the system is called the equivalent conductance (Λ).

Now suppose that 1 g equivalent of the electrolyte is dissolved in (v) ml of solution; then it follows that this will cover (v) cm^2 of electrode area. It follows that in the above equation (a) becomes (v) and (l) is unity and

$$\Lambda = \frac{1000\kappa}{c} \text{ ohms}^{-1}\text{cm}^2$$

where (c) is the electrolyte concentration in gram equivalents /liter.

or
$$\kappa = \frac{\Lambda c}{1000} \quad (4)$$

Substituting for (k) from (4) in (3) and rearranging,

$$C = \frac{1}{R} = \frac{\Lambda c a}{1000 L} \text{ ohms}^{-1} \quad (5)$$

It is seen that the resistance of the cell is inversely proportional to the electrolyte concentration (c), the equivalent conductance of the electrolyte and the geometry of the cell [(a) and (L)]. It follows that the output of the cell, which is, in fact proportional to the resistance change of the cell, must be modified to provide an output linearly related to electrolyte concentration.

The first conductivity detector was developed by Martin and Randall as long ago as 1951 [1]. Improved cell designs have been described by Harlan [2], Sjoberg [3] and more stable and sensitive electronic circuits have been Avinzonis and Fritz [4] and Berger [5]. Scott *et al.* [6] inserted electrodes in the wall of a column to monitor the progressive band dispersion along a packed LC column. Keller [7] described a bipolar electrical conductivity detector and Kourilova *et al.* [8] describe a electric conductivity sensor for use in LC having a volume of 0.1 µl.

The basic sensor, being so simple, has changed little over the years. The detecting volume has become much smaller but the sensor still consists of two electrodes (that can take a range of geometric shapes) situated in the column eluent, the resistance (or, as previously discussed, more strictly the impedance) of which is measured by a suitable electronic circuit. Figure 1 shows three simple forms of the conductivity sensor although many other geometric forms have been described in the literature.

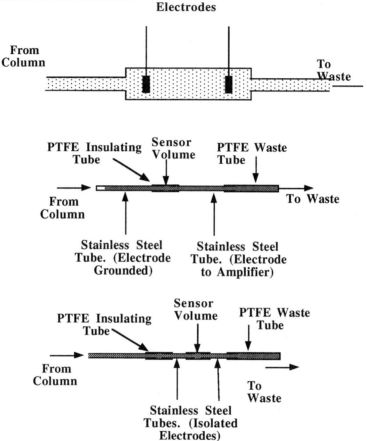

Figure 1   The Electrical Conductivity Detector

It is seen that the basic sensor is, indeed, very simple and the top design can be easily constructed to have an effective sensing volume of a few microliters. The lower sensors consist of short lengths of stainless steel tubing (about 0.020 in I.D. and 1/16 in O.D). The two tubes fit into a PTFE tube and the space between the two tubes constitutes the actual sensing volume. The lower sensor is basically the same, but two lengths of tube are employed, both of which are isolated from any other metal connection by PTFE tubing. As a consequence, the two electrodes are electrically isolated, which can be an advantage when they are used with certain types of electronic measuring circuits. It is clear that these types of sensor can have sensing volumes of only a few nanoliters. For this reason, geometric forms similar to these are often used as a detector in capillary electrophoresis. The total electrical conductivity detector system is shown in figure 2

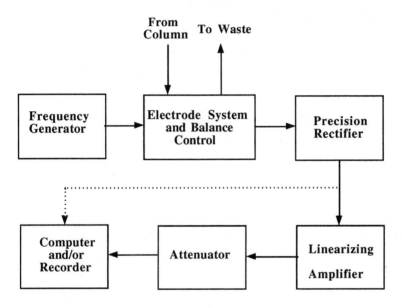

Figure 2 The Electrical Conductivity System

The Electrical Conductivity System consists of a frequency generator (the frequency is usually between 1,000 and 5,000 Hz) that provides an AC potential across the cell. As already discussed, an AC potential must be used to avoid electrode polarization. The sensor is usually placed in one arm of a Wheatstone bridge as shown in figure 3. The out-of-balance signal is then rectified with a precision rectifier and the DC signal either passed to nonlinear amplifier or a computer data acquisition system. The non-   linear amplifier modifies the signal so that the output is linearly related to ion concentration. Alternatively, if the output from the precision rectifier is passed directly to the computer, then the linearization can be carried out with appropriate software. If fed directly to the computer the chromatogram is presented on the computer printer.

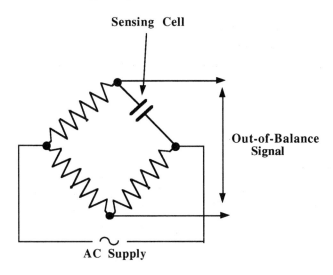

If a single sensor is employed, it is situated in one arm of the bridge; if, however, a reference sensor is also used, then the two sensors are situated in opposite arms of the bridge. In either case the out–of–balance signal can be passed to a precision rectifier and the output either handled by an analog circuit or sampled directly by the computer. Sometimes a variable resistance is situated in one of the other arms of the bridge and is used for zero adjustment to back-off any signal arising from ions contained by the mobile phase.

Figure 3 One Form of the Wheatstone Bridge Used with the Electrical Conductivity Detector

The conductivity detector is a bulk property detector and, as such, responds to any electrolytes present in the mobile phase (*e.g.* buffers etc.) as well as the solutes. It follows that the mobile phase must be arranged to be either non-conducting, which in many cases is difficult if not impossible to achieve, or the mobile phase buffer electrolytes must be removed prior to the detector. This technique of buffer ion removal is commonly called *ion suppression*. The first type of ion suppression that was developed involved the use of a second ion exchange column subsequent to the analytical column of an appropriate form for the removal of buffer cations, replacing them with hydrogen ions. Conversely if cations were being separated a suitable anion exchanger would replace the buffer anions with hydroxyl ions. This approach works well providing the ion suppresser (ion remover) does not cause too much peak dispersion and consequent loss of resolution.

Another alternative and far more efficient procedure is to employ an organic buffer, such as methane sulfonic acid and a short length of reversed phase column subsequent to the analytical column. After the separation has been achieved and the individual ions are eluted from the column, the methyl sulfonate that is eluted with them is absorbed, virtually irreversibly, on to the revered phase leaving the ions of interest to enter the detector in the absence of any buffer ions. As the suppresser column is, itself, a high efficiency LC column, little or no dispersion occurs and consequently no loss of resolution. If a basic buffer is required, then a tertiary butyl ammonium salt can be used which would be removed equally effectively by the reversed phase suppresser column.

It is obvious that this type of suppresser column will have a limited lifetime but can be regenerated rapidly and easily by passing acetonitrile through it, followed by water. An example of the use of this type of suppression system is given in figure 4. It is seen that an excellent separation of the cations was achieved and the separation was carried out with 20 nM of methane sulphonic acid present in the mobile phase which was completely removed by the ion suppresser column.

Although this technique of ion suppression is frequently used in ion exchange chromatography, depending on the nature of the separation, a specific type of ion suppression column or exchange membrane may be required that will be appropriate for the phase system that is chosen and the analyte ion involved.

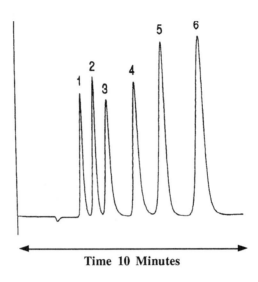

Courtesy of Dionex Inc.

A proprietary ion exchange column, IonPacCS12, was used and the mobile phase consisted of a 20 nM methanesulphonic acid solution in water. A flow rate of 1 ml/min was employed and the sample volume was 25 µl.

1. Lithium 2. Sodium 3. Ammonium
4. Potassium 5. Magnesium 6. Calcium

Figure 4 Determination of Alkali and Alkaline Earth Cations

A wide variety of different types of ion suppression columns are available for this purpose. It should be pointed out that any suppressor system introduced between the column and the detector that has a finite volume will cause some band dispersion. Consequently, the connecting tubes and suppression column must be very carefully designed to reduce this dispersion to an absolute minimum.

The electrical conductivity detector has a sensitivity (minimum detectable concentration) of about 5×10^{-9} g/ml and a linear dynamic range of 5×10^{-9} to 1×10^{-6} where $0.97 < r < 1.03$. It is used in probably over 95% of all analyses involving ion exchange procedures to separate inorganic and organic ions. It is simple, reliable and accurate besides being one of the least expensive detectors

The Electrochemical Detector

This detector responds to substances that are either oxidizable or reducible and the electrical output results from an electron flow caused by the reaction that takes place at the surface of the electrodes. If the reaction proceeds to completion, exhausting all the reactant, then the current becomes zero and the total charge that passes will be proportional to the total mass of material that has been reacted. For obvious reasons, this process is called coulometric detection. If, on the other hand, the mobile phase is flowing passt the electrodes, the reacting solute will be continuously replaced as the peak passes through the detector. While there is solute present between the electrodes, a current will be maintained, albeit varying in magnitude. Until relatively recently, this procedure was the most common employed in electrochemical detection and is called amperometric detection.

In the electrochemical detector, three electrodes are employed, the working electrode (where the oxidation or reduction takes place), the auxiliary electrode and the reference electrode (which compensates for any changes in the background conductivity of the mobile phase). The electrochemical processes that take place at the electrode surface can be very complex; however, the dominant reaction can be broadly described as follows. At the surface of the electrode the reaction of the solutes is extremely rapid and proceeds almost to completion. This results in the layer close to the electrode being virtually depleted of reactant. Consequently, a concentration gradient is established between the electrode surface and the bulk of the solution. This concentration gradient results in the solute diffusing into the depleted zone at a rate that is proportional to the concentration of the solute in the bulk of the mobile phase. As a result, the current generated at the surface of the electrode will be determined by the rate at which the solute reaches the electrode and consequently will exhibit a linear response with respect to solute concentration. The response of the detector or current (i) is described by the following equation,

$$i = nFAK_T cu^a \qquad (6)$$

where (n) is the number of electrons per molecule involved in the reaction,
- (F) is the Faraday Constant,
- (A) is the area of the working electrode,
- (K_T) is the limiting Mass Transfer Coefficient,
- (c) is the solute concentration,
- (u) is the linear velocity of the mobile phase over the surface of the electrode,

and (a) is a constant usually taking a value between 1/3 and 1/2.

It is seen that the current (i), and consequently the sensitivity of the detector, can be increased by either increasing the area of the electrodes, increasing the transfer coefficient or increasing the velocity of the mobile phase past the electrodes. It would appear that increasing the electrode area would be an attractive means of increasing detector sensitivity and, in fact, can change the nature of the detector function, which will be discussed later. However, the conventional increase in surface area while maintaining an amperometric response will also increase the noise, which is often so great that it results in an overall *reduction* in detector sensitivity. Weber and Purdy [9] and Hanekamp *et al.* [10] demonstrated that under certain conditions a reduction in the size of the sensor produces a significant increase in signal-to-noise and a consequent increase in sensitivity. By operating at higher flow rates, and thus increasing the rate of solute transfer rate, the sensitivity can also be increased and this would be an added advantage to miniaturization. However, as shown by equation (6), the sensor will be very flow sensitive and thus must be operated under constant flow conditions and can not be used with flow programming methods of chromatographic development. Miniaturization would also have the advantage of reducing peak dispersion in the sensor.

Electrode Configurations

The electrodes can take a number of different geometric forms which have been described in some detail by Poppe (11). Some examples of different electrode configurations are shown in figure 5.

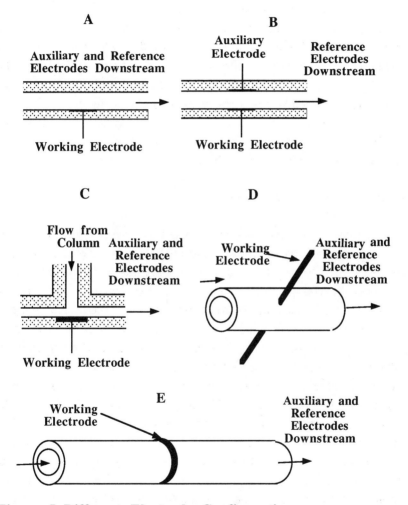

Figure 5 Different Electrode Configurations

Examples (A) and (B) are two common forms of thin-layer cells. (A) has the working electrode sealed into the wall of the cell with the reference and auxiliary electrodes situated down-stream to the working electrode. (B) is a similar type of cell with the auxiliary electrode sealed in the wall of the tube opposite the working electrode and the reference electrode down-stream. (C) is an example of a wall jet electrode where the column eluent is allowed to impinge directly onto the working electrode which is situated opposite the jet. This arrangement not only increases the value of (u), the velocity of the liquid passing over the electrode and thus the transfer coefficient (K_T), but also provides scrubbing action on the surface of the working electrode which reduces, a little, the need for frequent cleaning. (D) and (E) are two examples of cylindrical electrodes; in (D) the working electrode is in the form of a rod stretching across the diameter of the sensor cell and in (E) the working electrode comprises an annular ring set in the cell wall. In both cases the auxiliary and the reference electrodes are down-stream to the working electrode.

Electrode Construction

The choice of material that can be used for the construction of the working electrode is somewhat restricted owing to the need for it to be mechanically rugged and have long term stability. The most common material used earlier is a carbon paste made from a mixture of graphite and some suitable dielectric substance. The disadvantage of this material is its solubility in some solvents, although special waxes or polymers can be used as dielectric binders to contain the graphite which helped reduce the solubility problem. Vitreous or 'glassy' carbon is also an excellent material for electrode construction particularly if it is to be used with organic solvents and is probably the most popular contemporary electrode material. This material is produced by slowly baking a suitable resin at elevated temperatures until it is carbonized and then heating it to a very high temperature to cause vitriation. Vitreous carbon is relatively pure mechanically strong and has good electrical properties. It can be readily cleaned mechanically and performs particularly well relative to other alternative materials when operated at a negative potential. In general glassy carbon electrodes are to be preferred over carbon paste electrodes due to their inherent resistance to solvents but a number of other carbon types have been suggested [12,13]. Mercury was the material first used by Kermula [14] as early as 1952 for electrochemical detection and is still occasionally used in the form of amalgamated gold discs and amalgamated platinum wire [15] and as spools in capillary tubes. Joynes and Maggs [16] employed carbon impregnated silicone rubber membranes as working electrode material and Takato and Muto (17) examined the use of platinum and silver gauze.

Electrochemical detection can impose certain restrictions on both the type of chromatography that can be employed and the mobile phase that can be used. As the detecting system requires the mobile phase to be conducting, the mobile phase must contain water, which means that the majority of 'normal phase' systems are not usable. Furthermore, very high solvent concentrations in the mobile phase may render it insufficiently conducting. Reversed phase chromatography, however, is ideally suited to electrochemical detection. Nevertheless, certain stringent precautions must be taken for the effective use of the detector. The mobile phase must be completely free of oxygen, which can be removed by bubbling helium through the solvent reservoir. In addition, it is important to remove oxygen from the sample before an injection is made. The solvents must also be free of metal ions or a very unstable base line will result. Under some circumstances non-aqueous solvents such as pure acetonitrile can be employed but certain salts like tetrabutyl-ammoniumhexafluorophosphate must be added to render the solvent conducting.

Basic Electrochemical Detector Electronics

A simplified form of the circuit that is in general use is shown in figure 6. To simplify the function of the circuit, the amplifiers can best be considered as operational amplifiers such as 741 or perhaps a field effect operational amplifier such as the TLO81. The auxiliary electrode is held at a fixed potential by the first amplifier, the voltage being selected by the potentiometer (P) that is connected to a regulated power supply. The current flowing through the working electrode is processed by the second amplifier and the output fed to the recorder or data acquisition system.

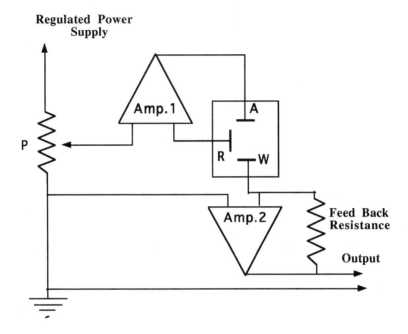

Figure 6 Basic Circuit Used with Electrochemical Detectors

The electrochemical detector in the form described above is extremely sensitive but suffers from a number of drawbacks. Firstly, the mobile phase must be extremely pure and in particular free of oxygen and metal ions. A more serious problem arises, however, from the adsorption of the oxidation or reduction products on the surface of the working electrode. The consequent electrode contamination requires that the electrode system must be frequently calibrated to ensure accurate quantitative analysis. Ultimately, the detector must be dissembled and cleaned, usually by a mechanical abrasion procedure. Much effort has been put into reducing this contamination problem but, although diminished, the problem has not been completely eliminated particularly in the amperometric form of operation. Due to potentially low sensing volume the detector is very suitable for use with small bore columns.

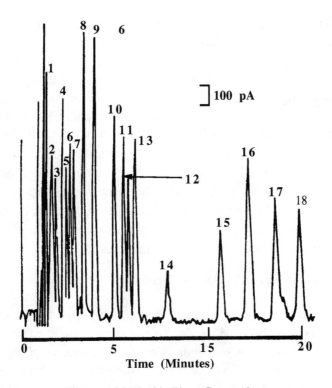

Courtesy of the Perkin Elmer Corporation

Column: HC-3 C18 (100 mm x 4.6 mm); mobile phase: aqueous solution of 100 nM formic acid , 0.35 nM octane sulphonic acid, 1.0 nM citric acid 0.10 nM EDTA, 5% acetonitrile, 0.25% v/v diethylamine, pH to 3.10 with potassium hydroxide; flow rate 1 ml/min; detection: oxidative amperometric with glassy carbon electrode at 100 mV potential *vs.* Ag/AgCl electrode.

1. 3,4 dihydroxymendelic acid 200 pg
2. L-dopa 600 pg
3. vanillymendelic acid 400 pg

4. norepinephrin 200 pg
5. α-methyl dopa 600 pg
6. 3-methoxy,4-hydroxyphenylglycol 400 pg
7. epinephrine 200 pg
8. 3,4-dihydroxybenzylamine 200 pg
9. normetanephrin 400 pg

10. dopamine 200 pg
11. metanephrine 400 pg
12. 3,4-dihydroxyphenylacetic acid 200 pg
13. N-methyl dopamine 400 pg
14. tyramine 1 ng
15. 5-hydroxyindole-3-acetic acid 200 pg
16. 3-methoxytyramine 400 pg
17. 5-hydroxytryptamine 200 pg
18. homovanillic acid 400 pg

Figure 7 The Separation of Some Catacholamines Monitored by an Electrochemical Detector

An example of the use of the electrochemical detector to monitor the separation of a series of catacholamines is shown in figure 7. It is seen that the detector operates at an extremely high sensitivity but the actual sensitivity in g/ml (which would allow some comparison with other detecting methods) can not be evaluated from the data. It should also be noted that the column appears to be packed with 3 micron particles which would provide an efficiency of *ca.* 16,000 theoretical plates. This high efficiency would means that the peaks occupied a very small volume of mobile phase, which, in turn, would mean that, even with the small peak masses, the peak concentrations may be relatively high. Quoting sensitivities in terms of the mass contained in each peak will always give an enhanced impression of detector sensitivity when a high efficiency column is used.

The Multi-Electrode Array Detector

The effective use of an electrode array as a liquid chromatography detector is largely possible due to the development of the porous carbon electrode. This electrode is made of porous graphitic carbon, which has a very high surface area, is mechanically robust and, more important, is permeable to the mobile phase. As a consequence, flow through electrodes can be constructed. The material commends itself to electrochemical detection in a number of ways. As the surface area is greatly in excess of that required for efficient electrochemical reaction, it can suffer excessive contamination before it fails to function. In fact up to 95% of the surface can be contaminated before it requires cleaning. Furthermore, should the electrode eventually become sufficiently contaminated to require cleaning (which, according to the manufacturers, may occur between one and three years of continual use), the contamination can be rapidly removed by flushing with nitric acid. This may be difficult with more than one electrode assembly but as its need is so infrequent the extra effort is not daunting.

The porous graphitic carbon electrode also facilitates the construction of electrode arrays. In use, the porous electrode offers such a large surface area to the solute that 100% of the material is reacted.

Consequently the electrochemical reaction is no longer amperometric, but now coulometric. This is an important difference and makes the array system practical. The electrode system is shown diagramatically in figure 8.

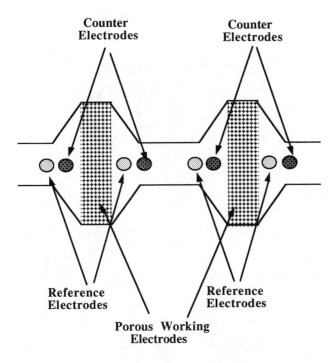

Courtesy ESA Inc.

Figure 8 The Coulometric Electrode System Employing Porous Graphitic Carbon Electrodes

Each electrode unit consists of a central porous carbon electrode, on either side of which is situated a reference electrode and a auxiliary electrode. As the pressure drop across the porous electrode is relatively small, these electrode units can be connected in series forming an array. Normally up to 16 units can be placed in series and these are commercially available. However, a sensor system that contains as many as 80 electrodes in the space of a few millimeters has also been constructed [18].

The array operates with a progressively greater potential being applied sequentially to the electrodes of each consecutive unit. This results in all the solutes migrating through the array until each reaches the unit that has the required potential to permit its oxidation or reduction.

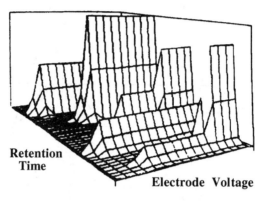

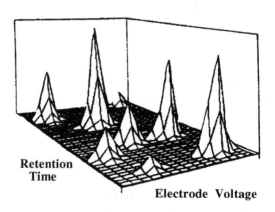

Courtesy of the Analyst.

Figure 9 Three Dimensional Graphs Demonstrating the Difference between Amperometric and Coulometric Detection Employing an Electrode Array

A comparison between amperometric and coulometric detection is shown in figure 9. In amperometric detection only a small part of the solute is reacted and so the remainder can proceed to the next electrode system and be detected again. In this way each of the units will detect all the solutes and the graph shown in the upper part of the figure is produced. It is seen that there is no discrimination by the different electrode voltages. Coulometric detection results in the total solute being reacted and thus it will be detected by that unit that has the required potential and not be sensed by other units. It follows that distinct peaks will be produced only at those units with the appropriate potentials. The result is a three dimensional graph similar to that shown in the lower part of the figure. It is seen that each unit produces a peak at a unit potential that is characteristic of the solute being detected.

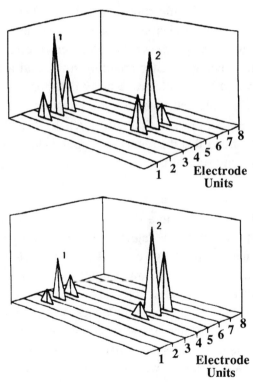

Courtesy of the Analyst.

Figure 10 Use of the Electrochemical Array to Confirm Compound Identity

In fact although the voltage increases progressively from one unit to the next, reaction is not completed at one unit only. This is due to the fact that reaction will take place, not at a specific potential, but over a relatively narrow range of potentials. In practice the signal is usually detected over three contiguous electrode units. For example, the first will oxidize a very small amount of the solute, the second unit will be the dominant unit and oxidize the majority of the solute, while the third unit will oxidize the small remaining quantity of solute. This produces a characteristic pattern of peaks for a particular solute. The ratio of peak height for the three contiguous electrode units that sense the substance will be different for different substances although they may be reacted at the same three electrodes. This is obviously a method of confirming the identity of the solute and is demonstrated in figure 10. The upper graph shows the reference chromatogram of two standards each showing specific retention times and a specific peak pattern as provided by the electrode array detector. The lower graph is for a similar sample and, although the retention times for the pair of solutes is very similar, the pattern given by the electrode array detector clearly shows that the second compound eluted is not the same as that of the second standard.

The electrode array detector also gives improved *apparent* chromatographic resolution in a similar way to that of the diode array detector or any spectroscopic detection system. Two peaks that have not been chromatographically resolved and are eluted together can still be shown as two peaks that are resolved electrochemically and can be quantitatively estimated. Another advantage is that high oxidation potentials can be used without the high background currents and noise that usually accompany such operating conditions. The electrodes that are operating at high voltages are "buffered" by the previous electrodes operating at lower voltages which results in *reduced* background currents and noise.

The advent of the porous carbon electrode made coulometric detection possible and thus opened the way for an effective electrode array detector. An example of the application of the detector to monitoring the separation of a number of neuroactive substances [19] is shown in figure 11.

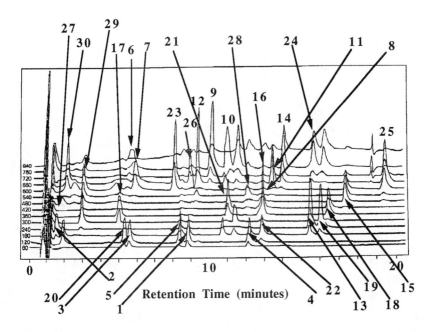

Mobile Phase 1% Methanol to 40% Methanol in a Phosphate (0.1 mol l^{-1} buffer with ion pairing (pH 3.4)

1. Dihydroxyphenylacetic acid
2. Dihydroxyphenylethylene glycol
3. L-Dopa
4. Dopamine
5. Epinephrine
6. Guanine
7. Guanosine
8. Homovanallic acid
9. Hydroxybenzoic acid
10. Hydroxyindoleacetic acid
11. Hydroxyphenylacetic acid
12. Hydroxyphenyllactic acid
13. Hydroxytryptophan
14. Kynurenine
15. Melatonin
16. Metenephrine
17. Methoxyhydroxyphenyl glycol
18. Methoxytyramine
19. N-methylserotonin
20. Norepinephrim
21. Normetenephrine
22. Salsolinol
23. Octopamine
24. Seratomin
25. Tryptophan
26. Tyrosine
27. Uric Acid
28. Vanillic acid
29. Vanylmandelic acid
30. Xanthine

Courtesy of the Analyst.

Figure 11 The Separation of 30 Neuroactive Substances Monitored by an Electrochemical Array

It is seen that for certain applications the electrochemical array detector can be extremely useful. Nevertheless, in order to use the detector, the solutes must be amenable to electrochemical reaction and capable of being separated using a mobile phase that will conduct an ion current

References

1. A. J. P. Martin and S. S. Randall, *Biochem J.*, **49**(1951)293.
2. H. D. Harlan, *Anal Chem.*, **54**(1965)89.
3. C. I. Sjoberg, *Acta Chem. Scand.*, **8**(1954)1161.
4. P. W. Avinzonis and F. Fritz, *Anal. Chem.*, **34**(1962)58.
5. D. Berger, Arkiv Kemie, 4(1952)401.
6. R. P. W. Scott, D. W. J. Blackburn and T. Wilkins, *J. Gas Chromatogr.*, 5(1967)183.
7. J. M. Keller, *Anal. Chem.*, **53(2)**(1981)344.
8. D. Kourilova, K. Slais and M. Krejce, *Inst. Anal. Chem., Czech Acad. Sci.*, **48(4)**(1983)1129.
9. S. G. Weber and W. C. Purdy, *Ind. Eng. Chem. Prod. Res. Dev.*, **20**(1981).
10 H. G. Hanekamp, P. Boss and R. W. Frei, *Trends in Anal. Chem*, **1**(1982)135.
11. H. Poppe, *Instrumentation for High-Performance Liquid Chromatography* (Ed. J. F. K. Huber) Elsevier, Amsterdam and New York (1978).
12.W. R. Heinaeman and P. T. Kissinger, *Anal. Chem.* **50**(1978)166R.
13. W. R. Heinaeman and P. T. Kissinger, *Anal. Chem.* **52**(1980)138R.
14. W. Kermula, *Roezniki Chem.* **26**(19520281.
15. E. S. Watson, *American Lab.*, **Sept** (1969)12.
16. P. L. Joynes and R. J. Maggs, *J. Chromatogr. Sci.*, **8**(1970)427.
17. Y. Takata and G. Muto, *Anal. Chem.*, **45**(1973)1864.
18. A. Atsushi, T. Matsue and I. Uchida, *Anal Chem.*, **64** (1992)44.
19. C. N. Svendsen, *Analyst,* **118(Feb)**(1993)123.

CHAPTER 11

THE REFRACTIVE INDEX DETECTOR AND ASSOCIATED DETECTORS

The Refractive Index Detector

The refractive index detector was one of the first on-line detectors to be developed and was described by Tiselius and Claesson [1] in 1942. It was also the first detector to be made commercially and at one time was the only on-line detector that was available for general use in chromatography. The refractive index detector is the least sensitive of all the commonly used detectors. It is very sensitive to changes in ambient temperature, pressure changes and flow-rate changes; furthermore, it can not be used for gradient elution. Nevertheless, this detector can be extremely useful for detecting those compounds that are nonionic, do not adsorb in the UV, and do not fluoresce.

Since the original model of Tiselius there have been many types of refractive index detectors introduced and a number of different optical systems utilized. However, only those in common use or having particular interest will be described here.

The Angle of Deviation Method

When a monochromatic ray of light passes from one isotropic medium, (A), to another, (B), it changes its wave velocity and direction. The change in direction is called the refraction and the relationship between the angle of incidence and the angle of refraction is given by Snell's law of refraction, namely,

$$n'_B = \frac{n_B}{n_A} = \frac{\sin(i)}{\sin(r)}$$

where (I) is the angle of incident light in medium (A),
  (r) is the angle of refractive light in medium (B),
  (n_A) is the refractive index of medium (A),
  (n_B) is the refractive index of medium (B),
and (n'_B) is the refractive index of medium (B) relative to that of medium (A).

The refractive index of a substance is a dimensionless constant that normally decreases with increasing temperature; values given in the literature are usually taken at 20º or 25ºC using the mean value taken for the two sodium lines. If a cell is constructed in the form of a hollow prism through which the mobile phase flows, a ray of light passing through the prism will be deviated from its original path and this can be focused onto a photocell. As the refractive index of the mobile phase changes, due to the presence of a solute, the angle of deviation of the transmitted light will also alter and the amount of light falling on the photocell will change. The angle of deviation method for refractive index monitoring has been used by a number of manufacturers in refractive index detector design.

The modern refractive index detector is the result of considerable research initiated by Zaukelies and Frost [2] and Vandenheuval and Sipas [3] and is now continued largely by the research and development laboratories of the major instrument companies. A diagram of a simple refractive index detector that is based on the angle of deviation method measurement is shown in figure 1.

The differential refractometer monitors the deflection of a light beam caused by the difference in refractive index between the contents of the sample cell and those of the reference cell. A beam of light from an incandescent lamp passes through an optical mask that confines the beam to the region of the cell. The lens collimates the light beam which passes through both the sample and reference cells to a plane mirror. The mirror reflects the beam back through the sample and reference cells to a lens which focuses it onto a photocell.

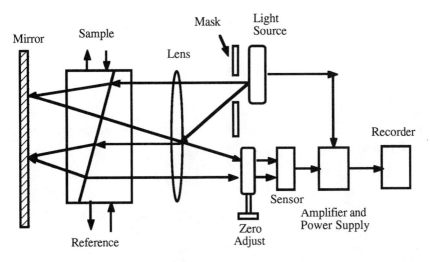

Courtesy of the Millipore Corporation

Figure 1 The Refractive Index Detector Based on the Angle of Deviation Method of Measurement

The location of the beam, rather than its intensity, is determined by the angular deflection of the beam caused by the refractive index difference between the contents of the two cells. As the beam changes its position of focus on the photoelectric cell, the output changes and the resulting difference signal is electronically modified to provide a signal proportional to the concentration of solute in the sample cell.

The Fresnel Method

The Fresnel method of refractive index measurement has also been used in the design of commercially available detectors. The different systems that are used provide comparable performance with respect to sensitivity and linearity and differ largely in the manufacturing techniques used to fabricate the different instruments.

The relationship between the reflectance from an interface between two transparent media and their respective refractive indices is given by Fresnel's equation,

$$R = \frac{1}{2}\left[\frac{\sin^2(i-r)}{\sin^2(i+r)} + \frac{\tan^2(i-r)}{\tan^2(i+r)}\right]$$

where (R) is the ratio of the intensity of the reflected light to that of the incident light and the other symbols have the meanings previously assigned to them.

Now,

$$\frac{\sin(i)}{\sin(r)} = \frac{n_1}{n_2}$$

where (n_1) is the refractive index of medium (1),
and (n_2) is the refractive index of medium (2).

Consequently, if medium (2) represents the liquid eluted from the column, then any change in (n_2) will result in a change in (R) and thus, the measurement of (R) could determine changes in (n_2) resulting from the presence of a solute. Conlon (4) utilized the principle to develop a practical refractive index detector. His device is now obsolete and can not be used with modern high efficiency columns but it illustrates the principle of the Fresnel method very simply. A diagram of Conlon's detector is shown in figure 2. The sensing element consists of a rod prism sealed into a tube through which the solvent flows. The rod prism is made from a glass rod 6.8 mm in diameter and 10 cm long, bent to the correct optical angle (just a little less than the critical angle) and an optical flat is ground on the apex of the bend as shown in figure 2. The optical flat is then sealed into the window of a suitable tube that acts as a flow-through cell. The photocell is arranged to be one arm of a Wheatstone bridge and a reference photocell (not shown) monitoring light direct from the cell, is situated in another arm of the bridge.

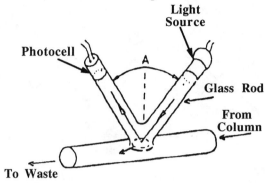

Figure 2 A Early Detector Based on the Fresnel Method of Refractive Index Measurement

This detector was never manufactured as it had too large a cell volume and limited sensitivity. However, it was the one of the first refractive index detector to work on the Fresnel method.

An example of a commercial refractive index detector also working on the Fresnel principle is shown in figure 3. Light from a tungsten lamp is directed through an IR filter to prevent heating the cell to a magnifying assembly that also splits the beam into two. The two beams are focused through the sample and reference cells respectively. Light refracted from the mobile phase/prism surface passes through the prism assembly and is then focused on two photocells. The prism assembly also reflects light to a user port where the surface of the prism can be observed. The output from the two photocells is electronically processed and either passed to a potentiometric recorder or a computer data acquisition system. The range of refractive index covered by the instrument for a given prism is limited and consequently three different prisms are made available to cover the refractive index ranges of 1.35–1.4, 1.31–1.44 and 1.40–1.55 respectively. An example of the separation of a series of polystyrene standards monitored by the detector is shown in figure 4.

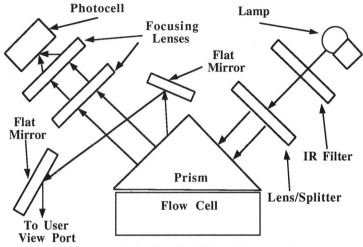

Courtesy of the Perkin Elmer Corporation

Figure 3 A Diagram of the Optical System of a Refractive Index Detector Operating on the Fresnel Method

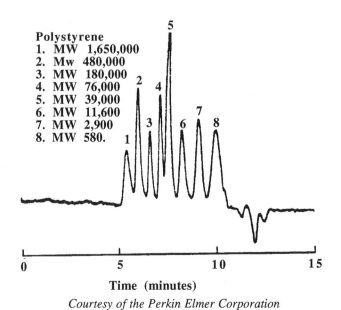

Courtesy of the Perkin Elmer Corporation

Figure 4 The Separation of Some Polystyrene Standards Using a RI Detector Operating on the Fresnel Method

The separation was carried out by size exclusion on a column packed with 5 µm particles operated at a flow rate 0.8 ml/min.

The Christiansen Effect Detector

This method of measuring refractive index arose from the work of Christiansen on crystal filters [5,6]. If a cell is packed with particulate material having the same refractive index as the mobile phase passing through it, light will pass through the cell with little or no refraction or scattering. If however, the refractive index of the mobile phase changes, there will be a refractive index difference between the mobile phase and that of the packing. This difference results in light being refracted away from the incident beam reducing the intensity of the transmitted light. If the transmitted light is focused on a photocell, and the refractive index of the packing and mobile phase initially matched, then any change in refractive index resulting from the elution of a peak will cause light scattering and a reduction in light falling on the sample photocell and thus provide a differential output.

In practice as the optical dispersions of the media are likely to differ, the refractive index will only match at one particular wavelength and thus the fully transmitted light will be largely monochromatic. Light of other wavelengths will be proportionally dispersed depending on the their difference from the wavelength at which the two media have the same optical dispersion. It follows that a change in refractive index of the mobile phase will change both the intensity of the transmitted light and its wavelength.

This device has been manufactured by GOW-MAC Inc., who claimed it had a sensitivity of 1×10^{-6} refractive index units. This would be equivalent to a sensitivity of 9×10^{-6} g/ml of benzene (refractive index 1.501) eluted in n-heptane (refractive index 1.388). The cell volume was kept to 8 µl, a little large for modern sensors but small enough to work well with normal 4.6 mm I.D. columns. Different cells packed with appropriate materials were necessary to cover the refractive index range of 1.31 to 1.60. A diagram of the Christiansen detector is shown in figure 5.

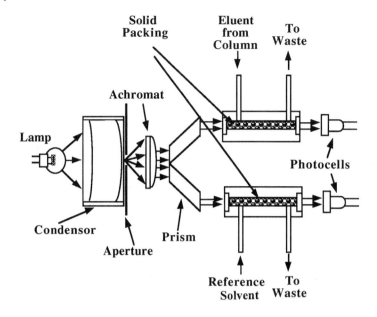

Figure 5 The Christiansen Effect Detector

The optical module contains the prefocused lamp, the voltage of which is adjustable to allow operation at low energy when the maximum sensitivity is not required. The condensing lens, aperture, achromat and beam splitting prisms are mounted in a single tube to prevent contamination from dust and permit easy optical alignment. The system has two cells, identical and interchangeable. The disadvantage of this detector is that the cells must be changed when alternative mobile phases are used in order to have a packing with the appropriate refractive index. Close matching of the refractive indices of the cell packing and the mobile phase can be achieved by using mixed solvents. Solvent mixing can usually be achieved without significantly affecting the chromatographic resolution, *e.g.* by replacing a small proportion of n-heptane in a mixture with either n-hexane or n-octane depending on whether the refractive index needs increasing or decreasing. However, this technique can only be used by someone with considerable knowledge of the effect of different solvents on solute retention. The limitations inherent in his type of refractive index measurement in combination with general disadvantages of the refractive index detector *per se* has not made the detector a very popular instrument.

The Interferometer Detector

The interferometer detector was first developed by Bakken and Stenberg [6] in 1971. The detector responds to the change in the effective path length of a beam of light passing through a cell when the refractive index of its contents changes due to the presence of an eluted solute. If the light transmitted through the cell is focused on a photocell coincident with a reference beam of light from the same source, interference fringes will be produced; the fringes will change as the path length of one light beam changes with reference to the other and consequently, as the concentration of solute increases in the sensor cell, a series of electrical pulses will be generated as each fringe passes the photocell.

The effective optical path length (d) depends on the change in refractive index (Δn), and the path length (l) of the sensor cell as follows,

$$d = \Delta n \, l$$

Further, it is to possible calculate the number of fringes (N) (sensitivity) which move past a given point (or the number of cyclic changes of the central portion of the fringe pattern) in relation to the change in refractive index by the equation,

$$N = \frac{2 \Delta n \, l}{\lambda}$$

where (λ) is the wavelength of the light employed.

The larger the value of (N) for a given (Δn), the more sensitive the detector will be. Therefore (l) should be made as large as possible but this is limited by the dead volume of the column and the dispersion that can be tolerated before chromatographic resolution is impaired. A diagram of the simple optical system originally employed by the authors is shown in figure 6.

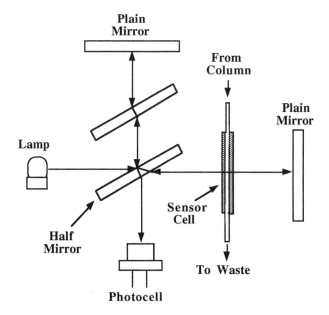

Figure 6 The Original Optical System Used by Bakken and Stenberg in Their Interferometer Detector

Light from a source strikes a half silvered mirror and is divided into two paths. Part of the beam is reflected by a plane mirror back along the same path and onto a photocell.

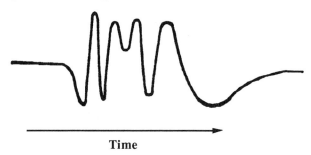

Figure 7 Chromatogram from the Bakken and Stenberg Interferometer Detector

The other part of the beam passes through the sensor cell to a plane mirror, where it is reflected back again through the sensor cell to the half silvered mirror that reflects it onto the photocell where interference takes place with the other half of the light beam. The trace resulting from the elution of 8 ml of dioxane through the cell is shown in figure 7. Each peak shown in figure 7 represents the passage of a fringe across the photocell and the combination of the four peaks represents a single chromatographic peak. The number of fringes will be directly proportional to the total change in refractive index, which will be proportional to the total amount of solute present. In this form the detector has limited use, but has been developed into a commercially viable instrument called the Optilab DSP by Wyatt Technology Inc. A diagram of the optical system of the Optilab interference detector is shown in figure 8.

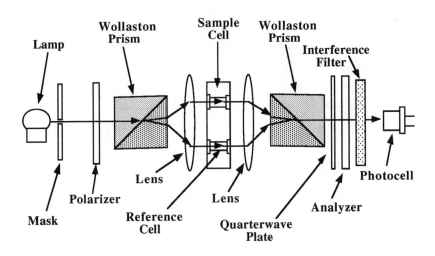

Courtesy of Wyatt Technology

Figure 8 The Optilab Interference Refractometer Detector

Light from the source is linearly polarized at -45° to the horizontal plane. Horizontal and vertical polarized light beams are produced and on passing through the Wollaston prism one passes through the sample cell and the other through the reference cell. The beam passing through the sample cell is horizontally polarized and that through the reference cell is vertically polarized. After passing through the cells, the beams are focused on a second Wollaston prism and then through a quarter-wave plate which has its fast axis set -45° to the horizontal plane. A beam that is linearly polarized in the fast axis plane will, after passing through the plate, lead another linearly polarized but orthogonal beam by a quarter of a wavelength. The phase difference results in a circularly polarized beam. It can be assumed that each of the beams focused on the Wollaston prism consists of two such perpendicular beams which, after the quarter wave plate, result in two circularly polarized beams of opposite rotation. These beams will interfere with each other to yield the original linearly polarized beam. A second polarizer is placed at an angle $(90 - \beta)$ to the first, allowing about 35% of the signal to reach the photocell. A filter transmitting light at 546 nm precedes the photocell.

If the sample cell contains a higher concentration of solute than the reference cell, in general, the refractive index will be higher and the interfering beams will be out of phase. The refractive index difference (Δn) and the phase difference (Δp) are related by

$$\Delta p = \frac{2\pi L \Delta n}{\lambda}$$

where (L) is the length of the cell,
and (l) is the wavelength of the light.

The circularly polarized beams, therefore, will interfere to yield a linearly polarized beam which is rotated $\frac{\Delta p}{2}$ radians, and the amplitude of the light striking the photocell (A_p) will be given by

$$A_p = A_o \cos\left(90 - \beta - \frac{\Delta p}{2}\right) = A_o \cos\left(\beta - \frac{\Delta p}{2}\right)$$

An extremely high sensitivity is claimed for this system but it is difficult to interpret the data in terms of minimum detectable concentration The smallest cell (1.4 µl) (a cell volume that would be suitable for use with microbore columns) is reported to give a sensitivity of about 2×10^{-7} RI units at a signal-to-noise ratio of two. Consequently, for benzene (RI = 1.501) sensed as a solute in n-heptane (RI=1.388) this sensitivity would represent a minimum detectable concentration of 5.6×10^{-5} g/ml. The alternative 7 µl cell would decrease the minimum detectable concentration to about 1×10^{-6} g/ml, similar to that obtained for other refractive index detectors. However, the cell volume is a little large for modern high efficiency columns.

Applications of the Refractive Index Detector

Bulk property detectors, and in particular, the refractive index detector, have an inherently limited sensitivity irrespective of the instrumental technique that is used. Consider a hypothetical bulk property detector that monitors, for example, the density of the eluent leaving the column. Assume it is required to detect the concentration of a dense material, such as carbon tetrachloride (specific gravity 1.595), at a level of 1 µg/ml in *n*-heptane (specific gravity 0.684).

This situation is typical for a bulk property detector and the sample chosen will be particularly favorable for this hypothetical detector, as the solute to be sensed exhibits a large difference in density from that of the mobile phase.

Let the change in density resulting from the presence of the solute at a concentration of 10^{-6} g/ml be (Δd). It follows that to a first approximation,

$$\Delta d = \frac{X_s(d_1 - d_2)}{d_1}$$

where (d_1) is the density of the solute, carbon tetrachloride,
(d_2) is the density of the mobile phase, *n*-heptane,
and (X_s) is the concentration of the solute to be detected.

Thus for the example given,

$$\Delta d = \frac{(1.595 - 0.684) \times 10^{-6}}{1.59}$$

$$= 5.71 \times 10^{-7}$$

Now the coefficient of cubical expansion of n-heptane is approximately 1.6×10^{-3} per °C. It is therefore possible to calculate the temperature ($\Delta\theta$) that would produce a change equivalent to the presence of carbon tetrachloride at a concentration of 10^{-6} g/ml.

Thus,
$$\Delta\theta = \frac{5.71 \times 10^{-7}}{1.6 \times 10^{-3}} \,°C$$

$$= 3.6 \times 10^{-4} \,°C$$

If it is assumed that a concentration of one part per million of carbon tetrachloride is just detectable (it provides a signal-to-noise ratio of two), then the temperature fluctuations must be maintained below 1.8×10^{-4} °C to realize this sensitivity. Such temperature stability can be extremely difficult to maintain in practice and thus the temperature control can place a severe limit on the sensitivity obtainable from such a detector. Even the heat of adsorption and desorption of the solute on the stationary phase can easily result in temperature changes of this order of magnitude.

In a similar way, the density of the contents of the cell will change with pressure and, if there is a significant pressure drop across the cell, also with flow rate. These arguments apply to all bulk property detectors and it must therefore be concluded that all bulk property detectors will have a limited sensitivity (probably, on average, and for most compounds, the maximum sensitivity that could be expected would be about 10^{-6} g/ml). Furthermore, to realize this sensitivity, the sensor must always be operated under very carefully controlled conditions. As a result of their limited sensitivity, bulk property detectors will also have a limited practical linear dynamic range for chromatographic purposes, usually about three orders of magnitude. In fact, the refractive index detector can have a much wider actual dynamic range but this extends into concentrations of 10^{-3} g/ml and above. Concentrations as high as 10^{-3} g/ml can not be used with chromatographic columns as the column loading capacity would be exceeded.

As a result of limited sensitivity and restricted linear dynamic range, the refractive index detector is often a "choice of last resort" and is used for those applications where, for one reason or another, all other detectors are inappropriate or impractical. However, the detector has one particular area of application for which it is unique and that is in the separation and analysis of polymers. In general, for those polymers that contain more than six monomer units, the refractive index is directly proportional to the concentration of the polymer and is practically independent of the molecular weight. Thus, a quantitative analysis of a polymer mixture can be obtained by the simple normalization of the peak areas in the chromatogram, there being no need for the use of individual response factors. The sensitivity of most RI detectors will be about 1×10^{-6} g/ml and the linear dynamic range around 1×10^{-6} to 2×10^{-4} g/ml with the response index (r) lying between 0.97 and 1.03.

A typical application of the RI detector is for carbohydrate analysis. Carbohydrates do not adsorb in the UV, do not ionize and although fluorescent derivatives can be made, the procedure is somewhat tedious. Consequently, the RI detector can be ideal for detecting such materials and an example of such an application is shown in figure 9.

The separation was completed in less than 20 minutes. These types of separation including other bio-monomers, dimers and polymers are frequently carried out employing refractive index detection. Two further examples of similar types of analyses using the RI detector are afforded by the separation of the products of β-cyclodextrin hydrolysis and the partial hydrolysis of galaction.

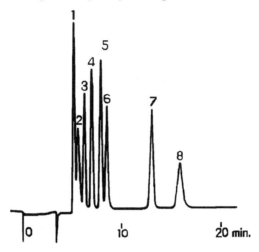

Courtesy of TOYO SODA Manufacturing Co. Ltd.

Column: TSKgel Amide-80 column 25 cm x 4.6 mm I.D at 80°C.; mobile phase: 80% acetonitrile 20% water; flow rate: 1 ml/min.

| 1. Rhamose | 2. Fucose | 3. Xylose | 4. Fructose |
| 5. Mannose | 6. Glucose | 7. Sucrose | 8. Maltose |

Figure 9 The Separation of Some Carbohydrates Monitored by the Refractive Index Detector

A chromatogram demonstrating the separation of the hydrolysis products of β-cyclodextrin is shown in figure 10. The TSKgel packing used in this separation is a vinyl polymer based material suitable for separation by size exclusion using aqueous solvents. There are a number of grades of this product available that are suitable for separations covering a wide range of molecular weights. It is seen that the products of the hydrolysis are well separated and almost all of the oligomers are resolved. The separation of the hydrolysates of galaction provides a further example of the use of the RI detector for monitoring the separation of organic polymers which is again achieved by the use of size exclusion. The chromatogram is shown in figure 11.

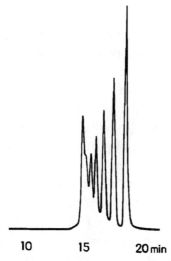

Courtesy of TOYO SODA Manufacturing Co. Ltd.

Column: TSKgel G-Oligo-PW   30 cm x 7.8 mm I.D. at 60°C :Flow rate of 1 ml/min

Figure 10 The Separation of Hydrolyzed β–Cyclodextrin Monitored by a Refractive Index Detector

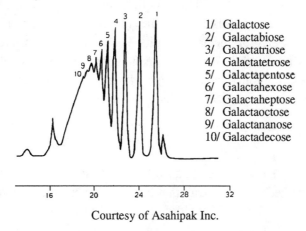

1/ Galactose
2/ Galactabiose
3/ Galactatriose
4/ Galactatetrose
5/ Galactapentose
6/ Galactahexose
7/ Galactaheptose
8/ Galactaoctose
9/ Galactananose
10/ Galactadecose

Courtesy of Asahipak Inc.

Figure 11 A Chromatogram of the Partial Hydrolysis of Galaction Monitored by the Refractive Index Detector

Two columns were used both GS 220, x 2, 50 cm x 7.6 mm I.D. The two long, wide columns were necessary as, in size exclusion chromatography, the total separation must be achieved in the pore volume of the column or about 40% of the dead volume. This is an interesting example of the practical use of large volume columns to provide adequate peak capacity. As is usual with many size exclusion separations, the mobile phase was pure water. To improve the column efficiency an elevated column temperature of 60°C was used with a mobile phase flow rate of 1 ml/min. For a column of 7.6 mm I.D. a flow rate of 1 ml/min would provide a linear mobile phase velocity close to the column optimum velocity. The conditions represent the result of a struggle to achieve adequate resolution by arranging for a large peak capacity to accommodate all the peaks (a large column volume) and an attempt to achieve the highest possible efficiency by operating close to the optimum velocity. Besides being an interesting example of the use of the refractive index detector, this separation is also an excellent example of how the operating conditions of a chromatographic system can be adjusted to perform a particular and difficult analysis.

Detectors Associated with Refractive Index Measurement

There are a number of LC detectors that have been developed that are either based on refractive index measurement or function on some physical property of the mobile phase system that is related to the refractive index. Most of these are not commercially available but nevertheless demonstrate the range of sensing techniques that have been investigated as possible methods of detection.

The Thermal Lens Detector

When a laser is focused on an absorbing substance the refractive index may be affected in such a way that the medium behaves as a lens. This effect was first reported by Gorden *et al.* (7,8) in 1964 and since that time it has been investigated by a number of workers. Thermal lens formation results from the absorption of laser light which may be extremely weak. The excited-state molecules subsequently decay back to ground state and as a result, localized temperature increases occur in the sample. Since the refractive index of the medium depends on the temperature, the resulting spatial variation of refractive index produces an effect which appears equivalent to the formation of a lens within the medium.

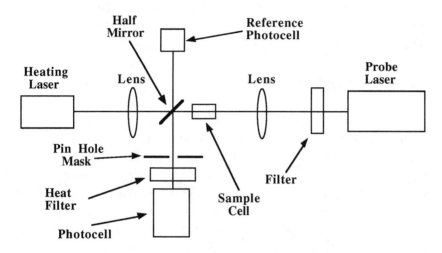

Figure 12 The Layout of a Thermal Lens Detector

The temperature coefficient of refractive index is, for most liquids, negative, consequently, the insertion of a liquid in the laser beam produces a concave lens that results in beam divergence.

The thermal lens effect has been used by Buffet and Momis [9] to develop a small volume detector. A diagram of their system is shown in figure 12. It consists of a *heating* laser, the light from which is passed directly through the sample via two lens and a half mirror. Another laser, the *probe* laser, passes light in the opposite direction through one lens, through the sample to the half mirror where the light is reflected onto a photocell. Between the mirror and the photocell is a filter to remove the heating laser light and a pinhole. When an absorbing solute arrives in the cell, a thermal lens is produced which causes the probe light to diverge, and consequently the intensity of the light passing through the pinhole and on to the photocell is reduced. The cell can be made a few microliters in volume and would thus be suitable for use with microbore columns. A sensitivity of 10^{-6} AU was claimed for the detector and a linear dynamic range of about three orders of magnitude. The use of two laser's adds significantly to the cost of the device. Basically, the thermal lens detector is a special form of the refractive index detector and, as a consequence can be considered as a type of universal detector. However, it can not be used with gradient elution or flow programming and its sensitivity is no better, if as good as, other refractive index detectors.

The Dielectric Constant Detector

Under the influence of small fields, electrons move quite freely through conductors, whereas in insulators or dielectrics these fields displace the electrons only slightly from their equilibrium positions. As an electric field acting on a dielectric causes a separation of positive and negative charges, it is said to polarize the dielectric. The polarization can occur as a result of two effects: the induction effect and the orientation effect. The electric field always induces dipoles in molecules on which it is acting, whether or not they originally contain them. If the dielectric does contains molecules that have permanent dipoles, the field tends to align these dipoles along its own direction. As a result of the induction or orientation it is found experimentally that when a dielectric is introduced between the plates of a capacitor, the capacitance is increased by a factor (ε) called the dielectric constant. Thus if (C_o) is the capacitance in vacuum the capacitance with a dielectric between its plates will be (C) where

$$C = \varepsilon C_o$$

In this way the dielectric constant of a substance can be defined. Due to the electromagnetic nature of light, its transmission is also affected by the dielectric constant of the medium through which it passes. It follows that the refractive index of a substance is a complementary property to the dielectric constant and in some circumstances is a direct function of it. For example for non-polar substances, the relationship between dielectric constant (ε) and refractive index (n) is given by

$$\varepsilon = n^2$$

For semi-polar substances or mixtures of semi-polar substances and non-polar substances the Lorentz-Lorenz equation applies

$$\frac{\varepsilon - 1}{\varepsilon + 2} = \frac{n^2 - 1}{n^2 + 2}$$

For polar substances and mixtures of polar and semi-polar substances the relationship breaks down and no simple functions describe dielectric constant in terms of refractive index.

In general, the more polar the substance, the larger is its dielectric constant. This is always true for substances having single functional groups and generally true for substances having two or more functional groups but there are some exceptions. For example the molecule of dioxane contains two ether groups, but has a fairly low dielectric constant although it is a very polar solvent. The low value for the dielectric constant results from the fact that the two dipoles formed by the ether groups are electrically in opposition and thus partially neutralize the effect of each others' charge. This effect should be considered when choosing the mobile phase for use with a dielectric constant detector.

In *normal* chromatography the mobile phase is usually less polar than the solutes being eluted as they need to be retained on the column by polar forces to achieve the separation. Thus, the presence of a solute in the mobile will, in general, *increase* the dielectric constant of the mobile phase. In *reversed phase* chromatography, however, the solute is usually less polar than the solute and, consequently, the dielectric constant of the mobile phase is *reduced* by the presence of a solute. It follows that if a device is situated at the end of the column, which responds to changes in dielectric constant, such a device would act as a chromatography detector. In practice the sensor often takes the form of a cylindrical or parallel plate condenser. The volume of the sensor must be as small as possible to ensure the separation achieved by the column is not impaired. Thus as the sensitivity of the device is directly related to the electrical capacity of the sensor, the plates of the capacitor must be very close together.

The capacity (C) of a parallel plate condenser is given by

$$C = \frac{0.080(N-1)A\varepsilon}{d}$$

where (N) is the number of plates in the sensor,
(A) is the area of each plate in cm^2,
and (d) is the distance between the plates in cm.

The capacity of a cylindrical condenser is given by

$$C = \frac{0.2416\,\varepsilon\,l}{\log r_1 - \log r_2}$$

where (l) is the length of the cylinder,
(r_1) is the radius of the outer cylinder,
and (r_2) is the radius of the inner cylinder.

The impedance (ϕ) of a capacitor to an applied AC potential is

$$\phi = \frac{1}{2\pi f C}$$

where (f) is the frequency of the applied AC potential.

It follows that as the capacity (C) is directly proportional to the dielectric constant of the material between its plates, then the electrical impedance of the cell will vary inversely as the dielectric constant. The most appropriate circuit to use in dielectric constant measurement is an appropriate electrical "bridge", the detector cell being situated in one arm of the bridge. If the cell can be designed to have a capacity greater than 100 pF, then a Wein bridge can be used; however such a cell may well have a fairly large volume. For cells having smaller capacities the Schering bridge is more appropriate and a diagram of the circuit of a Schering bridge is shown in figure 13.

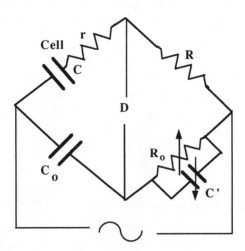

Figure 13 The Schering Bridge for the Measurement of Small Capacities

No capacitor is ideal, in that it will have some inductance and resistance in addition to its capacity. In fact, because the plates of the capacitor are situated in the mobile phase, if uninsulated, it will certainly have a significant resistance component. As already discussed the current though the resistive component of a conductor is in phase with the applied voltage and the capacity component lags the applied voltage by 90°.

It follows that there are two components to be balanced before the output of the bridge (across (D)) can be used to monitor the elution of a solute. Balance in the Schering bridge is achieved by the iterative adjustment of (R_o) and (C'). At balance the following relationships will hold:

$$\frac{C}{R_o} = \frac{C_o}{R} \text{ and } \frac{r}{R} = \frac{C'}{C_o}$$

The resistance-component of the cell across the sensor reduces the bridge sensitivity to changes in capacity and thus, wherever possible, the plates should be well insulated to prevent conductivity through the mobile phase. Alternatively, the capacity of the sensor can be measured by making it one component of a resistance/capacity or an inductance/capacity oscillator. The frequency will depend, among other things, on the capacity of the sensor. If the frequency is heterodyned against a reference oscillator, then the frequency difference will be proportional to the change in capacity and hence the dielectric constant of the mobile phase.

One of the early dielectric constant detectors was that designed by Grant [10] but the detector cell had a volume of 2-3 ml. Poppe and Kunysten (11) described a dielectric constant detector which included a reference cell for temperature compensation. The cell consisted of two stainless steel plates 2 cm x 1 cm x 1 mm separated by a gasket 50 µm thick. The two cells were identical and clamped back to back, sharing a common electrode.

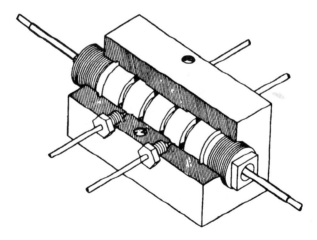

Figure 14 The Sensor of a Dielectric Constant Detector

The device had a sensitivity of 10^{-6} g/ml for chloroform ($\varepsilon = 4.81$) in n-octane. Unfortunately, it was found to be very sensitive to pressure changes in the cell (thought due to deformation of the plates) even when constant flow pumps were employed. The first dielectric constant detector became commercially available in 1979 [12] and was described by Benningfield and Mowery [13]. Several applications were reported by Bade *et al.* [14]. A diagram of the sensor is shown in figure 14. Each cell consists of a concentric cylinder (inner electrode) inside a larger cylinder (the outer electrode) which forms the outer wall of the cell. Both electrodes are made of stainless steel. The two cylinders are electrically isolated with a cylindrical flow path through the cell.

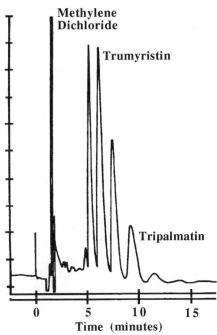

Figure 15 The Separation of Some Triglycerides Monitored by a Dielectric Constant Detector

The inner cylindrical electrodes are 1.26 cm in diameter and 0.625 cm long with a separation distance to the inner cylinder of about 0.009 cm. The linear dynamic range of the detector was reported to be 3.5 x 10^4 but this range probably included concentrations that are too high for practical column operation. The sensitivity was quoted as about 1 x 10^{-7} g/ml, which would be close to the theoretical limit for bulk property detectors. An example of the use of the dielectric constant detector to monitor a separation of triglycerides is shown in figure 15.

Bulk property detectors generally have neither the sensitivity nor the linear dynamic range of solute property detectors and, as a consequence, are less frequently used in modern LC analyses. Furthermore, none can be used with gradient elution, flow programming or temperature programming and so they place considerable restrictions on the choice of chromatographic system. They do, however, have certain unique areas of application, some of which have already been mentioned, but their use probably represents less than 5% of all LC analyses.

References

1. A. Tiselius and D. Claesson, *Ark. Kemi. Mineral. Geol.* **15B(No 18)**(1942).
2. D. Zaukelies and A. A. Frost, *Anal. Chem.* **21**(1949)743.
3. F. A. Vandenheuval and E. Sipas, *Anal. Chem.*, **33**(1961)286.
4. R. D. Conlon, Rev. Sci. Instrum., **34**(1961)1418.
5. C. Christiansen, *Ann. Phys. Chem.*, **3**(1884)298.
6. C. Christiansen, *Handbook of Chemical Microscopy* (Ed. E. M. Chanot and C. W. Mason), John Wiley and Sons, New York, **Vol. 1**((1958)101 and 189.
6. M. Bakken and V. J. Stenberg, *J. Chromatogr. Sci.*, **9**(1971)603.
7. J. P. Gorden, R. C. C. Leite, R. S. Moore, S. P. S. Posto, J. R. Whinnery, *Bull. Am. Phys. Soc.,* **(2) 9**(1964)501.
8. J. P. Gorden, R. C. C. Leite, R. S. Moore, S. P. S. Posto, J. R. Whinnery, *J. Appl. Phys.*, **36**(1965)3.
9. C. E. Buffet and M. D. Momis, *Anal. Chem.* **54**(1982)1824.
10. R. A. Grant, *J. Appl. Chem.* **8**(1959)136.
11. M. Poppe and J. Kunysten, *J. Chromatogr. Sci.* **10**(1972)16A.
12. L. V. Benningfield Jr., *Pittsburgh Conference on Analytical Chemistry and Applied Spectroscopy,* **March 5-9** (1979)paper 123.
13. L. V. Benningfield Jr. and R. A. Mowery Jr., *J. Chromatogr. Sci.,* **19**(1981)115.
14. R. K. Bade, L. V. Benningfield Jr., R. A. Mowery and E. N. Fuller, *Am. Lab.***13(10)**(19810130.

CHAPTER 12

MULTIFUNCTIONAL DETECTORS AND TRANSPORT DETECTORS

Multifunctional Detectors

Contemporary liquid chromatographic procedures are now well developed, accurate and reliable and both high speed [1-3] and high resolution columns [2,4,5] have been available for some years. There are, at present, a veritable plethora of detectors available to the practicing chromatographer. In spite of this however, as already stated, the vast majority of contemporary LC analyses are carried out employing one of four detectors: namely, either some form of UV detector, the fluorescence detector, the electrical conductivity detector or, to a significantly lesser degree, the refractive index detector. There remains, notwithstanding, the need for a detector having greater versatility; a detector, for example, that concurrently monitors more than one property of the solute and thus increases the range of solutes types that can be sensed. In addition, such a detector could help confirm the identity of the eluted solute. The latter is often achieved by the use of "tandem" systems where an appropriate spectrometer is coupled directly to the chromatograph which provides spectra of each eluted solute. Such combinations can provide unambiguous solute identification and many systems are commercially available but can be rather expensive. Tandem systems will be discussed in a subsequent chapter. A less sophisticated solution is to use a diode array UV absorption detector to provide UV spectra of each solute or to employ a series of different detectors in sequence, each monitoring a different physical or chemical property of the eluent. The former has limitations due to the restricted information provided by most UV spectra and, unfortunately, the latter can cause serious peak dispersion that may arise in the extensive connecting tubing that is necessary and also in each sensor cell. The latter system, nevertheless, has been used successfully for solute identification (6) where the chromatographic resolution is sufficient to tolerate the inevitable extracolumn dispersion that will occur. However, excess chromatographic resolution is not normally available, and thus the dispersion arising from the connection of a number of detectors in series or parallel is usually unacceptable and furthermore can also be very expensive.

The problem was solved by the development of *multifunctional* detectors, where more than one property of the solute is concurrently measured while it is situated in a single sensor cell. This arrangement reduces both the cost of the detector and also the extracolumn dispersion as only one cell is employed and only the normal column detector connection is necessary. The first multifunctional detector was developed by DuPont and was a bifunctional detector that simultaneously measured UV absorption and solute fluorescence.

Bifunctional Detectors

The DuPont UV/Fluorescence Detector

The UV/fluorescence detector designed and manufactured by DuPont is no longer commercially available but is an interesting example of one of the early bifunctional detectors. A diagram of the DuPont UV/Fluorescence detector is shown in figure 1. Light from an appropriate UV source is collimated by means of a quartz lens through a cylindrical sensor cell and the transmitted light is focused by means of another quartz lens onto a photodiode. The output is processed by suitable electronic circuitry to provide a signal that is proportional to the concentration of solute in the mobile phase. The body of the cell is made of an appropriate transparent material and any fluorescent light resulting from the excitation of the solute molecules by the incident UV light is focused by means of another lens onto a second photo diode. This in turn is processed by suitable electronics to provide an output that is proportional to solute concentration.

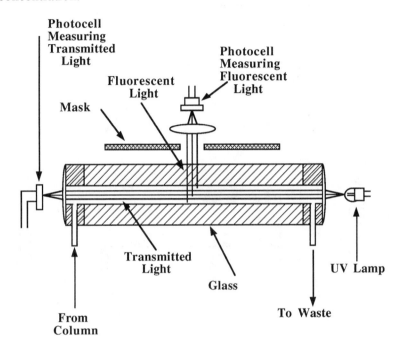

Figure 1 The DuPont UV/Fluorescence Detector

The results can be presented on a dual pen recorder or the raw data from the detectors acquired by a computer and the chromatograms from each sensor presented separately. The bifunctional system can give a response proportional to the absorbed UV light, the emitted fluorescent light thus helping to confirm solute identification. The DuPont instrument was the pioneer of multifunctional detection but, due to a number of problems that arose in its operation, it did not prove to be a technical or economic success. At the time of its design, the importance of low dispersion cells and conduits was not well established and, as a consequence, the device was not designed to contain dispersion and the system exhibited serious peak spreading

The UV/Conductivity Detector

In 1984 Baba and Housako (7) described a second bifunctional detector based on the concurrent measurement of UV absorption and electrical conductivity. A diagram of the detector is shown in figure 2.

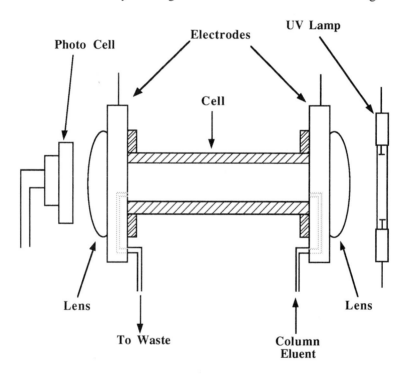

Figure 2 UV/Conductivity Detector

The UV absorption system is very similar to that of the DuPont bifunctional detector. Light from a UV lamp is collimated through the cell by a quartz lens that serves as one end of the sensor cell, and is then focused by another quartz lens at the other end of the cell onto a photodiode. The output from the photodiode is processed electronically in the usual manner to provide an output that changes linearly with solute concentration. The ends of the cell, between the cell body and the quartz lens, where the mobile phase from the column enters the cell and where the mobile phase, after passing through the cell, flows to waste, are made of stainless steel. These spacers serve as the electrodes of an electrical conductivity sensor. An AC potential is applied across the electrodes that can form one arm of an AC bridge (*e.g.* the Wein bridge). As the impedance of the cell changes with the presence of an ionic solute between the sensor electrodes, the out of balance signal is electronically processed and the output fed either to a potentiometric recorder or to an A/D converter and thence to a computer. Consequently, as the mobile phase flow rate is constant, the bifunctional detector can provide, simultaneously, chromatograms that represent the change in UV absorption and the change in electrical conductivity of the sensor contents with time. This device, although a very effective bifunctional detector, does not appear to have had great commercial success despite its being basically a simple system that could be manufactured at a relatively low cost.

The UV/Refractive Index Detector

Another bifunctional detector that was developed by Knauer and made commercially available in Europe operated on the simultaneous measurement of UV absorption and refractive index. This detector is particularly interesting as it combines a solute property detector with a bulk property detector and consequently should have a wide field of application.

A diagram of the instrument is shown in figure 3. There are two light sources, a low pressure mercury vapor lamp and a tungsten filament lamp. Light from the low pressure mercury lamp passes through a filter to remove light of unwanted wavelengths, through the cuvette carrying the mobile phase and onto a photocell. As with the previous devices, the output of the photocell is electronically modified to provide an output linearly related to solute concentration. Light from the filament lamp passes to a half silvered mirror that reflects the light through the cuvette onto a concave mirror that in turn reflects the light back through the cuvette through a glass plate and focuses it on the apex of a prism. The light splits into two and is reflected onto two photocells. By moving the glass plate, the relative amount of light falling on the two photocells can be adjusted and is used to provide an optical and thus electrical zero.

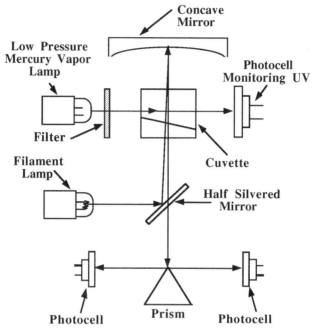

Figure 3 The UV/Refractive Index Detector

The output from the refractive index sensor is also electronically modified to provide a output linearly related to solute concentration.

Specifications for this detector are not very explicit. It would appear that the sensitivity of the refractive index detector was about 10^{-6} refractive index units and the UV detector about 10^{-3} absorption units. The cell volume was about 12 µl, which is too large for use with modern high efficiency LC microbore columns. Nevertheless, the detecting system is an interesting association as the combined performance of the UV absorption detector and the refractive index detector approaches that of the universal detector.

There is another bifunctional detecting system being developed at the time of writing this book, but unfortunately details are not readily available The detector simultaneously measures the UV absorbance and the optical activity of the column eluent. Considering the great interest that has arisen recently in the relative physiological activity of different enantiomorphs of the same compound the device should be widely accepted. This detector will be discussed in the chapter on chiral detectors.

Trifunctional Detectors

Trifunctional detectors are the natural extension of the multifunctional concept but, to date, there appears to be only one trifunctional detecting system that has been fully developed and become commercially successful. In fact, there is no reason why four, five or even more sensing devices could not be incorporated together in an effective multifunctional detector. However, it appears that, so far, the TriDet, a *trifunctional* detector has the maximum number of different sensors of any commercially available detector.

The TriDet Detector

It was the popularity of the UV detector, the electrical conductivity detector and the fluorescence detector that motivated Schmidt and Scott (8,9) to develop a trifunctional detector that detected solutes by all three methods simultaneously in a single low volume cell. A diagram of their detector is shown in figure 4. The UV adsorption system consists of a low pressure mercury lamp emitting light at 254 nm and a solid state photocell with quartz windows allowing the photo-cell to respond to light in the UV region.

The detector cell is 3 mm long terminated at one end by a cylindrical quartz window and at the other by a plano-convex quartz lens that disperses the transmitted light over a significant portion of the light sensitive area to the photocell. The total volume of the sensor cell was about 2.5 µl. Next to the quartz windows are two stainless steel discs separated by a 3 mm length of Pyrex tube. The mobile phase enters and leaves the detector cell through radial holes in the stainless steel discs that connect to the central orifice of the disc. The stainless steel discs also act as the electrodes for conductivity detection. The inlet tube to the cell is electrically connected to the column by the stainless steel union and is consequently at ground potential.

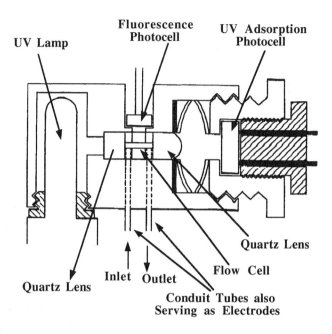

Courtesy of Bacharach Inc.

Figure 4   The TriDet Trifunctional Detector

The outlet tube is insulated by means of a short length of PTFE tubing interposed between the cell and the outlet union. At right angles to the short length of Pyrex tube between the disc electrodes is situated another photocell that receives any fluorescent light emitted normal to the incident UV excitation light. The output from each sensor passes to an appropriate amplifier to provide an output that is linearly related to solute concentration. Consequently, the column eluent is continuously and simultaneously monitored by UV adsorption, fluorescence and electrical conductivity. A block diagram of the electronic system is shown in figure 5. The UV and fluorescence amplifiers are very similar and consist of an impedance converter as the first stage which is provided with coarse and fine zero controls.

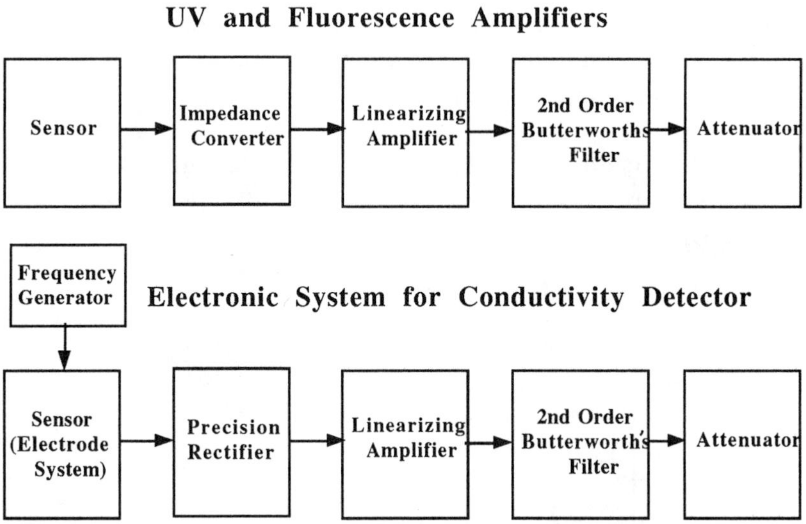

UV and Fluorescence Amplifiers

Sensor → Impedance Converter → Linearizing Amplifier → 2nd Order Butterworth's Filter → Attenuator

Electronic System for Conductivity Detector

Frequency Generator ↓

Sensor (Electrode System) → Precision Rectifier → Linearizing Amplifier → 2nd Order Butterworth's Filter → Attenuator

Figure 5   Schematic Diagram of the Electronic System of the TriDet

The second stage is a signal modifying amplifier that renders the output signal linearly related to the solute concentration. This is followed by a second order Butterworth's filter that removes any electrical noise that might be generated in the circuitry and finally the output is controlled by a simple six stage binary attenuator providing attenuation in steps of two up to a maximum of 32. The only difference between the UV detector electronic system and that of the fluorescence detector is in the signal modifying amplifier which must be designed to suit the response of the sensor.

The electronic system of the electrical conductivity detector is comprised of a 1 kHz frequency generator, the output of which is fed *via* a suitable impedance to the detector electrodes. The voltage across the electrodes is fed to a precision rectifier to provide a DC signal that is related to the conductivity of the fluid between the sensor plates. The DC signal is then passed to a signal modifying amplifier to provide an output linearly related to the ion concentration between the electrodes and then to a filter and attenuator similar to those used in the UV and fluorescence amplifiers.

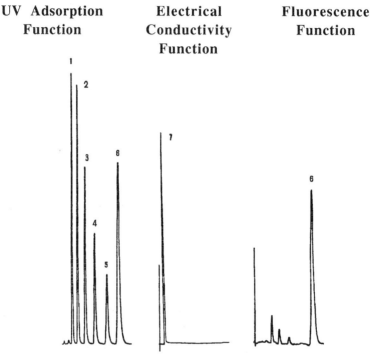

The column used was a Pecosphere™ 3 mm in diameter and 3 cm long carrying a C18 stationary phase. The mobile phase was a mixture of methanol (75%) and water (25%) at a flow rate of 2 ml/min. The solutes were 1, benzene; 2, toluene; 3, ethyl benzene; 4, isopropyl benzene; 5, t-butylbenzene; 6, anthracene; and 7, sodium chloride.

Courtesy of Bacharach Inc.

Figure 6 Chromatograms Demonstrating the Simultaneous Monitoring of a Mixture by All Three Detector Functions

The sensitivity of the UV absorption function is about 1.7×10^{-7} g/ml for toluene with a linear dynamic range of about 1.5×10^3. These specifications compare well with those expected for a fixed wavelength UV detector. The fluorescence function provides a sensitivity of about

2.5 × 10^{-8} g/ml and a linear dynamic range of 1.2 × 10^3.

Column-Pecosphere	Column Pecosphere	Column Pecosphere 3
Size 3 mm x 3 cm C18	Size 4.6 mm x 15 cm C18	Size 3 mm x 3 cm C18
17% Methanol/Water	90% Acetonitrile/Water	1nMtetrabutyl-ammonium hydroxide and buffer
Flow-Rate 3.0 ml/min.	Flow-Rate 2 ml/min.	Flow-Rate 1.5 ml/min.

UV Detector   **Fluorescence Detector**   **Conductivity Detector**

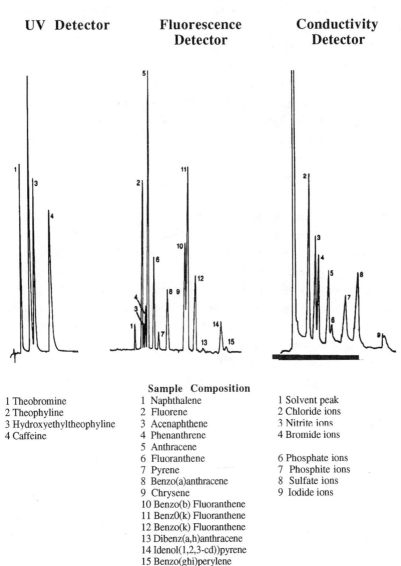

1 Theobromine	Sample Composition	1 Solvent peak
2 Theophyline	1 Naphthalene	2 Chloride ions
3 Hydroxyethyltheophyline	2 Fluorene	3 Nitrite ions
4 Caffeine	3 Acenaphthene	4 Bromide ions
	4 Phenanthrene	
	5 Anthracene	
	6 Fluoranthene	6 Phosphate ions
	7 Pyrene	7 Phosphite ions
	8 Benzo(a)anthracene	8 Sulfate ions
	9 Chrysene	9 Iodide ions
	10 Benzo(b) Fluoranthene	
	11 BenzO(k) Fluoranthene	
	12 Benzo(k) Fluoranthene	
	13 Dibenz(a,h)anthracene	
	14 Idenol(1,2,3-cd))pyrene	
	15 Benzo(ghi)perylene	

Figure 7 The Use of the Different TriDet Functions to Monitor the Separation of Diverse Sample Types

Finally the sensitivity of the conductivity function to sodium chloride is 5×10^{-8} g/ml with a linear dynamic range of 3×10^3. This detector, is obviously very versatile and, perhaps a little surprisingly, is relatively inexpensive. The detector together with a column, sample valve, pump and recorder, which constitutes a basic liquid chromatograph, costs about $10,000.

Chromatograms demonstrating the simultaneous use of all three detector functions are shown in figure 6. It is seen that the anthracene is clearly picked out from the mixture of aromatics by the fluorescence detector and the chloride ion, not shown at all by the UV adsorption or fluorescence detectors, is clearly shown by the electrical conductivity detector.

However, it is not only the simultaneous use of all detector functions that makes this detector so useful. An equally important advantage of the trifunctional detector is that it allows the analyst a choice of the three most useful detector functions in one detecting system. Furthermore, any or all of the three functions can be chosen at the touch of a switch and without any changes in hardware. Examples of the use of the three individual detector functions in the analyses of three quite different types of sample are shown in figure 7. The separations clearly demonstrate the flexibility of the TriDet and its value for use in widely diverse analyses.

Transport Detectors

Transport detectors are a unique type of solute property detector in that the signal from the sensor is entirely independent of the solvent that is used as the mobile phase. Various forms of transport detectors have been commercially available over the years past but, due to certain deficiencies in the early models, they did not become popular and (to the author's knowledge) none are currently being manufactured. Nevertheless, the transport detector has the potential qualities that are inherent in the ideal detector, *i.e.* universal detection, high sensitivity, wide linearity and unaffected by the nature of the mobile phase, providing the solvents are volatile. For this reason transport detectors will be discussed in some detail.

A transport detector consists of a carrier that can be, for example, a metal chain, wire or disc that continuously passes through the column eluent taking a sample with it as a thin film of mobile phase adhering to its surface. The mobile phase is then removed by evaporation leaving the solute originally contained in the mobile phase as a coating on the carrier. The carrier is then examined by a suitable sensing procedure to monitor the solute alone. If, for example, a flame ionization detector is employed to monitor the pyrolysis products of the solute produced by heating the carrier, all pyrolysis products containing carbon will be detected. The only restrictions to the system is that the solutes must be *involatile* and the constituents of the mobile phase must be *volatile*. The former is almost always true in liquid chromatography; otherwise the separation would be carried out by gas chromatography. The latter is easy to arrange as there is a wide choice of solvents that can be used and that are readily available. The system seems ideal but the early models had some disadvantages. The instruments were bulky and expensive and incorporated a ^{90}strontium source that was unpopular. As a result of the basic design, the anticipated high sensitivity was not realized and the apparatus was clumsy and difficult to operate. Nevertheless, due to it being a universal detector and it being unaffected by the solvents used, it was readily accepted by the control laboratories of the soap and cosmetic industry. The first transport detector to be discussed will be that developed by James *et al.* [10] in 1964.

The Moving Wire Detector

The wire transport detector developed by James *et al.* [10] was subsequently manufactured and marketed by Pye Unicam and a diagram of the detector is shown in figure 8. Wire from a spool passes through a cleaning oven maintained at 750°C to burn off any lubricants remaining on the wire after drawing. The wire then passes round a pulley and through a coating block where the eluent from the column passes over the wire, coating it with a thin film of mobile phase. Subsequent to the coating block, the wire passes through an evaporator oven held at about 105°C or at a temperature appropriate for the volatilization of the solvents in the mobile phase. The evaporation is aided by a nitrogen stream flowing counter-current to the movement of the wire. The wire, with the solvents removed and any solute remaining as a residue on the wire surface, enters a pyrolysis tube *via* restriction that is held at about 500°C. The pyrolysis tube has a restriction and a nitrogen supply entering at either end. The nitrogen sweeps the tube from either end and any pyrolysis products that are formed pass out through a center tube into the FID.

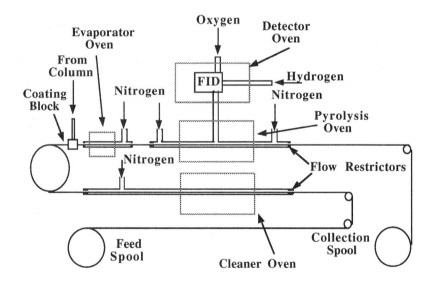

Figure 8   The Pye Unicam Moving Wire Detector

The FID was a standard detector used in gas chromatography and the output was processed in the normal manner. The detector functioned well in respect to operating independent of the nature of the mobile phase but the sensitivity realized was disappointing, being little better than the average refractive index detector *viz.* 5×10^{-6} g/ml. The poor sensitivity appeared to result from excessive noise, not a weak signal, which probably arose from the presence of high boiling impurities in the solvents, irregularities in the nitrogen flow and irregularities in the pyrolysis process. In addition the linear dynamic range was found to be less than two orders of magnitude. It did, however, unambiguously establish the transport system as a viable method of LC detection. Another disadvantage of the early detectors was their cost, which was excessive compared with the price of other liquid chromatography detectors that were commercially available.

The Chain Detector

About the same time as the development of the wire transport detector Haahti and Nikkari [11] described a similar device, more simple in design, that employed a chain loop in place of the wire transport system. A diagram of their apparatus is shown in figure 9.

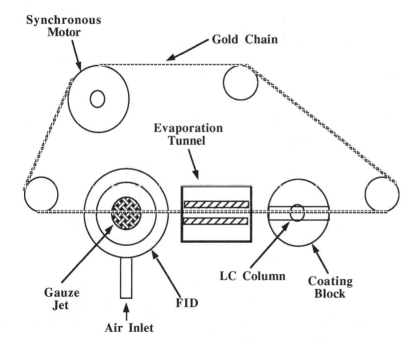

Figure 9   The Chain Detector

A gold chain driven by a synchronous motor passes over a coating block where the chain is wetted with the column eluent. The wetted chain then passes into an evaporator tunnel where the chain is heated and the solvent volatilized leaving any solute present in the eluent deposited on the chain. The chain passes out of the tunnel into the flame of an FID. During combustion of the solute, ions are produced in the expected manner and the ion current monitored by the detector electronics. Unfortunately, due to the occlusion of local, high concentrations of solute between the links of the chain, the detector output is extremely noisy and the overall system exhibits a relatively poor sensitivity. A chromatogram obtained with the chain detector is shown in figure 10.

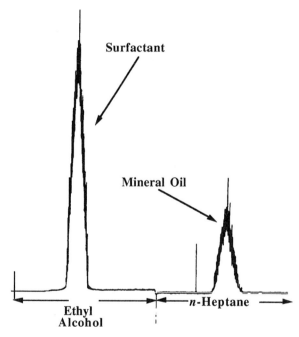

Sample: mineral oil and a surfactant, solvent: n-heptane/ethyl alcohol, column: 2 x 300 mm, column packing: silica gel, flow rate: 0.7 ml/min, chart speed: 24 cm/min., evaporator temperature: 150°C, nitrogen flow: 30 ml/min., hydrogen flow rate: 25 ml/min, oxygen flow rate: 30 ml/min.

Figure 10 Chromatogram Obtained from the Chain Detector

The noise spikes on each peak are clearly seen, which, besides affecting the overall sensitivity of the detector, also renders quantitative analysis approximate and difficult.

The Modified Moving Wire Detector

In the early 1960s, many workers in the field attempted to improve the performance of the transport detector and in 1966 Karmen [12] introduced an aspirating system to draw the pyrolysis products into the hydrogen flame detector. In 1970, Scott and Lawrence [13] developed the system of Karmen further and introduced a modified form of the original moving wire detector. The full sensitivity of the moving wire detector employing the original pyrolysis system was only realized for certain compounds, for example, high boiling hydrocarbons, such as squalane or long chain fatty acids such as stearic acid. For highly oxygenated solutes such as carbohydrates, polyglycols, etc., the sensitivity of the detector could be reduced by as much as an order of magnitude.

The sensitivity of the original wire transport detector, besides being degraded by the high noise level, was also determined by the quantity of pyrolysis product that could find its way to the FID. Excluding synthetic polymers, which often quantitatively produce monomers on pyrolysis, many compounds yield only a few percent of volatile compounds and the higher boiling components of these often condense in the conduits and never actually reach the FID. Thus the FID may only sense a very small fraction of the products from the solute deposited on the wire.

However, if, instead of pyrolysing the solutes, they were completely combusted in an oxygen or air stream then all the carbon in the solute would be converted to carbon dioxide. Further if the carbon dioxide was then reduced to methane by mixing it with hydrogen and passing it over a nickel catalyst, the carbon dioxide would be quantitatively converted to methane which could be detected by the FID. Such a system would increase the sensitivity of the detector to substances that gave poor yields of volatile compounds on pyrolysis and, furthermore, potentially provide a much wider linear dynamic range and possibly a predictable response.

A diagram of the original moving wire detector modified in this way is shown in figure 11. The FID was modified by enlarging the hydrogen lines in the detector body to reduce the flow impedance and permit the satisfactory operation of the aspirator. The detector was connected to a 2 in. length of 1/2 in I.D. thin walled stainless steel tube and then to the aspirator. The wide tube was closed with a loose plug of quartz wool and filled with about 2 g of nickel catalyst.

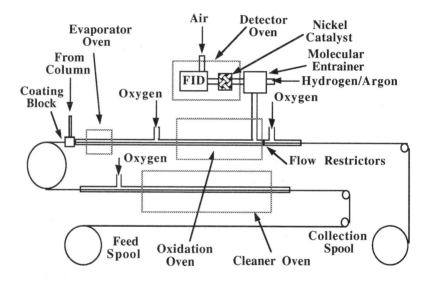

Figure 11 The Modified Moving Wire Detector

The nickel catalyst was prepared by absorbing a saturated solution of nickel nitrate onto 20/40 BS mesh brick dust, decomposing the nitrate by heating to 500°C for 3 hours followed by reduction of the nickel oxide to metal in a stream of hydrogen at 250°C. The aspirator consisted of a jet and venturi and was placed in line with the hydrogen flow to the detector. The gas used was a mixture of hydrogen and argon to improve the aspirating efficiency, and the passage of the mixture through the jet resulted in a pressure drop around the venturi allowing the combustion gasses to be continuously sucked into the hydrogen stream. It is seen in figure 11 that the reduced pressure side of the aspirator was connected to the side limb of the oxidation tube and the two tube system used in the original moving wire detector was replaced by a single tube. The oxygen or air is fed into the center of the tube providing both the evaporator flow and the oxidation flow. All tubes were constructed of quartz. The linear dynamic range of the system was shown to be about four orders of magnitude as indicated by the curve in figure 12.

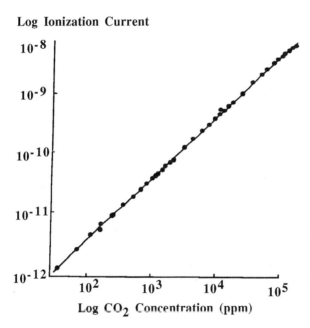

Figure 12 Linearity Curve for the Modified Moving Wire Detector

The response index for a series of compounds of different chemical types ranged from 0.96 to 1.04. The response of the detector was found to be proportional to the carbon content of the solutes tested, which would be expected. However, due to the limited number of compounds that were tested this relationship should be assumed only with caution. A chromatogram of blood lipids obtained by incremental gradient elution and monitored by the modified detector is shown in figure 13. As incremental gradient elution involves a program of 12 solvents ranging from hydrocarbons, chlorinated hydrocarbons, nitroparaffins, esters, ketones and alcohols. This separation illustrates the versatility that is provided by this detector for solvent selection.

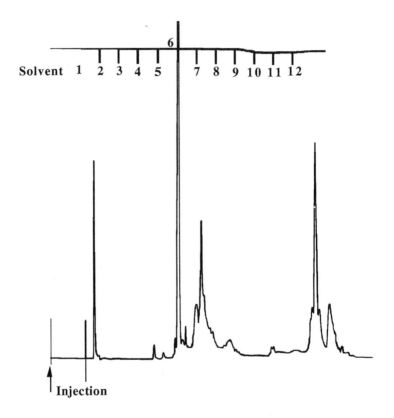

Figure 13 The Separation of Blood Liquids Employing Incremental Gradient Elution and Monitored by the Modified Moving Wire Detector

Van Dijk [14] developed a spray procedure for coating the wire in an attempt to improve the sensitivity of the detector. The column eluent passed directly to an atomizer, the nozzle of which was situated directly above the wire and 1-2 mm above it. The effect of the spray was firstly, to concentrate the solute in the mobile phase by partial evaporation during atomization and secondly, to increase the load on the wire. A linear dynamic range of about 3×10^3 was obtained from the system. The author claimed a sensitivity increase of 50 over the original wire transport detector. It is difficult to access the exact sensitivity of the device from the publication but from the calibration curves given it would appear to be about 3×10^{-6} g/ml.

Yang *et al.* [15] also developed a thermal spray procedure for coating the wire and also claimed an increased sensitivity. The authors employed a heated chamber above a moving stainless steel belt through which the conduit from the column passed. The solvent was rapidly brought to its boiling point resulting in spray leaving the exit of the conduit and coating the belt. The authors also employed a photo-ionization detector and an electron capture detector as alternatives to the FID.

Compton and Purdy [16] refashioned the FID of the Pye Unicam Modified moving wire detector by inserting a rubidium silicate glass bead above the flame and thus made its response specific and changed it into a nitrogen phosphorus detector.

Stolyhwo *et al.* [17] attempted to improve the sensitivity of the detector by using metal spirals wound on wire and stranded wire to increase the surface area of the carrier and thus increase the proportion of the column eluent taken into the detector. The authors claimed a minimum detectable mass of 100 ng of triolein. However, again the exact volume of mobile phase in which the mass of solute was contained was not clear from the publication. If the solute was eluted in a peak 1 ml wide at the base, the concentration at the peak maximum would be twice the average concentration *i.e.*, 2×10^{-7} g/ml, which, for a transport detector, would be a greatly improved sensitivity. If, however, the same mass was eluted as an early peak in the chromatogram with a band width of only 50 µl, then the sensitivity would be 4×10^{-6} g/ml, which would be no better than the previously developed transport detectors. This confusion emphasizes the importance of specifying sensitivity in terms of concentration, which allows the direct comparison of one detector with another.

Pretorious and Van Rensburg [18] attempted to increase the quantity of column eluent taken on the carrier by coating the wire with sodium silicate, kaolin and copper kaolin. Again, sensitivity was not reported in terms of minimum detectable concentration, so the precise change in sensitivity that resulted is not known. From the publication, however, a significant improvement appears to have been realized. The introduction of the wire coating procedure complicates an already complex instrument. It would appear that this approach might lead to serious instrumental problems arising from the dust produced by the disruption of the coating as the wire passed over the different pulleys.

Slais and Krejei [19] replaced the normal FID with the NPD detector and used it to detect chlorine compounds. They also used a combustion technique, mixed the combustion gases with hydrogen and passed the mixture directly into the NPD. At a column flow rate of 0.37 ml/min, the sensitivity of the detector was stated to be about 3×10^{-7} g/sec, which is equivalent, in concentration units, to about 1.6×10^{-6} g/ml. The moving wire detector has also been modified to produce radioactive detection by Dugger [20] for monitoring tritium or ^{14}carbon labeled compounds. To detect ^{14}carbon compounds, the solute on the wire was oxidized to carbon dioxide and the radioactive gas passed to a Geiger-Muller tube. To detect tritium, the tritiated water produced on combustion was passed over heated iron to reduce it to hydrogen and tritium, which was then also passed to a Geiger-Muller tube.

The wire transport system has many attractive characteristics as an LC detector, but for general use, its sensitivity needs to be increased by at least an order of magnitude. Furthermore, the overall system needs to be simplified to render it more reliable, less expensive and easier to operate.

The Disc Detector

The disc detector originally developed by Dubsky [21] employed a rotating gauze disc as a carrier and a diagram of the device is shown in figure 14. It consists of a rotating disc the perimeter of which is made of wire gauze. The column exit is situated just above the gauze and the eluent flows through the gauze coating it with a film of mobile phase. The excess of mobile phase is collected in a suitable container situated below the disc. A little ahead of the point of coating, in the direction of rotation of the disc, is situated an infrared lamp or some other appropriate heater which evaporates the solvent and leaves the solute coated on the gauze. Diametrically opposite to the point of coating is situated the FID. The flame jet is situated beneath the gauze, such that the flame itself is in contact with the gauze.

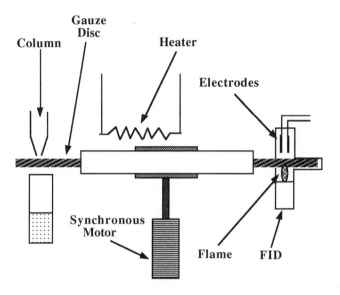

Figure 14 The Disc Detector

The electrodes of the FID are place just above the gauze directly over the flame. The current resulting from the ions collected by the electrode are amplified in the usual way and fed to a recorder. The system has the advantage of being simple compared with the conventional wire transport detector but although the author claimed a tenfold increase in sensitivity this is difficult to confirm from the data given. Szakusito and Robinson [22] claimed that the metal gauze disc carrier produced excessive noise and, in particular, "spikes" that resulted from local concentrations of solute accumulating at the intersections of the wire mesh during evaporation. They replaced the wire disc with an alumina disc 4.5 in diameter with the edge tapered to 0.25 mm thick. The thin edge was used for coating and detection. It was claimed that a significant reduction in noise was achieved but again, sensitivities were not given terms that would allow comparison with other detectors. The life of the alumina disc also seems a little questionable as, in continued use, the pores of the alumina would eventually become blocked with residue from incompletely combusted solutes or mobile phase components. The disc does appear to be a simpler transport system than the wire or chain but its reliability and sensitivity remain to be established.

References

1. I. Halasz, R. Endele and J. Asshauer, *J Chromatogr.*, **112**(1957)37.
2. R. P. W. Scott and P. Kucera, *J. Chromatogr.*, **169**(1979)51.
3. E. Katz and R. P. W. Scott, *J. Chromatogr.*, **253**(1982)159
4. R. P. W. Scott and P. Kucera, *J. Chromatogr.*, **125**(1976)251.
5. M. Verzeli and C. Dewaele, *J. HRC & CC* **5**(1982)286.
6. H. Yoshida, S. Kito, M. Akinoto and T. Nakajima, *J. Chromatogr.*, **240**(1982)493.
7. N. Baba and K. Housako US Patent 4,462.962, July 31 (1984).
8. R. P. W. Scott US Patent. No 4,555,936.
9. G. J. Schmidt and R. P. W. Scott. *Analyst,* **110**(1985)757.
10. A. T. James, J. R. Ravenhill and R. P. W. Scott, **Chem. Ind.** (1964)746.
11. E. O. A. Haahti and T. Nikkari, *Acta. Chem. Scand.*,**17**(1963)2565.
12. A. Karmen, *Anal. Chem.* **38**(1966)286.
13. R. P. W. Scott and J. F. Lawrence, *J. Chromatogr. Sci.*,**8**(1970)65.
14. L. M. Van Dijk, *J. Chromatogr. Sci.* **10**(1972)31.
15. L. Yang, G. J. Ferguson and M. L. Vertal, *Anal. Chem.*,**56**(1984)2632.
16. B. J. Compton and W. C. Purdy, *J. Chromatogr.*, **169**(1979)39.
17. A. Stolyhwo, O. S. Privet and W. L. Erdahl, *J. Chromatogr. Sci.* **8**(1970)65.
18. V. Pretorius and J. F. van Rensburg, *J Chromatogr. Sci.*, **11**(1973)263.
19. K. Slais and M Krejci, *J. Chromatogr.*, **91**(1974)181.
20. N. A. Dugger, US Patent 528,343.
21. H. Dubsky, J. Chromatogr.,71(1972)395.
22. J. J. Szakusito and R. E. Robinson, *Anal. Chem.*, 46(1974)1648.

CHAPTER 13

CHIRAL DETECTORS

Over the last decade there has been a growing interest in *chiral chromatography,* a term given to the separation of optically active compounds by chromatographic techniques. This interest has arisen largely as a result of the recognition that the majority of physiologically active substances exist in chiral form. Furthermore, the different enantiomers of a drug can exhibit widely different physiological activity in both degree and nature. For example, the hormonal activity of the two enantiomers of adrenaline differ very significantly, one being many times more active than the other. Similarly, one enantiomer of ephedrine not only has no physiological activity, but degrades the activity of the other enantiomer when also present as a mixture. A particularly sad example of the contrasting physiological effect of different enantiomers is the drug Thalidomide. This drug was originally marketed as a racemic mixture of N-phthalylglutamic acid imide. The desired pharmaceutical activity resides in the R-(+)-isomer and it was not found, until too late, that the corresponding S-enantiomer, was strongly teratogenic and its presence in the racemate caused serious fetal malformations. Largely as a result of this tragedy, the chiral purity of drugs has become a very important aspect of drug assay. Furthermore, in many countries, it is now law that if a drug can exist in different chiral structures, the relative physiological activity of the different enantiomers must be determined. It follows that the separation and identification of enantiomers is now a very important analytical problem and chiral chromatography is the natural technique to apply to the resolution of such mixtures. Chemically, chirality can result from a chiral center such as a carbon atom with different groups attached to each bond, from structural helicity as in the case of proteins, from planar chirality as demonstrated in some substituted cyclic compounds, from axial chirality as in the spiranes and from torsional chirality as shown in substituted allenes. Furthermore, other atoms such as sulfur, nitrogen, phosphorus and boron can also exhibit chirality. Simple chirality with one chiral center can be represented as follows.

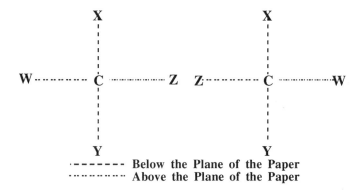

The two structures are seen to be non-superimposable images and are said to be enantiomeric.

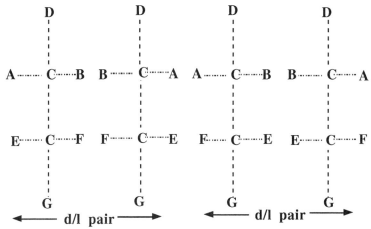

This rotation of polarized light is an important physical property that is used in chiral detection. If a molecule has two asymmetric (chiral) carbon atoms in its structure, then it is possible to have two pairs of optically active molecules and the pairs are said to be diastereoisomers as in the second diagram. If the two asymmetric carbons have identical substitution, again there will be two pairs of diastereoisomers formed, but one pair will possess a plane of symmetry and thus be optically *inactive*. The inactive pair are called *meso diastereoisomers*.

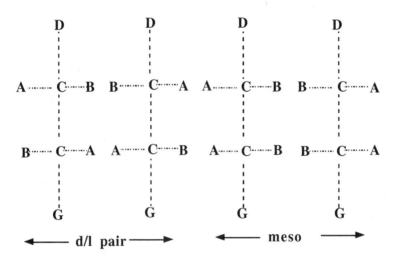

Enantiomers possess identical physical and chemical properties but diastereoisomers can *differ* and these differences can be used in appropriate physical chemical procedures to separate enantiomers. The enantiomer mixture is reacted with an appropriate chiral reagent and the diastereoisomers so formed are then separated by classical means.

One of the first chromatographic separation of enantiomers was by Gil-Av, who used gas chromatography to separate the trifluoroacetyl derivatives of some amino acids on an optically active stationary phase consisting of N–trifluoroacetyl-L-phenylalanine cyclohexyl ester. It was noted that the L isomers eluted last on the L stationary phase but first on the D stationary phase. Gas chromatography has not proved the ideal technique for separating optically active isomers as all solutes must be volatile and that restricts the molecular weight of the solutes that can be eluted and also their polarity, despite the possibility of derivatization.

In contrast, liquid chromatography lends itself to chiral separations and there are two basic procedures for separating optically active solutes. Firstly, reversed phase chromatography can be employed and a chiral substance can be added at low concentrations in the mobile phase. The chiral additive will be absorbed onto the surface of the reversed phase and act as an adsorbed stationary phase having chiral activity. This approach makes chiral detection more difficult, as it provides a background of optical activity in the mobile phase that will be many orders greater than that from the chirally active solutes. This is inevitably accompanied by a *high noise level* and consequent *poor sensitivity*. The second approach is to employ specific chirally active materials that are bonded to a silica or polymer surface to provide chirally specific interactions with the solutes.

Among the most effective chiral agents are the cyclodextrins The cyclodextrins are produced by the partial degradation of starch followed by the enzymatic coupling of the glucose units into crystalline, homogeneous toroidal structures of different molecular size. Three of the most widely characterized are the *alpha-, beta-* and *gamma*-cyclodextrins which contain 6, 7 and 8 glucose units respectively. Cyclodextrins are, consequently, chiral structures and *beta*-cyclodextrin has 35 stereogenic centers. These polymers can be bonded to silica gel by an aliphatic carbon chain that can range from 5 to 10 methylene units long. The chiral activity can be further augmented by bonding other chirally active groups onto the hydroxyl groups of the cyclodextrin. An example of a cyclodextrin phase that has an enantiomer of naphthylethyl carbamate bonded to it is shown in figure 1. The cyclodextrin is depicted as an open cone that is bonded directly to the silica base. There are a number of other optically active materials that can be bonded to silica or some polymeric material but the cyclodextrins have become particularly popular.

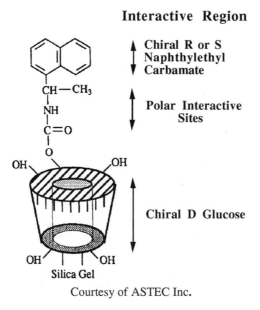

Courtesy of ASTEC Inc.

Figure 1 The Naphthylethyl Carbamate Derivative of a Cyclodextrin Bonded Phase

This latter method of achieving chiral selectivity renders chiral detection far more straightforward. The chiral material is strongly bonded to the stationary phase and thus none is present in the column eluent. Consequently, the only optical activity that can arise in the mobile phase flowing through the detector will be due to the presence of optically active solutes. This makes detection much simpler and allows a much higher sensitivity to be realized.

The Production and Properties of Polarized Light

As already discussed earlier in this book, light consists of a sinusoidally changing electric field normal to, and in phase with, a sinusoidally changing magnetic field. It is the electric vector of the electromagnetic wave that affects matter and thus can be sensed and measured. The plane of the electric vector in normal light takes no particular orientation, but in plane polarized light the electric vector is either vertically or horizontally polarized. If the electric vector transcribes an helical path, either to the right or left, the light is said to be *circularly polarized*. A linearly polarized beam of light can be considered to be the resultant of two equal-intensity in-phase components, one left and the other right circularly polarized, or two orthogonal linear components at ± 45°.

The differential absorption of these two ±45° linear components in a medium is known as *linear dichroism*; if there is a differential velocity between the two ±45° linear components when they pass through a medium (*i.e.* the refractive indicies of the medium and the two light components differ), this is known as *linear birefringence*. In an analogous manner, the difference in the adsorption characteristics of a medium to left and right circularly polarized light is termed *circular dichroism* (CD); and it follows, that the difference in refractive index of a substance to the two light components is called *optical rotary dispersion* (ORD), sometimes reported as *specific optical rotation*.

CD spectra are usually measured as the differential absorption of left and right circularly polarized light, *i.e.* (A_L–A_R) and are usually reported as the differential molar extinction coefficient ($\Delta\varepsilon$),

$$\Delta\varepsilon = (\varepsilon_L - \varepsilon_R) = \frac{(A_L - A_R)}{cl}$$

where (l) is the length of the cell,
and (c) is the molarity of solute.

The basic apparatus for measuring circular dichroism is shown in figure 2. It consists of a light source, a linear polarizer, a Fresnel rhomb that converts the linear polarized light to circularly polarized light, a sample cell and finally an appropriate light intensity measuring device. The rotation of the linear polarizer ±45° to the appropriate Fresnel rhomb axis induces the generation of left or right polarized light. A combination of a quartz Rochon prism and a silica Fresnel prism works well throughout the UV and visible light region. Below

220 nm, magnesium fluoride, calcium fluoride or lithium fluoride optics are required. In the infrared region a wire grid polarizer and sodium chloride rhomb are satisfactory.

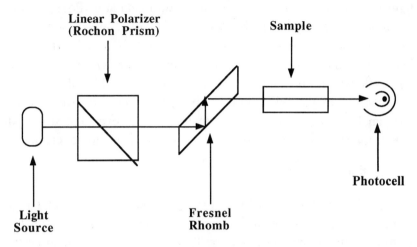

Figure 2 The Basic Apparatus for Measuring Circular Dichroism

The modern form of the CD spectrometer, on which the various types of chiral detectors are based, is shown in figure 3. It consists of a broad emission light source (*e.g.* xenon lamp) which is required because only small light intensities reach the sensor. Consequently, a high intensity light source is necessary and if a wide wavelength range is not required, then laser light sources can often be used.

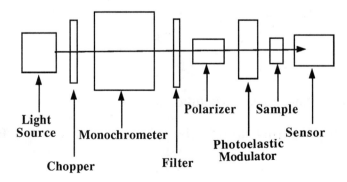

Figure 3 A Modern CD Spectrometer

The light passes through a chopper and then to a monochromator that allows light of a selected wavelength to be passed through a filter to the polarizer. The polarizer can be a Rochon prism. The polarized light is then passed through a photoelastic modulator (Pockel's cell) the function of which will be described below. The selected left or right circularly polarized light is then passed through the cell and the intensity of the transmitted light monitored by the sensor. In a sense, this could be considered as a very large volume LC chiral detector and it only requires the cell volume to be appropriately reduced for use with an LC column. This size reduction is, however, not simple in practice, particularly if the sensitivity of the device is to be retained.

Polarization Modulation

Polarization modulation, which is the alternate production of left or right circularly polarized light, is an essential process for CD measurement. A linearly polarized light beam can be said to be the resultant of two orthogonal, in phase, linear light beams. Consider a block of isotropic fused silica that is rendered birefringent by pressure exerted along the (x) or (y) axis as shown in figure 4.

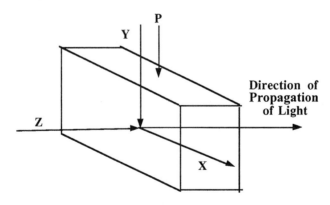

Figure 4 Polarization Modulation

Under these condition the refractive indexes (n_x and n_y) will differ,

$$\text{and} \quad n_x \neq n_y$$

If the light beam passing through the block is oriented with its resultant axis at 45° to the pressure axis, one of the components will travel faster through the medium than the other. If $n_x > n_y$, then the (x) component of the light beam will travel more slowly. If the retardation is exactly $(\frac{\lambda}{4})$ the emergent light beam will be right circularly polarized. If the retardation is $(-\frac{\lambda}{4}$ or $-\frac{3\lambda}{4}, -\frac{7\lambda}{4},$ etc.) then the emergent light beam will be left circularly polarized. This an example of the general principle of photoelasticity as first described by Brewster [1]. Electro-optic modulizers that are used in some modern instruments operate in the same way.

Electro-optic Modulators

The Pockel's cell used by Grosjean and Legrand [2] is based on the above principle and is composed of a **Z** cut section of a single crystal of ammonium dihydrogen phosphate. A diagram of a Pockel's cell is shown in figure 5.

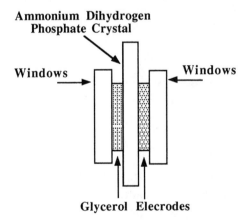

Figure 5 The Pockel's Cell

When transparent electrodes (glycol windows) are placed on either side of the crystal of ammonium dihydrogen phosphate and linearly polarized light, oriented at 45° to the crystallographic axis, is propagated down this unique axis, the device can act as a retardation plate. The retardation, at a particular wavelength, is a function of the voltage applied across the crystal faces. The voltage (V) required to maintain the quarter-wavelength condition for DC measurements at a wavelength (λ nm) is given by

$$V = 1350 + 10.6(\lambda - 220)$$

It is clear that fairly high voltages must be employed. As the crystal section is only 2 mm thick, voltage breakdown is possible and the maximum potential that can be used is about 5400 V with a corresponding wavelength maximum of 600 nm. The minimum wavelength is set by the transmission of the crystal which is normally about 185 nm. Another disadvantage of the above system is that light propagated along any axis, other than the isotropic unique axis of the ammonium dihydrogen phosphate crystal, will be birefringent. As a consequence the acceptance angle is critical (±1-2°) and so the incident light must be well collimated. Another problem associated with the above equipment is the vulnerability of the ammonium dihydrogen phosphate crystal to degradation if operated at high voltages continuously in the visible region or in conditions of relatively high humidity.

Piezo-Optical Modulators

Mechanical engineers have used polarized light to detect stress patterns for many years and more recently a number of workers [3-6] have explored the possibilities of photoelasticity for the fabrication of polarization modulators. The systems have been given several names, photoacoustic modulators, photoelastic modulators and stress modulators, the term photoelastic modulator will be used in this chapter. There are basically two types of photoelastic modulator, composite resonators and matched element resonators. Diagrams of the composite resonator and matched element resonator are shown in figures 6 and 7 respectively. The original piezo-optical devices [7] were composite resonators composed of a central block of optical material, (e.g. silica or germanium) with ceramic piezoelectric plates at either end. Blocks of brass are usually joined to the plates to "anchor" the motion. Each plate serves a different purpose: one drives the system and the other senses the modulation and controls and stabilizes it by an electrical feedback loop.

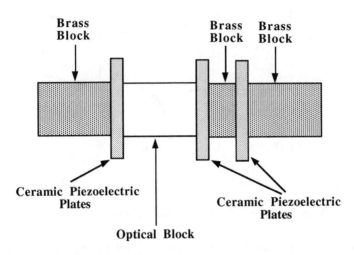

Figure 6 The Composite Resonator

The devices are reasonably simple to construct as their dimensions are not critical. Unfortunately the maximum oscillation will occur at the junction of the optical element and the piezoelectric plate. This can cause adhesion problems leading to static strain in the optical element and joint disruption during modulation.

The matched element resonators function using the sympathetic oscillation in an optical element when it is in contact with a piezoelectric driver, the frequency of which is set at the natural frequency of the optical element. An example of a matched two element resonator would consist of a quartz block, x-cut and gold plated, that is cemented to a block of optical material such as silica or calcium fluoride. The dimensions of the optical block are cut to match the natural frequency of the gold plated quartz block. The system is arranged such that during modulation a standing wave is set up in the device with a *node* situated at the junction between the two blocks.

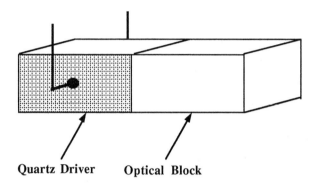

Figure 7 The Matched Element Resonator

This ensures minimum disruption of the joint during operation. The optical element of both types of resonator should be chosen to be isotropic, which will provide a relatively wide acceptance angle for light beams. The devices can operate at fairly high energy input, but there is a limit and if this limit is exceeded the optical unit may shatter.

Practical Chiral Detectors

There are two basic types of chiral detectors for LC, those that measure optical rotation and those that measure circular dichroism. At the time of writing this book, the only commercially available chiral detectors are those that measure optical rotation. Nevertheless, a detector that measures circular dichroism and utilizes a diode array sensor system is thought to be in the design stage and will be briefly described later.

The successful development of a chiral detector based on optical rotation measurement hinges on the use of the Faraday effect. If a plane polarized beam of light passes through a medium that is subjected to a strong magnetic field, the plane of the polarized beam is rotated a small angle (α), where (α) is given by

$$\alpha = VdH$$

where (V) is the Verdet Constant,
(d) is the path length
and (H) is the magnetic field strength.

This relationship is the Faraday effect. The magnetic field is generated in an air core coil inside of which is a rod made from glass having a high Verdet constant. Now,

$$H = iN$$

where (i) is the current through the coil,
and (N) is the number of turns in the coil.

Thus, $\alpha = VdiN$

Thus (a) can be controlled by varying (i).

The rotational resolution ($\Delta\alpha$) can be as little as 10^{-5} but due to heat losses in the coil the maximum value of ($\Delta\alpha$) is about $\pm 2°$. A diagram of the detector is shown in figure 8.

Light from a tungsten lamp passes through two condenser lenses to a polarizer. Plane polarized light from the polarizer passes through a temperature controlled cell to a Faraday modulator and thence through an analyzer and onto a photomultiplier. The modulator is supplied with a crystal controlled AC component. If an optically active sample is present, the intensity of the light falling on the photomultiplier changes. By the use of a phase sensitive amplifier and electrical feed-back, the current though the modulator is automatically adjusted until no AC component appears on the photomultiplier output.

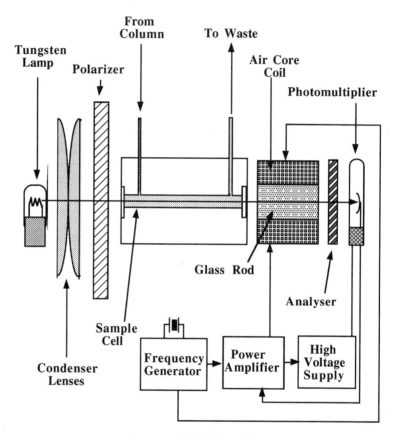

Courtesy of JM Science Inc.

Figure 8 A Chiral Detector Monitoring Optical Rotation

Unfortunately the sensor has a volume of about 40 µl, which is extremely large for modern LC columns. However, the urgent need for chiral detection is such that the large cell volume is accepted, and is accommodated by the use of large diameter columns. The device is rigidly supported and due to its stability can measure a rotation (α) of $10^{-4°}$. The separation of some carbohydrates monitored by the detector is shown in figure 9. The separation was carried out on a column 12.5 cm long, 4.6 mm in diameter packed with Hypersil APS 1. The mobile phase was acetonitrile/ water (8:2), at a flow rate of 0.5 ml/min. It is seen that the resolution of the mixture is maintained to satisfactory level despite the very large cell volume.

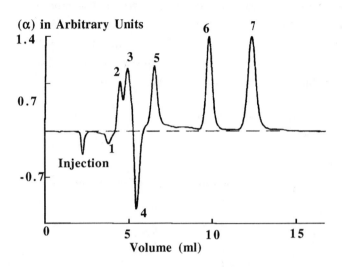

Courtesy of JM Science Inc.

(1) L-Rhamnose	0.02 mg	(2) D-(+)-Xylose	0.04 mg
(3) L-(+)-Arabinose	0.04 mg	(4) D-(-)-Fructose	0.02 mg
(5) D-(+)-Glucose	0.02 mg	(6) Sucrose	0.02 mg
(7) Maltose	0.2 mg		

Figure 9 The Separation of Some Carbohydrates Monitored by a Chiral Detector

It is also seen that the direction of rotation is clearly and unambiguously indicated by the direction of the peak. The peaks represent between 20 and 40 µg of material, which also indicates reasonable sensitivity. From approximate calculations made from the chromatogram the sensitivity for peak (6) appeared to be about 1.4 x 10^{-7} g/ml, which is only a factor of 2-4 less than that obtainable from the diode array UV detector. However it is extremely difficult to estimate the noise from the chromatogram shown in figure 9. A more accurate estimation of the detector sensitivity can be made from the chromatogram of some essential oil components shown in figure 10.

The separation shown in figure 10 was carried out on a column 20 cm long and 8 mm I.D. packed with Zorbax silica. The mobile phase was n-hexane/chloroform (4:1) and the flow rate was 0.5 ml/min. The width of the second peak((-)-α-terpinyl acetate) is about 0.16 mm and from the flow rate axis, 1.06 cm is equivalent to a volume of 5 ml. Thus the peak width at the base is about 0.75 ml.

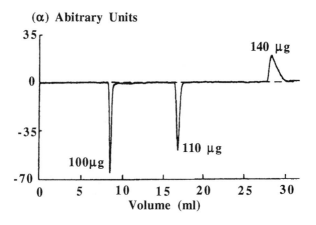

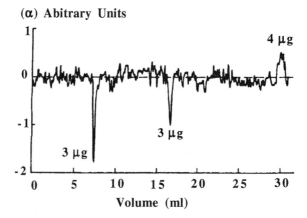

Figure 10 The Separation of Three Optically Active Fragrance Compounds Monitored by a Chiral Detector

The peak represented 3 µm of material and, taking the concentration at the peak maximum as twice the peak average concentration, the peak maximum concentration was about 8 µm/ml. The peak height appears to be about three times the noise and so the sensitivity (that concentration that will give a signal equivalent to twice the noise) is about 5.3 µm/ml or in more standard terms 5.3×10^{-6} g/ml. As the chiral detector is a bulk property detector, a sensitivity of 5.3×10^{-6} g/ml seems more realistic. Nevertheless, this sensitivity is a great improvement on many chiral detectors previously described.

The value of a chiral detector in the analysis of physiologically active materials is clear, but the methods so far used have been found somewhat insensitive. A more encouraging procedure would be the measurement of circular dichroism and such instrumentation employing diode array detection is presently under development. Details of the device are difficult to obtain due to patent applications pending and particulars are not available. The basic arrangement, however, is thought to be similar to that depicted in figure 11.

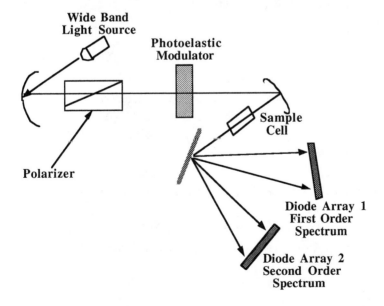

Figure 11 Projected Diode Array Chiral Detector

Light from a wide band source such as a xenon lamp passes through a polarizer and photoelastic modulator producing left and right polarized light of all wavelengths generated by the lamp. The light then passes through the sample cell and the transmitted light falls onto a diffraction grating. The dispersed light is focused onto two diode array sensors. One array receives the zero order spectrum and the second the first order spectrum. It is claimed, that the output from these two diode arrays can be processed to give both the optical rotary dispersion and the circular dichroism for each group of wavelengths monitored by each diode of the arrays. This approach is a great advance on the simpler detector presently available and hopefully more details will be forthcoming in the not too distant future. Fundamentally, the system is a *bifunctional* detector that provides both chiral and absorption measurements on the column eluent.

It is recommended that those further interested in chiral measurements read a review paper by Drake [7] which gives an excellent survey of the subject and a paper by Drake and Jonas [8] that describes techniques for measuring optical activity.

References

1. D. Brewster, *Philos. Trans. R. Soc. London,* 156(1816).
2. M. Grosjean and M. Legrand, *C. R. Acad. Sci.,* **251**(1960)2150.
3. J. C. Kemp, *J. Opt. Soc. Am.,* **59**(1969)950.
4. S. N. Jasperson and S. E. Schatternley, *Rev. Sci. Instrum.,* **40**(1969)761.
5. L. F. Mollenauer, D. Downie, H. Sengstrom and W. B. Grant, *Appl. Opt.,* **8**(1969)661.
6. J. C. Cheng, L. A. Nafie, S. D. Allen and A. I. Braunstein, *Appl. Opt.,* **158**(1976)1960.
7. A. F. Drake, *J. Phys E. Sci. Instrum.* **19**(1986)170.
8. A. F. Drake and G. D. Jonas, *Chromatogr. Anal.,* **February**(1989)11.

CHAPTER 14

THE RADIOACTIVITY DETECTOR AND SOME LESSER KNOWN DETECTORS

The Radioactivity Detector

The use of radioactive tracers in the study of reaction mechanisms has been steadily increasing over the last two decades. Tracer techniques have been used to elucidate the mechanism of complex reactions throughout the field of biotechnology and, in particular, to follow metabolic pathways, both synthetic and natural in both plants and animals. Just as the analytical problems that arose in such work educed the development of GC radioactivity detectors, so have they also evoked the development of LC radioactivity detectors. In LC, however, as the mobile phase is a liquid, radioactivity detection is more difficult and the procedures used for radioactivity detection in GC cannot be used.

The most useful radioactive tracers are compounds containing tritium (^{3}H) or ^{14}carbon, both of which are relatively low energy β emitters. The first problem is to convert the β energy into light photons that can be detected by a photo multiplier and provide an electrically countable pulse. This can be achieved by either mixing a suitable scintillating reagent with the column eluent prior to detection, or by passing the eluent through a transparent tube packed with a scintillator that is closely associated with a photomultiplier. Both methods are effective and both are used, the choice depending somewhat on the nature of the sample.

The sample must be counted for a finite time and so the sensor that is exposed to the photocell must have a finite volume. Thus, at given flow rate, the solute will reside for a given period (the residence time) in the vicinity of the photomultiplier.

The residence time (T_R) is given by

$$T_R = \frac{V_c}{Q}$$

where (T_R) is the volume of the sensor next to the photomultiplier,
and (Q) is the flow rate.

Thus the counts per minute (CPM) will be given by

$$CPM = \frac{N_T}{T_R}$$

where (N_T) is the total number of counts,
and (T_R) is the residence time of the solute in the sensor.

The efficiency of a radioactivity detector is the percentage of counts that are detected expressed as a proportion of the total number of radioactive events that actually occur in the sample during the counting period. The sensitivity of a radioactivity detector has a different meaning from that used in other parts of this book. The *sensitivity* of a *radioactivity detector* is a measure of the detector's ability to identify a given number of disintegrations above any background signals. The background signals can arise from radioactive contamination but are mostly due to background counts from cosmic rays. For a given combination of detector, radioactive nuclei, chromatography eluent and solute concentration, the counting efficiency is constant. It can be determined by counting a calibrated standard and comparing the number of disintegrations recorded to the calibrated activity of the standard. The following are the approximate counting efficiencies achievable under normal mobile phase flow conditions.

Nucleus	Liquid Scintillator	Solid Scintillator
^3H	>55%	up to 10%
^{14}C	>95%	up to 90%

Under isocratic development the counting efficiency remains sensibly constant but with gradient elution, the counting efficiency will change as the mobile phase composition changes and an efficiency calibration curve may need to be obtained. The calibration curve is often called a *quench* curve.

The liquid scintillation process involves the conversion of radioactive particle energy to photons of light. Both the sample and the scintillator are in direct contact, which allows the energy to be transferred. Any process that reduces the efficiency of the energy transfer is called *quenching*. There are two types of quenching, *chemical* quenching and *color* quenching. Chemical quenching occurs during the transfer of energy from the solvent to the scintillator; color quenching results from the absorption of the light produced, before it is sensed by the photomultiplier. The two types of quenching are depicted as follows.

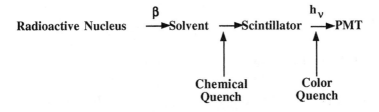

Quench calibration curves may need to be constructed from data obtained from samples of standard activity and counting efficiency.

Such curves are necessary for quench correction when, for the example given, the counting efficiency changes during gradient elution. If standard sensors and conditions are used, appropriate quench corrections are usually made available with the instrument. If not, standards are usually provided, and specific calibration instructions are given for each type of instrument.

The minimum detectable activity (MDA), which is analogous to the minimum detectable concentration, is given by

$$\text{MDA} = \frac{BWE}{T_R}$$

where (B) is the background count,
(W) is the base width of the peak,
and (E) is the efficient expressed as % efficiency/100.

The sensitivity, analysis time and resolution are obviously all related. Slowing the flow rate will increase the residence time and increase the counting efficiency, but it will also increase the analysis time. Conversely, increasing the flow rate will reduce the analysis time, but reduce the counting efficiency. Increasing the volume of the sensor will increase the counting efficiency but, as already discussed, may also impair the chromatographic resolution. Reducing the sensor volume will allow very high efficiency columns to be used, but the counting efficiency will be reduced and the minimum detectable activity increased. As a general rule, the volume of a sensor should be between a quarter and half the volume of the smallest peak to be monitored.

One of the first LC flow through cells for the continuous monitoring of radioactive material in liquid streams was designed by Schram and Lombaert [1] in 1961. They achieved this by scintillation counting, utilizing anthracene powder as the scintillation agent. In fact, although other scintillation agents are now used, the modern radioactivity detectors are all based on the same system as that devised by Schram and his coworkers. The counting cell was made from 60 cm of polyethylene tubing 2.2 mm I.D. and 3.2 mm O.D., which was filled with anthracene powder. The anthracene was packed into the tube, using a slurry packing technique in conjunction with a suspending solvent composed of 30% water in ethanol. The packed tube was then coiled into a flat spiral and placed in a flat Lucite vial containing silicone oil. The size of the anthracene particles was found to affect both the flow impedance and the counting efficiency of the tube. The authors compromised on the particle size and employed particles 300 µm in diameter for counting ^{14}carbon and 150 µm in diameter for counting tritium. The cell volume was about 1 ml, which, although the inherent band dispersion was significantly reduced due to the presence of the anthracene packing, would be far too great for modern high efficiency small bore columns. The tube was situated in a light-tight box and counting was achieved by using two refrigerated photo-multipliers. The counting efficiency was about 2% for tritium and 55% for ^{14}carbon. The background gave 60 counts/min. and thus at a signal to noise ratio of two, 1.5 nC/ml of tritium and 0.5 nC/ml of ^{14}carbon could be detected.

In 1964, Sjoberg and Agren [2] described a dual in-line UV absorption and radioactivity detector combination. This type of system can be very useful as the output from the absorption detector discloses the position and relative proportion of all UV absorbing species present in the mixture, whereas the radioactivity detector indicates which of those substances are radioactive. The sensor cell consisted of a commercially available plastic scintillation tube in the form of a flat spiral, 1.75 in diameter, composed of a single tube, 0.7 mm I.D. and 1.5 mm O.D. having a volume of about 300 µl. Their report contains an excellent example of the use of the system in monitoring the separation of some soluble nucleotides extracted from diploid tumor ascilllis cell 30 minutes after incubation with an inorganic radioactive phosphate.

The detecting systems devised by Schram and Lombaert [1] and Sjoberg and Agren [2] were reasonably efficient for counting

^{14}carbon and isotopes having β-particle emission of higher energy but their efficiency for counting tritium was relatively poor. Moreover, some compounds could be adsorbed by the anthracene, causing a build-up of background noise and consequent loss of material. In addition, the cell could only be used with solvents that did not affect the cell material or the anthracene. Scharpenseel and Menke [3] attempted to improve the efficiency of the tritium count by employing a toluene based liquid scintillation agent. The column eluent, or a portion thereof, is mixed with the reagent and then passed to the scintillation cell. The counting efficiency was increased a little (2-3%) but the system was extremely sensitive to the presence of salts even when the eluent-to-scintillator ratio was low. Hunt [4] attempted to improve this system by replacing the toluene based scintillator with a solution of naphthalene 2,5–diphenyloxazole and (1,4-*bis*-2(3-methyl-5-phenyl oxazole) benzene) in carefully purified dioxane.

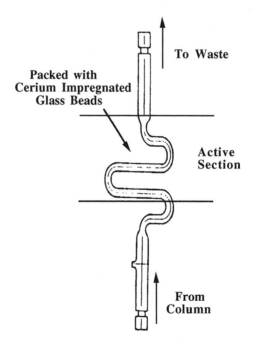

Figure 1 The Packed Scintillation Sensor of Schutte

The cell Hunt used was a coiled tube similar to that of Schram and Lombaert [1] so that peak dispersion would still be a problem with high efficiency columns, but nevertheless he achieved a counting efficiency for tritium of 14% and about 70% for ^{14}carbon. Schutte [5] examined the methods of Hunt and Sjoberg and developed a dual system of absorption and radioactivity detection. Schutte also introduced a heterogeneous scintillation system employing cerium activated lithium glass beads as the scintillation agent. A diagram of the cell employed by Schutte is shown in figure 1.

Unfortunately the counting efficiency of the system was relatively poor, 0.2% for tritium and 17% for ^{14}carbon. However, the advantage of this method is that due to the cell being packed with beads, it would have little flow resistance and limited peak dispersion and thus if used in conjunction with suitable low dispersion connecting tubes, it could be used with relatively high efficiency columns. As a consequence, many modern commercial radioactivity detectors are designed on the same principle, but with more efficient scintillators and more efficiently designed sensors.

As already mentioned under transport detectors, Dugger [6] modified the moving wire detector to detect tritium and ^{14}carbon. To detect ^{14}carbon, the solute coated on the wire after evaporation of the solvent was oxidized to carbon dioxide and water. The radioactive carbon dioxide was passed to a Geiger counter and detected in the same manner as that described by James and Piper [7] which was discussed under GC radioactivity detectors in an earlier chapter. Tritium could be detected by passing the water vapor from the oxidation process over heated iron to reduce it to hydrogen and tritium, which was then also passed through a Geiger counter.

The continuous flow monitoring of radioactivity in columns eluents, which usually involves a scintillation counting technique, can be classed as either homogeneous or heterogeneous counting systems. In homogeneous counting systems the column eluent is mixed with a liquid scintillation reagent before passing through the counting cell situated between the photo multiplier tubes of a photomultiplier. In heterogeneous counting systems the column eluent is passed directly through a flow cell that is packed with particles of a suitable scintillating agent such as anthracene or cerium activated lithium glass beads. Heterogeneous counting systems are usually free from chemical quenching effects and the sample can be easily recovered from the eluent leaving the sensor.

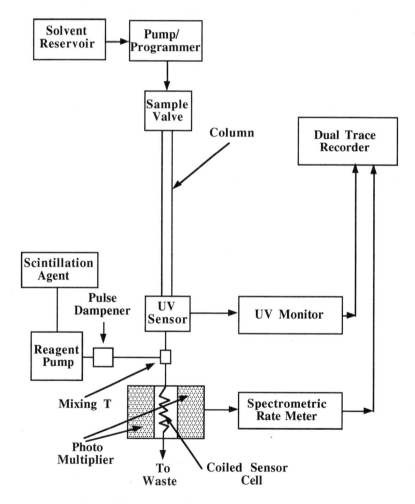

Figure 2 A Liquid Chromatography System Incorporating UV and Radioactivity Monitors

However, the counting efficiency can be low and the nature of the scintillator can restrict the choice of mobile phase. As previously mentioned, solutes can be adsorbed on the surface of the scintillator producing an increasing background noise and reduced counting efficiency.

Heterogeneous counting systems can be advantageous in preparative chromatography where solute concentration and radioactivity are relatively high. Homogeneous radioactivity detectors are generally to be preferred in analytical LC where the recovery of the sample is not important but sensitivity and versatility are essential. Nevertheless, the coupling of a homogeneous radioactivity monitor to a liquid chromatography will still require some compromise between the sensitivity of the monitor and the speed and resolution obtained from the liquid chromatograph.

Reeve and Crozier [8] assessed the optimum arrangement for a homogeneous radioactivity detecting system from a theoretical point of view and the design they recommended is shown in figure 2. The eluent from the column first passes through the sensor of a UV absorption detector and is then blended with the scintillation agent in a small-volume mixing T. The reagent is delivered by a micro metering-pump fitted with a pulse dampener. The authors recommended a dioxane/naphthalene mixture as the scintillation reagent for normal phase chromatography (i.e. silica gel and other polar stationary phases). However, the optimum mixture had to be determined by experiment.

Mackey *et al.* [9] described an optimized heterogeneous radioactivity detector cell based on a packed tube similar in principle to that of Schutte [5] employing cerium impregnated glass powder as the scintillator. A diagram of their cell is shown in figure 3. The glass they used was NE901 and NE913 supplied by Nuclear Enterprises, Edinburgh, Scotland, UK. They claimed that the optimized detector provided a counting efficiency greater than 70% for ^{14}carbon, which is the kind of efficiency normally expected from standard liquid scintillation techniques.

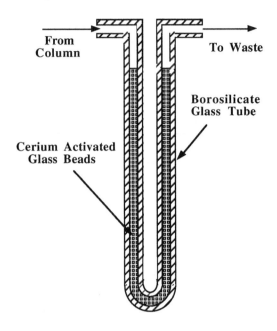

Figure 3 Diagram of a Heterogeneous Radioactivity Sensor

In order to achieve this counting efficiency, however, special electronic techniques were necessary which will be discussed shortly.

An example of a contemporary LC system for monitoring radioactivity is afforded by the ΔRadiomatics Beta Model A-280 manufactured by ΔPackard Inc., a diagram of which is shown in figure 4. The eluent from the column passes to a valve that can direct the flow to waste or to the radioactivity counter. The eluent from the valve then passes to a splitter where a selected portion can be passed to the mixing T for subsequent counting and the remainder passed to a fraction collector. The scintillation agent is transferred from a reservoir by an appropriate pump to the mixing T where it blends with the column eluent and passes to the counting sensor.

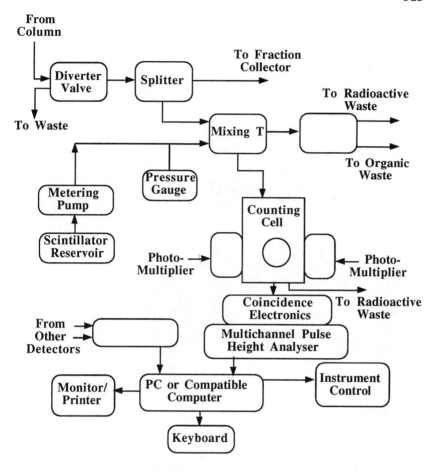

Courtesy of ΔPackard Inc.

Figure 4   Block Diagram of the Radiometric LC Scintillation Detecting System

A diagram of the counting sensor is shown in figure 5. From the counting sensor the residue passes to radioactive waste. The counting cell contains a coiled tube of a defined volume situated in a transparent case. On either side of the sensor case are situated two photomultipliers the outputs of which are processed by coincidence electronics. Much of the noise that produces the background count arises from the electronics and the power supplies. If two photomultipliers are employed, noise spikes arising from the power supplies will coincidentally activate both photomultipliers. The purpose of the coincidence electronics is to reject any signal that arises simultaneously from both photomultipliers, thus eliminating power supply noise spikes. In this way the background noise is significantly reduced, which increases the overall sensitivity of the detector.

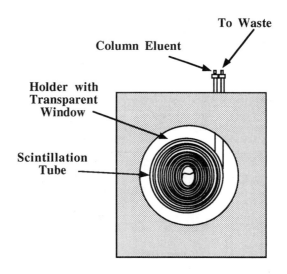

Courtesy of ΔPackard Inc.

Figure 5 The Radiomatic LC Scintillation Sensor

The output from the coincidence electronics passes to a multichannel pulse height analyzer and thence to a computer. The computer can also take data from other detectors that are in line with the column eluent in addition to instrument control. Data is processed and then presented on a monitor or printer. Cells for both homogeneous and heterogeneous counting are available. Homogeneous counting cells provide efficiencies of 55% for tritium and at least 95% for ^{14}carbon, whereas the heterogeneous cells provide about 10% for tritium and about 90% for ^{14}carbon.

An example of the use of the radioactivity detector to monitor some alkylethoxylate urinary metabolites (9) is shown in figure 6.

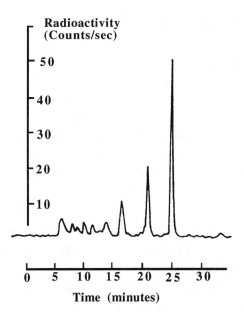

Figure 6 An Example of the Use of the Radioactivity Detector to Monitor Some Urinary Metabolite Derivatives

The separation was achieved with the aid of a reversed phase column with an acetonitrile–water mobile phase buffered with acetic acid. The radioactive metabolites are clearly seen.

Some Lesser Known Detectors

Over the last thirty years of chromatography research and development a large number of gas and liquid chromatography detectors have been devised, fabricated and evaluated. Of these, only a relatively small number are now in general use and even less have become commercially available. Nevertheless, many of these detectors are interesting examples of innovative thinking and, furthermore, many have specific types of response that may have use in applications other than chromatography. Some of the lesser known GC detectors have already been discussed in this book and in this chapter some of the more remote LC detectors will also be described.

The Heat of Adsorption Detector

During the passage of a solute band through a chromatographic column, the solute is continuously adsorbed into the stationary phase in the front portion of the peak and progressively desorbed into the mobile phase at the rear portion of the peak. The adsorption process arises from the continuously increasing solute concentration in the front of the peak causing more and more solute to be *adsorbed* until the peak maximum is reached. Subsequent to the peak maximum, in the rear portion of the peak where the solute concentration is continuously falling, the converse occurs and the solute is *desorbed* from the stationary phase. As a result of the heat of adsorption of a solute in the stationary phase, the adsorption/desorption process is accompanied by heat evolution and heat absorption. Concurrent with this heat change is a corresponding temperature change. Thus, in the front of the peak where the heat of adsorption is evolved, there is a rise in temperature and at the rear of the peak, where the solute is desorbed and the heat of adsorption is absorbed, there is a fall in temperature. In 1959, this temperature effect was employed by Claxton [10] as a means of detection in LC. In fact, this effect can also be observed in GC columns but the temperature changes are much smaller.

The heat of adsorption detector, devised by Claxton, consists of a small plug of adsorbent, usually silica gel, through which the chromatographic eluent passes subsequent to leaving the column. Embedded in the silica gel is either a thermocouple or a thermistor that continuously measures the temperature of the adsorbent and mobile phase.

When a solute is eluted from the column, it is adsorbed by the silica gel, heat is evolved and the temperature rises. As the peak passes, and the solute is desorbed from the silica gel, heat is adsorbed and the temperature falls.

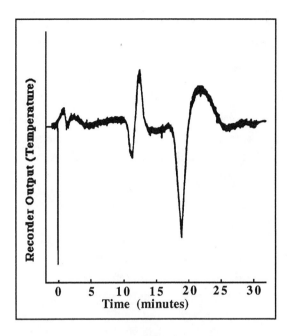

Figure 7 A Chromatogram Showing Two Peaks Monitored by the Heat of Adsorption Detector

Thus, the curve relating temperature with time appears as an **S**-shaped curve for each solute band that is eluted. An example of a chromatogram containing two peaks monitored by the heat of adsorption detector is shown in figure 7.

The heat of adsorption detector was investigated by a number of workers [11-13] and although one model was manufactured, it was not a commercial success and consequently this detector has been little used in LC analysis. One reason for this is the curious and apparently unpredictable shape of the temperature-time curve that results from the detection of the usual Gaussian or Poisson concentration peak profile. The shape of the curve changes with detector geometry, the operating conditions of the chromatograph, the retention volume of the solute and for closely eluted peaks, it produces a complex curve that is extremely difficult to interpret. Much work has gone into the modification of the temperature profile, both by sensor design and by electronic signal shaping, but no change has been sufficiently successful to make the detector a practical eluent monitoring device. There is good reason for this, as the function that describes the temperature profile is very complex. Nevertheless, it would appear that there are certain conditions where the integral of the temperature time curve can give an accurate representation of the concentration profile of the peak although these conditions are extremely difficult to meet in practice. The function describing the temperature-time curve can be derived theoretically.

A Theoretical Treatment of the Heat of Absorption Detector

Consider a small cell, containing the adsorbent, situated at the end of the column through which the column eluent passes. It is assumed that the eluent is brought to a constant temperature by a suitable heat exchanger situated between the cell and the column and that the exchanger does not contribute significantly to band dispersion. Let the cell have internal and external radii of (r_1) and (r_2) respectively and length (l). The following postulates must be made:

1/ The flow of mobile phase through the cell is constant.

2/ The temperature of the cell surroundings is constant.

3/ The cell is of sufficient size that solute equilibrium between the two phases can be assumed.

4/ The cell itself does not contribute significantly to band dispersion.

Consider the heat balance of the cell,

(Cell Heat Capacity)x(Temperature Change)=(Heat Evolved in Cell)

−(Heat Convected from Cell by Mobile Phase) − (Heat Conducted from Cell Through the Cell Walls)

The volume flow of mobile phase will be measured in 'plate volumes' [14] (v) of the attached column.

The plate volume is define as $(v_m + Kv_l)$ which for the column will be designated as (c_a) for solute (a). (v_m) and (v_l) are the volumes of mobile phase and stationary phase per plate respectively and (K) is the distribution coefficient of the solute between the two phases.

Let a volume (dv) of mobile phase pass through the cell carrying solute that is absorbed onto the silica with the evolution of heat, and let the resulting temperature change be (dø).

Then assuming the heat capacity of the solute is negligible,

Hdø = (Heat Evolved in Cell) - (Heat Convected from Cell)
                        - (Heat Conducted from Cell)   (1)

where (H) is the heat capacity of the cell,

i.e. $\quad\quad H = V_m d_m S_m + V_s d_s S_s + V_g d_g S_g$

and (V_m), (V_s), (V_g) are the volume of mobile phase in the cell, the volume of absorbent in the cell and the wall volume respectively, (d_m), (d_s), (d_g) are the densities of the mobile phase, absorbent and cell walls respectively and (S_m), (S_s), (S_g) are the specific heats of the mobile phase, absorbent and cell walls respectively.

Note the italic form, (V), is used to distinguish it from volume, (V), and plate volume, (v).

Let a volume (dv) of mobile phase, equivalent to $(c_a dv)$ ml, enter the cell, and let the concentration of solute (a) in the incremental volume be (X_n).

Let an equivalent volume $(c_a dv)$ of mobile phase be displaced from the cell, and the solute concentration in the cell prior to the introduction of volume (dv) be (X_m).

Now the net change of mass of solute (dm) in the cell will be

$$dm = (X_n - X_m)c_a dv \tag{2}$$

As equilibrium is assumed to occur in the detector cell the introduction of the mass of solute (dm) will result in a change in concentration of solute in the mobile phase and adsorbent of (dX_m) and (dX_s) respectively, where (X_s) is the concentration of solute in the absorbent.

Thus, $\quad dm = V_s dX_s + V_m dX_m$

and as $\quad dX_s = K dX_m$

$$dm = V_s K dX_m + V_m dX_m$$

$$= (V_s K + V_m) dX_m \tag{3}$$

Equating equations (2) and (3) and rearranging,

$$\left(\frac{V_s K + V_m}{c_a}\right)\frac{dX_m}{dv} + X_m = X_n$$

Now, $V_s K + V_m$ is the "effective cell volume" in much the same way that (c_a) is the column "plate volume".

Let, $\dfrac{(KV_s + V_m)}{c_a} = C_a = \dfrac{\text{"effective cell plate volume"}}{\text{"column plate volume"}}$ for solute (a)

Thus, $\quad C_a \dfrac{dX_m}{dv} + X_m = X_n \tag{4}$

Multiplying throughout by $(e^{\frac{v}{C_a}})$,

$$C_a e^{\frac{v}{C_a}} \frac{dX_m}{dv} + X_m e^{\frac{v}{C_a}} = X_n e^{\frac{v}{C_a}}$$

or
$$\frac{d\left(C_a e^{\frac{v}{C_a}} X_m\right)}{dv} = X_n e^{\frac{v}{C_a}}$$

Integrating, $\quad C_a e^{\frac{v}{C_a}} X_m = \int X_n e^{\frac{v}{C_a}} dv + R$

Now, when $v = 0$, the solute has not moved from the point of injection on the column.

Thus, when $v = 0$, $X_m = X_n = 0$ and consequently, $R = 0$ and

$$X_m = \left(\frac{e^{-\frac{v}{C_a}}}{C_a}\right) \int_0^V X_n e^{\frac{v}{C_a}} dv \qquad (5)$$

Equation (24) provides an expression for (X_m). Continuing; if the change in mass of solute on the absorbent due to a volume flow of mobile phase $c_a dv$, is (dm_s), then the consequent heat evolved dG in the cell will be given by

$$dG = g\left(\frac{dm_s}{dv}\right) dv$$

where (g) is the heat of adsorption of the solute in calories per gram of solute.

Hence, $\quad dG = gV_s\left(\dfrac{dX_s}{dv}\right) dv \quad$ and, as $dX_s = KdX_m$

$$dG = KgV_s\left(\frac{dX_m}{dv}\right) dv$$

From equation (4),
$$\frac{dX_m}{dv} = \left(\frac{X_n - X_m}{C_a}\right)$$

Substituting for $\left(\dfrac{dX_m}{dv}\right)$, $\quad dG = KgV_s\left(\dfrac{X_n - X_m}{C_a}\right)dv$

Thus, substituting for (X_m) from equation (5) and rearranging,

$$dG = \left(\dfrac{KgV_s}{C_a}\right)\left(X_n - \left(\dfrac{1}{C_a}\right)e^{-\frac{v}{C_a}}\int e^{-\frac{v}{C_a}} X_n\, dv\right) dv$$

Now, from the Plate Theory [15] $\quad X_n = X_o \dfrac{e^{-v} v^n}{n!}$

Therefore,

$$dG = \dfrac{KgV_s}{C_a}\left(X_o \dfrac{e^{-v} v^n}{n!} - \left(\dfrac{X_o}{C_a}\right) e^{\frac{-v}{C_a}}\int e^{\frac{-v}{C_a}}\left(\dfrac{e^{-v} v^n}{n!}\right) dv\right) dv$$

Let

$$X_o \dfrac{e^{-v} v^n}{n!} - \left(\dfrac{X_o}{C_a}\right) e^{\frac{-v}{C_a}}\int e^{\frac{-v}{C_a}}\left(\dfrac{e^{-v} v^n}{n!}\right) dv = f(v)$$

Hence, $\quad dG = \dfrac{KgV_s}{C_a} f(v)\, dv \quad\quad (6)$

Now, the heat conducted from the cell will be considered to be controlled by the radial conductivity of the total cell contents and not by the cell walls alone. Furthermore, the axial conductivity of the cell will be ignored as its contribution to heat loss will be several orders of magnitude less than that lost by axial convection.

Consequently, as the cell is cylindrical, the heat conducted radially from the cell has been shown to be [16]

$$\frac{2\pi l E \theta \, dt}{\text{Log} \frac{r_1}{r_t}}$$

where (E) is the thermal conductivity of the cell contents,
   (q) is the excess cell temperature above its surroundings,
   (r_t) is the radius of the sensing element,
and (dt) is the time for a volume $c_a dv$ of mobile phase to pass through the cell.

Now (dt) refers to the time interval during the introduction of the volume ($c_a dv$) of mobile phase and thus, if the flow rate is (Q),

$$\frac{dv}{dt} = \frac{Q}{c_a} \quad \text{or} \quad dt = \frac{c_a dv}{Q}$$

Thus, heat conducted from the cell is $\dfrac{2\pi l E c_a \theta \, dv}{Q \, \text{Log}_e \frac{r_1}{r_t}}$

and the heat convected from the cell is $d_m S_m \theta c_a dv$.

Thus, inserting the above expressions for the heat conducted from the cell and the heat convected from the cell, together with the heat evolved from the cell, from equation (6) in the heat balance equation,

$$H \, d\theta = \frac{KgV_s}{C_a} f(v) \, dv - \frac{2\pi l E c_a \theta \, dv}{Q \, \text{Log}_e \frac{r_1}{r_t}} - d_m S_m \theta c_a \, dv$$

or

$$\frac{d\theta}{dv} = \frac{A}{H} f(v) - \frac{\beta c_a \theta}{H} \qquad (7)$$

where $A = \dfrac{KgV_s}{C_a}$ and $\beta = \dfrac{2\pi l E c_a \, dv}{Q \, \text{Log}_e \frac{r_1}{r_t}} - d_m S_m$

Multiplying equation (7) throughout by $e^{\beta c_a \frac{v}{H}}$ and integrating,

$$e^{\beta c_a \frac{v}{H}} \theta = \int e^{\beta c_a \frac{v}{H}} \frac{A}{H} f(v)\, dv + R$$

Now, on injection of the sample, when $v = 0$, $q = 0$ and $f(v) = 0$ and consequently, $R = 0$.

Therefore,
$$\theta = e^{-\beta c_a \frac{v}{H}} \theta \int_0^V e^{\beta c_a \frac{v}{H}} \frac{A}{H} f(v)\, dv \tag{8}$$

Letting
$$\frac{A}{H} = \varphi \quad \text{and} \quad \frac{\beta c_a}{H} = \phi$$

Then,
$$\theta_v = \varphi\, e^{-\phi v} \int_0^V e^{\phi v} f(v)\, dv \tag{9}$$

It is seen from equation (9) that the constant (φ) merely affects the magnitude of (θ) but the constant (ϕ) and $f(v)$ condition the shape of the temperature profile and produces the curiously shaped peaks recorded by the detector. The constant (ϕ) can be considered the heat loss factor of the cell. It should be noted that the magnitude of $f(v)$ will depend on the value of (C_a) the ratio of the "effective volume of the cell" to the "plate volume" of the column.

Inserting the full expression for $f(v)$ in equation (9),

$$\theta_v = \varphi\, e^{-\phi v} \int_0^V e^{\phi v} \left(X_0 \frac{e^{-v} v^n}{n!} - \left(\frac{X_0}{C_a}\right) e^{\frac{-v}{C_a}} \int_0^V e^{-\frac{v}{C_a}} \left(\frac{e^{-v} v^n}{n!}\right) dv \right) dv \tag{10}$$

Equation (10) is the explicit equation for the temperature of a detector and can be used to synthesis the different shaped curves that the detector can produce. Employing a computer in the manner of Smuts *et al.* [17] ,Scott [18] calculated the relative values of (θ) for (v= 74 to 160 for a column of 100 theoretical plates, and for (C_a) ranging from 0.25 to 4 and (φ) ranging from 0.01 to 1.25. The curves obtained are shown in figure 8.

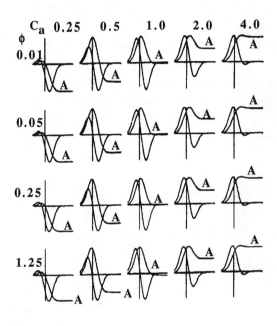

(A) Integral Curves

Figure 8   Theoretical Temperature Curves and the Integral Temperature Curves Obtained from the Heat of Absorption Detector

The twenty curves shown are graphs of (θ) versus (v) and the integral of (θ) versus (v) for different values of (C_a) and (φ) and are all normalized to the same peak height. The curves cover the practical range of heat loss factors that might be expected from a heat of adsorption detector cell. They demonstrate the effect on the shape of the (θ) versus (v) curves of changes in "detector cell capacity"/"column plate capacity" ratios that would result from different cell designs but when detecting a peak of constant width. The curves for different values of (C_a) would also represent the effect on peak shape of solutes of different retention, and consequently, having different peak widths passing through a cell of fixed dimensions.

It is seen that the major effect on peak shape is the "detector cell capacity"/"column plate capacity" ratio, (C_a). When the capacity of the detector cell is less than the plate capacity of the column, ($C_a<1$), the negative part of the signal dominates, and when the detector cell capacity exceeds the column plate capacity, ($C_a>1$), the positive part of the signal dominates. For this reason when ($C_a>1$), the integral curve rises to a maximum but does not return to the baseline. Conversely, when ($C_a<1$), the integral curve first rises and then falls below the baseline and does not return.

Only when ($C_a=1$) does the detector signal simulate the differential form of the elution curve and consequently the integral curve describes a true Gaussian peak.

The effect of the heat loss factor (ϕ) on peak shape is small but the magnitude of the signal varies inversely as (ϕ), although this is not apparent in figure (8), as all curves are normalized. It should also be noted that for low values of (β), where the maximum sensitivity is realized the peak maximum is displaced. However, for large values of (β), the maximum of the integral curves for ($C_a=1$) is almost coincident with the maximum of the elution curve.

It follows that if the detector was to be effective and produce the true Gaussian form of the eluted peak, then (C_a) must at all times be unity and consequently the detector must have the same plate capacity as that of the column. This means that the detector must employ the same absorbent, have the same geometry and be packed to give the same plate height as the column. It is obvious that to accomplish this, the column must also be the detecting cell and the temperature sensing element must be placed in the column packing itself. This situation can also be considered theoretically.

Restating the expression for f(v),

$$f(v) = X_0 \frac{e^{-v}v^n}{n!} - \left(\frac{X_0}{C_a}\right) e^{\frac{-v}{C_a}} \int e^{\frac{v}{C_a}} \left(\frac{e^{-v}v^n}{n!}\right) dv$$

Now, if the sensor is in the column packing, $C_a=1$.

Consequently, $$f(v) = X_0 \frac{e^{-v}v^n}{n!} - X_0 e^{-v} \int e^v \left(\frac{e^{-v}v^n}{n!}\right) dv$$

$$= X_0 \frac{e^{-v}v^n}{n!} - X_0 \frac{e^{-v}}{n!} \int v^n \, dv$$

Thus, $$f(v) = X_0 \frac{e^{-v}v^n}{n!} - X_0 \frac{e^{-v}}{n!} \left[\frac{v^{n+1}}{n+1}\right]$$

$$= X_0 \frac{e^{-v}v^n}{n!} - X_0 \frac{e^{-v}v^{(n+1)}}{(n+1)!}$$

$$= X_n - X_{(n+1)}$$

Now it has been shown [15] that $X_{(n+1)} = X_0 \frac{e^{-v}v^{(n+1)}}{(n+1)!}$

Thus, $$f(v) = \frac{dX_{(n+1)}}{dv} \qquad (11)$$

Now, as there is no (n+1) plate that constitutes the detector, the sensing element can be considered to be placed in the (n)th plate.

Thus, $$f(v) = \frac{dX_n}{dv} \qquad (12)$$

If the (n)th plate of the column acts as the detecting cell, there can be no heat exchanger between the (n -1)th plate and the (n)th plate of the column. Thus, there will be a further convective term in the differential equation that will take into account the heat brought into the (n)th plate from the (n -1)th plate by the flow of mobile phase (dv).

Thus the heat convected from the (n -1)th plate to plate (n) by (dv) will be

$$d_m S_m q(n-1) c_a dv \tag{13}$$

Substituting for f(v) from equation (12) ion equation (7) and inserting the extra convection term from (13),

$$\frac{d\theta_n}{dv} = \left(\frac{A_p}{H_p}\right)\frac{dX_n}{dv} - \left(\frac{\beta_p c_a}{H_p}\right)\theta_n + \left(\frac{d_m S_m c_a}{H_p}\right)\theta_{(n-1)}$$

where the subscript (p) accounts for the change from the already defined physical characteristics of the detecting cell to those of the last plate of the column.

Thus, $$\frac{d\theta_n}{dv} = \alpha \frac{dX_n}{dv} - B\theta_n + \gamma\, \theta_{(n-1)} \tag{14}$$

where $\alpha = \dfrac{A_p}{H_p}$ $\quad B = \dfrac{\beta_p c_a}{H_p}\quad$ $\gamma = \dfrac{d_m S_m c_a}{H_p}$

A solution to the differential equation (14) is given by

$$\theta_n = \frac{\alpha}{(B-1)} \sum_{r=0}^{r=n}\left[\frac{(\gamma-1)}{(B-1)}\right]^r \frac{dX_{(n-r)}}{dv} \tag{15}$$

The validity of this solution can be confirmed by differentiation as follows:

$$\frac{d\theta_n}{dv} = \frac{\alpha}{(B-1)} \sum_{r=0}^{r=n} \left[\frac{(\gamma-1)}{(B-1)}\right]^r \frac{d^2 X_{(n-r)}}{dv^2} \quad (16)$$

Now if
$$X_n = X_0 \frac{e^{-v} v^n}{n!}$$

Then,
$$\frac{dX_n}{dv} = X_0 \frac{e^{-v} v^{(n-1)}}{(n-1)!} - X_0 \frac{e^{-v} v^n}{n!} = X_{(n-1)} - X_n$$

Thus,
$$\frac{d^2 X_n}{dv} = \frac{dX_{(n-1)}}{dv} - \frac{dX_n}{dv}$$

Substituting for $\dfrac{d^2 X_{(n-r)}}{dv^2}$ in equation (16)

$$\frac{d\theta_n}{dv} = \frac{\alpha}{(B-1)} \sum_{r=0}^{r=n} \left[\frac{(\gamma-1)}{(B-1)}\right]^r \left(\frac{dX_{(n-1-r)}}{dv} - \frac{dX_{(n-r)}}{dv}\right)$$

or

$$\frac{d\theta_n}{dv} = \frac{\alpha}{(B-1)} \sum_{r=0}^{r=n} \left[\frac{(\gamma-1)}{(B-1)}\right]^r \frac{dX_{(n-1-r)}}{dv} - \frac{\alpha}{(B-1)} \sum_{r=0}^{r=n} \left[\frac{(\gamma-1)}{(B-1)}\right]^r \frac{dX_{(n-r)}}{dv}$$

Now, for the series, $\dfrac{dX_{(-1)}}{dv} = 0$

Thus,
$$\frac{d\theta_n}{dv} = \theta_{(n-1)} - \theta_n$$

and
$$\frac{d\theta_n}{dv} + B\theta_n - \gamma \theta_{(n-1)} = \theta_{(n-1)} - \lambda \theta_{(n-1)} - \theta_n + \beta \theta_n$$
$$= (1-\gamma)\theta_{(n-1)} + (B-1)\theta_n \quad (17)$$

From equation (15) and using its expanded form,

$$(B-1)\,\theta_n = \alpha\frac{dX_n}{dv} + \frac{\alpha(\gamma-1)}{(B-1)}\frac{dX_{(n-1)}}{dv} + \alpha\left[\frac{(\gamma-)}{(B-1)}\right]^2\frac{dX_{(n-2)}}{dv} + \ldots$$

and $(1-\gamma)\theta_{(n-1)} = -\frac{\alpha(\gamma-1)}{(B-1)}\frac{dX_{(n-1)}}{dv} - \alpha\left[\frac{(\gamma-1)}{(B-1)}\right]^2\frac{dX_{(n-2)}}{dv} - \ldots$

Thus,
$$(1-\gamma)\,\theta_{(n-1)} + (B-1)\,\theta_n = \alpha\frac{dX_n}{dv} \qquad (18)$$

Substituting $\alpha\dfrac{dX_n}{dv}$ from equation (18) for $(1-\gamma)\theta_{(n-1)} + (B-1)\theta_n$ in equation (17) and rearranging,

$$\frac{d\theta_n}{dv} = \alpha\frac{dX_n}{dv} - B\theta_n + \gamma\,\theta_{(n-1)} \qquad (19)$$

It is seen that equations (17) and (18) are identical, which substantiates the validity of equation (16) for (θ_n). It is also seen from equation (16) that (α) only affects the magnitude of the curve while (γ) and (B) affect its shape as well as its magnitude. Scott [18] assumed practical values for the various physical parameters of the system and calculated the temperature and integral temperature curves for a series of different practical values of (B) and (γ). The results are shown in figure 9. In figure 3 the values of (γ) are represented as(G).

It is seen from the curves in figure 3 that the heat convected into the detector cell or plate also distorts the curves. It is apparent that unless the heat lost *radially* is *extremely high*, so that little heat is convected to the sensor, symmetrical integral peaks will not be obtained. This heat loss will be difficult to achieve in practice and for normal columns, the heat of absorption detector does not seem viable for LC. However, adequate radial heat loss might be possible with very small bore columns, and consequently the possibilities of the heat of adsorption detector might well be worth re-examining for use with microbore columns.

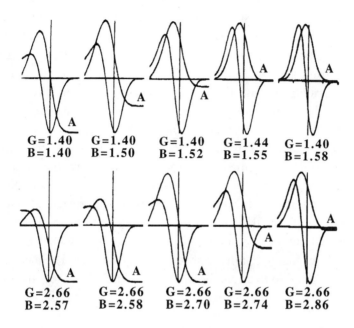

Figure 9 Theoretical Temperature and Integral Temperature Curves Generated by a Temperature Sensor Situated in the Column Packing

The Spray Impact Detector

The production of electrical charges during the disruption of a liquid to produce an aerosol has been known for many years. As far back as 1892 Lenard [19] noted the ionization of air at the base of a waterfall where the water splashed against the rocks. Christianson [20] investigated the electric charges produced when water was sprayed onto a solid plate, and in 1958 Loeb [21] examined the effect in greater detail and termed the phenomenon *spray electrification*. The charges are thought to be generated by the rapid break-up of a liquid surface either by a gas stream or by impact with a surface. Various mechanisms have been suggested to explain the production of electrical charges but the one generally accepted is that given by

Mattison [22]. Mattison suggests that any ions present in the liquid reside just below the surface and are not actually situated in the surface layer. If a new surface is formed by the break-up of the original surface, the ions in the new surface are pulled inward by attraction from the bulk liquid, the force of which will depend on both the field associated with the ion and its mobility. The surface thus becomes richer in the slower, larger species of ion. For example, with water, the new surface would become richer in hydroxyl ions than hydrogen ions and thus assume a net negative charge. In the author's opinion there is still some uncertainty as to the exact mechanism of charge formation and those readers interested in the subject are recommended to read Mattison's original paper. An alternative mechanism would be that caused by drop evaporation, something akin to that which takes place in *electro-spray ionization,* a technique used as a liquid chromatograph/mass spectrometer interface. The mechanism of electro-spray ionization, however, will be discussed in a later chapter.

Mowery and Juvet [23] have employed the spray electrification effect in a novel form as an LC detector, primarily for use in reversed phase chromatography. The column eluent passes through a central jet, where an annular flow of air around the jet ruptures the liquid into droplets. The spray of droplets is directed onto a target electrode which acts as an impact wall and shatters the droplets into a fine spray. The potential of the electrode is continuously monitored. A diagram of a Mowery and Juvet spray impact detector is shown in figure 10. Two types of electrodes were investigated, one composed of glassy carbon and the other of gold plated platinum. Impact between the eluent drops and the electrode resulted in the spray becoming positively charged and the target electrode negatively charged. Small quantities of solute in the mobile phase were found to produce profound changes in the electrode potential. For example, at a standing potential of 2,000 V, 1.5×10^{-7} g of nitrophenol resulted in a change in potential of about 400 V. The detector system was situated in a grounded metal box through which filtered laboratory air was passed. This was necessary to reduce noise from air contaminants and reduce the effect of charged droplets that tended to accumulate in the neighborhood of the electrode. The performance characteristics of the system are summarized in tables 1, 2 and 3.

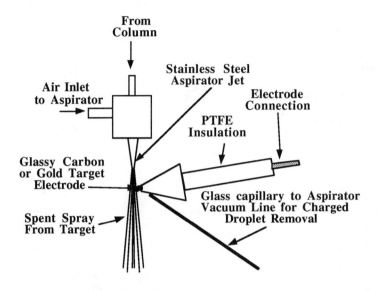

Figure 10   The Spray Impact Detector

In table 1 the detection limit and linear dynamic range of the detector for a series of organic compounds are given, employing water as the mobile phase and two different types of electrode. Generally, it was found that the glassy carbon electrode gave a higher sensitivity and, furthermore, required no prior conditioning before use with aqueous mobile phases. It is also seen that very high sensitivities were obtained with linear dynamic ranges of 3 to 4 orders of magnitude for most substances. In table 2 the sensitivity and linear dynamic range is given for a number of inorganic salts again employing water as the mobile phase. Very high sensitivities were once more obtained but there appeared to be two linear portions of the response curve with different slopes. In table 3 the sensitivity is given for a number of substances employing a range of different mobile phases.

Table 1 Performance Characteristics of the Spray Impact Detector
Mobile Phase, Boiled Distilled Water

Compound	Det.Limit g/s×10^{-11}	Lin.Dynam Range (log)	Target Material
n-Octanoic acid	7	4	gold
n-Nonanoic acid	8	4	gold
n-Decanoic acid	5	4	gold
Trifluoroacetylacetone(a)	30	4	gold
Trifluoroacetylacetone(b)	20	3	gold
Ammonium 8-anilino-1-naphthalene sulphonate	90	3.5	gold
o-Nitrophenol	90	3.5	gold
o-Nitrophenol	20	3.5	carbon
Sodium dodecylsulfate	50	3.5	gold
Sodium dodecylsulfate	20	3.5	carbon
Sodium dodecylsulphonate	20	–	carbon
Sodium tridecylsulfate	50	3.3	carbon
Ethylamine	200	2.5	carbon
n-Octanol	8,000	–	carbon
n-Nonanol	8,000	–	carbon
Ethyl Butyrate	10,000	>2	carbon
2-Heptanone	20,000	–	carbon

(a) Eluent flow rate 3.1 ml/min. (b) Eluent flow rate 8.6 ml/min.

Table 2 Performance Characteristics of the Spray Impact Detector for Inorganic Salts
Mobile Phase, Boiled Distilled Water

Compound	Det. Limit g/s × 10^{-11}	Linear Dynamic Range (log)
Lithium Nitrate	100	2.7 + 1.7 (a)
Potassium Nitrate	90	2.8 + 1.5 (a)
Lanthanum Nitrate	40	3.0 + 1.5 (a)
Thorium Nitrate	50	3.0 + 1.5 (a)

(a) Two linear regions were shown present having different slopes with a discontinuity occurring at zero current.

Table 3   Performance Characteristics of the Spray Impact Detector Employing Organic and Mixed Organic/Water Mobile Phases   Glassy Carbon Electrodes

Compound	Det. Limit (g/s)	Polarity	Mobile Phase
Glycine	5×10^{-11}	+	acetonitrile (a)
Stearic Acid	7×10^{-10}	+	acetonitrile (a)
Sodium Tridecylsulphonate	3×10^{-10}	+	12.5% acetonitrile/water
Sodium Tridecylsulphonate	8×10^{-11}	+	25% acetonitrile/water
o-Nitrophenol	2×10^{-7}	+	acetonitrile (b)
o-Nitrophenol	1×10^{-10}	+	15% acetonitrile/water
Sodium Cholate	3×10^{-11}	+	15% acetonitrile/water
p-Ethylphenol	1×10^{-7}	+	15% acetonitrile/water
n-Heptanol	3×10^{-7}	+	15% acetonitrile/water
Thorium Nitrate	4×10^{-8}	−	1mg/ml sodium dodecylsulfate/water
Water	3×10^{-6}	+	methyl ethyl ketone
Dimethylformamide	1×10^{-6}	+	methyl ethyl ketone
Ethyl Acetate	4×10^{-5}	−	methyl ethyl ketone
n-Heptane	7×10^{-6}	−	methyl ethyl ketone
Toluene	3×10^{-5}	−	methyl ethyl ketone
n-Octaldehyde	1×10^{-6}	−	methyl ethyl ketone
n-Octanol	2×10^{-8}	−	methyl ethyl ketone
o-Nitrophenol	2×10^{-7}	+	methyl ethyl ketone
Tetrachloroethylene	7×10^{-6}	−	methyl ethyl ketone
2-Heptanone	1×10^{-5}		methyl ethyl ketone

(a) Target first conditioned by wetting with water. (b) Target not conditioned with water.

It appears that the nature of the solvent can have a profound effect on the detector sensitivity. Employing methyl ethyl ketone as an alternative to acetonitrile or an acetonitrile water mixture as the solvent reduces the sensitivity of the detector by two or three orders of magnitude. In a similar way the sensitivity also varies with the type of solute. For example sensitivities of 10^{-10} g/s are realized for o-nitrophenol contained in a 15% solution of acetonitrile in water but, in the same solvent, the sensitivity of the detector to *p*–ethylphenol is only 10^{-7} g/s.

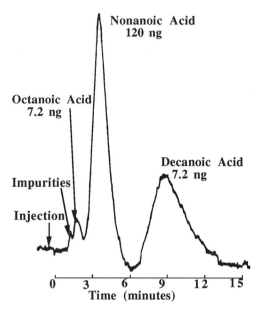

Column: 60 cm x 6 mm, C18 Corasil, target, gold, mobile phase water at 5.9 ml/min., air flow 4550 ml/min.

Figure 11   The Separation of the n-C_8 —n-C_{10} Fatty Acids Monitored by the Spray Impact Detector

A chromatogram showing the separation of some fatty acids monitored with the spray impact detector is shown in figure 11. The high sensitivity is clearly indicated. This detector has not been developed commercially but remains an interesting detecting principle based on a well-established natural phenomenon.

The Electron Capture Detector

The electron capture detector only gives a significant response to electron capturing substances and, therefore, even when used as a GC detector, it will only sense certain classes of compounds. A significant proportion of the solvents used in LC do not give a response with the electron capture detector and, consequently, such solvents can be volatilized along with any solute that is present and passed through the detector. The detector will then only respond to the solute provided it has electron capturing properties. This procedure was examined by Nota and Palombari [24] as a possible LC detector. Such a detector will be very selective, but as a number of important classes of compounds (e.g. pesticides, herbicides, carcinogens, etc.) are electron capturing in nature, this detector could be helpful in pollution studies and food analysis. The sensor examined by Nota and Palombari is shown schematically in figure 12.

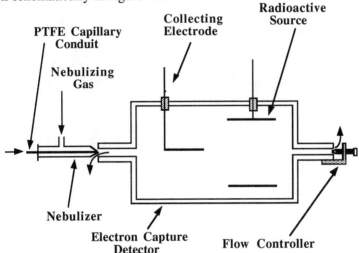

Figure 12 The Electron Capture LC Detector as Designed by Nota and Palombari

The column eluent passes directly into a nebulizer, the outlet tube from the column terminating at the nebulizer nozzle. A portion of the atomized eluent passes directly into the electron capture detector and then out to waste. The detector is operated under the same conditions as those that would be used if it were being used as a GC detector. The work of Nota and Palombari established the system as viable and demonstrated that the solvents, benzene, hexane, cyclohexane, pyridine, methanol, ethanol, diethyl ether, and acetone could all be employed as components of the mobile phase without significantly affecting the standing current of the detector. The disadvantages of this system included the need for a large quantity of nebulizing gas and the fact that the solutes eluted from the GC column had to be relatively volatile to provide a vapor to detect. It follow that the solutes would probably be more effectively separated by GC.

Willmott and Dolphin [25] developed an improved form of the detector and a diagram of their detecting system is shown in figure 13.

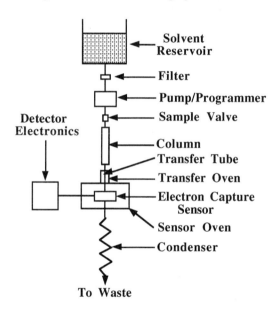

Figure 13 The Pye Unicam LC Electron Capture Detector

It is seen that the eluent is vaporized directly into the detector in an atomized form by means of a heated transfer tube situated in an oven held at 300°C. On entering the oven, the mobile phase and eluent are immediately vaporized, and the increase in volume resulting from the vaporization forces the vapor into the electron capture detector sensor (utilizing a ^{63}nickel radioactive source). The sensor is also held at 300°C and a purge of nitrogen gas sweeps the vapor through the detector and out through a condenser coil. The electron capture detector can be operated in either the constant current or pulsed mode. The sensitivity to electron capturing substances is extremely high (1.2 x 10^{-10} g/ml) but the linear dynamic range is only about two orders of magnitude.

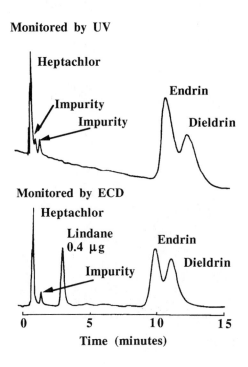

Figure 14 The Comparative Performance of the UV and Electron Capture Detectors

Chromatograms of pesticides obtained by Wilmott and Dolphin monitored with a standard UV absorption detector and the electron capture detector are shown in figure 14. It is seen that 3.2 x 10^{-8} g of lindane is clearly detected by the electron capture detector whereas it is not discernible on the trace from the UV detector. However, for the system to work the solutes must be relatively volatile and, as already discussed, the mixture would probably be better separated and analyzed employing a GC technique.

The Density Detector

The density of the mobile phase leaving the column will be changed when a dissolved solute is present and, thus, the continuous measurement of solvent density could be a possible means of detection. Such a method has been described by Fornstedt and Porath [26] and a diagram of their sensor is shown in figure 15.

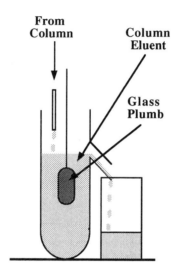

Figure 15   The Density Detector

It consists of a spherical glass plumb that is supported from the arm of an automatic, continuously recording balance and is totally immersed in the liquid leaving the column. The plumb is about 5 mm in diameter and about 0.4 g in weight. The solvent flows around the plumb, into the overflow tube and then to waste. In order to maintain stability, the chamber containing the float is situated in a thermostat and the mobile phase is brought to the temperature of the thermostat by means of a small heat exchanger prior to entering the float chamber. Unfortunately, the arrangement is inherently very sensitive to mechanical and electrical noise. Mechanical vibrations can be reduced by mounting the balance on a very heavy support. Electrical disturbances must be minimized by careful grounding. Temperature gradients within or around the apparatus can seriously disturb the system and consequently must also be minimized. The entire system needs to be thermally insulated so that its temperature does not vary more than +/− 1°C. The greatest disadvantage of the system is the enormous dead volume of the sensor which would seriously impair the resolution obtained from modern high efficiency columns.

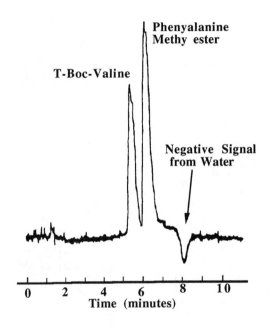

Figure 16 A Chromatogram Monitored by the Density Detector

An example of the separation of a mixture of tert-Boc-valine and phenylanaline methyl ester from a Sephadex LH–20 column monitored by the density detector is shown in figure 16. Each peak represents 50 mg of solute and thus the sensitivity is extremely low. Although it is a bulk property detector, and thus will detect all substances that have a density that differs from that of the mobile phase, it will obviously not tolerate gradient elution. Density measurement may be a basis for LC detection, and, in fact, this work has proved its validity. Nevertheless, the procedure suggested above is not practical considering the requisites of modern high efficiency columns.

The Density Bridge Detector

The density bridge detector is a novel application of the Gow-Mac gas density bridge designed for use in LC and was investigated by Quillet [27]. The GC model has already been discussed and it will be recalled that the balance functions by measuring the differential flow across the base of two columns of gas, one containing pure reference gas and the other the column eluent. The resulting differential flow of gas is measured by appropriately positioned thermistors. A diagram of the gas density bridge used in the LC mode is shown in figure 17.

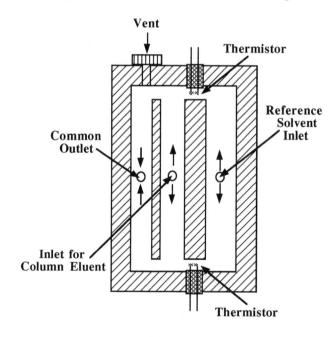

Figure 17 The Density Bridge Detector

The density bridge was situated in an appropriate thermostat and a reference flow of solvent and the column flow connected to the complementary ports used for GC detection. A vent was fitted to the top of the sensor to allow the air to be displaced by the respective liquid flows when the sensor is initially set up. Once the bridge was full of liquid, the vent was closed. The same ancillary electronic equipment was used, and a chromatogram showing the elution of 10 µl of *n*-cetane and 10 µl of decaline is shown in figure 18.

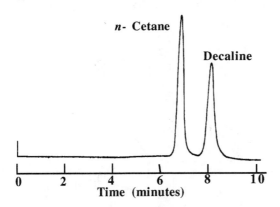

Figure 18 The Separation of *n*-Cetane and Decaline Monitored by the Density Bridge Detector

According to Quillet, the sensitivity of the detector was 10^{-4}–10^{-5} g/ml and had a linear dynamic range of about one order of magnitude. In the author's opinion it would seem highly unlikely that the detector was responding to changes in density but rather to changes in the thermal properties of the mobile phase or even changes in viscosity. The sensitivity of the detector is seen to be very poor but it might find use as a detector in preparative chromatography where high sensitivity is not required.

The Thermal Conductivity Detector

The presence of a solute in the mobile phase will change the thermal properties of the mobile phase and, thus, any device that can monitor the thermal properties of the solvent could potentially perform as a LC detector. The GC katherometer detector acts on this principle and it follows that a similar device situated in an LC column eluent might respond in the same way. The thermal properties that change in the presence of the solute, as previously discussed, are thermal conductivity and specific heat. Ohzeki *et al.* [28] have described, what they term a thermal conductivity LC detector. It appears, however, from their paper that the sensor is responding more to changes in specific heat than thermal conductivity. A diagram of their detector is shown in figure 19.

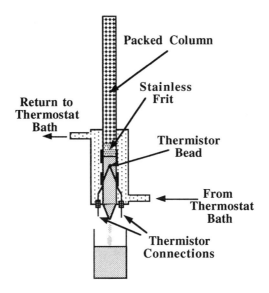

Figure 19 The Thermal Conductivity LC Detector

A thermistor, heated by an appropriate current, is situated in the mobile phase close to the end of the column. The thermistor is situated in one arm of a Wheatstone bridge. When equilibrium has been established, the bead reaches a constant temperature where the heat lost to the mobile phase is equal to the heat ohmically generated in the bead. As a consequence, the resistance of the bead is also constant and the output of the bridge can be balanced to zero. In the presence of a solute, the thermal equilibrium is destroyed and the heat lost from the bead changes as a result of either a change in specific heat or a change in the thermal conductivity of the mobile phase. Thus, the temperature changes and consequently the bead resistance also changes. The bridge is thus unbalanced and the differential output from the bridge is amplified and fed to either a recorder or a computer.

An example of some elution curves obtained from the LC thermal conductivity detector is shown in figure 20.

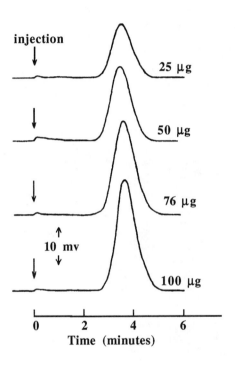

Figure 20 Elution Curves Monitored by the LC Thermal Conductivity Detector

The peaks are for blue Dextran separated on a Sephadex G–25 column. This detector will, unfortunately, suffer from the disadvantages of all bulk property detectors: An extremely constant flow rate must be maintained, thermal equilibrium must be carefully controlled and temperature programming or gradient elution will not be possible.

The Mass Detector

The mass detector [29] functions by allowing the column eluent to drop continuously into a heated container that is suspended from a recording balance. Providing the temperature of the container is high enough, the mobile phase will *flash evaporate* leaving the solute as a residue in the container. The trace recorded by the balance will be an integram, relating the mass of solute eluted, to time.

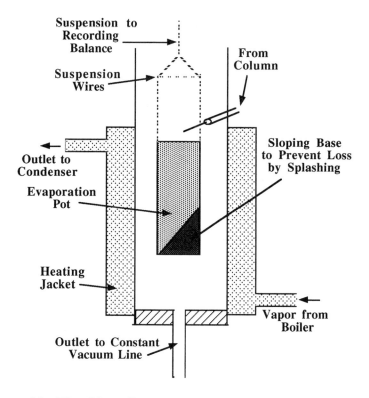

Figure 21   The Mass Detector

The elution of each component will be represented as a step on the integram. A diagram of the detecting system is shown in figure 21. The evaporation pot is made from aluminum and weighs about 100 g. It is constructed with an internal sloping base to prevent solute being thrown vertically out of the pot during the flash heating of the mobile phase. The boiling point of the vapor jacket is chosen to be about 50° to 100°C above the boiling point of the solvent to ensure adequate heat for flash heating. A constant suction is applied to the internal tube of the jacket. This suction satisfies a dual role; it withdraws vapor from the system and exerts a constant downward force on the pot. The constant downward force increases the apparent weight of the pot, accentuates the inertia of the balance and thus increases the stability of the signal obtained. An example of an integram obtained from this detector is shown in figure 22. It is seen that the detector has relatively low sensitivity but, at the same time, is surprisingly stable considering the dynamics of the evaporation procedure.

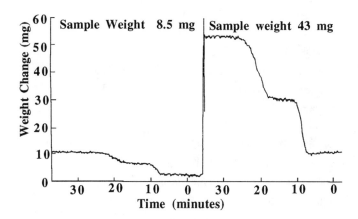

Figure 22 The Separation of Squalane and Methyl Stearate on Silica Gel Monitored by the Mass Detector

The level portions between the peaks are unambiguous so reasonable quantitative accuracy could be expected. Although a highly unlikely candidate for general analytical use, the mass detector represents one of the few chromatography detectors that has an *absolute* response.

Many of the detectors developed in the early days of liquid chromatography were rejected for many different reasons. Common shortcomings were poor linearity, high cost and, probably the most common, inadequate sensitivity. Materials, fabrication techniques and electronic systems have all changed and improved radically in the intervening years. In addition demands are being made on LC detectors for new and different applications. It might well be worthwhile re-examining some of the early "detector rejects" as some might well prove useful and effective if redesigned using contemporary materials and techniques.

References

1. E. Schram and R. Lombaert, *Anal. Biochem.*, **3**(1962)68.
2. C. J. Sjoberg and G. Agren, *Anal. Chem.* **36**(1964)1017.
3. N. W. Scharpenseel and K. M. Menke, *Tritium Ply. Biol. Sci. Proc. Symp.*, (1962)281.
4. J. A. Hunt, *Anal. Biochem.*, **23**(1968)289.
5. F. Shutte, *J Chromatogr.*, **72**(1972)303.
6. N. A. Dugger, US patent **528343**.
7. A. T. James and E. A. Piper, *J. Chromatogr.*, **5**(1961)265.
8. D. R. Reeve and S. Crozier, *J. Chromatogr.*, **137**(1977)271.
9. L. N. Mackey, P. E. Rodriguez and F. B. Schroeder, *J. Chromatpgr.*, **208**(1981)1.
10. G. Claxton, *J. Chromatogr.*, **2**(1959)136.
11. A. J. Groszek, *Nature,* **182**(1958)1152.
12. K. P. Hupe and E. J. Bayer, *J. Gas Chromatogr.*, April(1967)197.
13. J. I. Cashaw, R. Sigura and A. Zlatkis, *J. Chromatogr. Sci.*, **8**(1970)363.
14. R. P. W. Scott, *Liquid Chromatography Column Theory*, John Wiley and Sons, Chichester-New York (1992)18.
15. R. P. W. Scott, *Liquid Chromatography Column Theory*, John Wiley and Sons, Chichester-New York (1992)15.
16. R. H. Perry, C. H. Chilton and S. D. Kirkpatrick, *Chem. Eng. Handbook,* McGraw Hill, (1975)10.
17. T. W. Smuts, P.W. Rechter and V. Pretorious, *J. Chromatogr.Sci.*,**9**(1971)457.
18. R. P. W. Scott, *J. Chromatogr. Sci.* **11**(1973)349.
19. P. Lenard, *Ann. Phys.(Leipzig)* **46**(1892)584.
20. C. Christianson, *Ann. Phys. (Leipzig),* **4,40**(1913)107.
21. L. B. Loeb, *Static Electrification,* Springer Verlag, Berlin (1958)
22. M. J. Mattison, *J. Colloid Interface Sci.* **37**(1971)879.
23. R. A. Mowery and R. S. Juvet, *J. Chromatogr. Sci.*, **12**(1974)687.
24. G. Nota and R. Palombari, *J. Chromatogr. Sci.*, **162**(1971)153.

25. F. W. Willmott and R. J. Dolphin, *J. Chromatogr. Sci.*, **12**(1974)695.
26. N. Fornstedt and J. Porath, *J. Chromatogr.*, **42**(1969)376.
27. R. Quillet, *J. Chromatogr. Sci.*, **8**(1970)405.
28. K. Ohzeki, T. Kambara and K. Saitoh, *J. Chromatogr.*, **38**(1968)393.
29. R. P. W. Scott and J. G. Lawrence, *J. Chromatigr. Sci.*, **8**(1970)65.

CHAPTER 15

DETECTION IN THIN LAYER CHROMATOGRAPHY

The very first detector used in chromatography by Tswett was the human eye and, to this day, the eye is still the most frequently used detector in thin layer chromatography. In Tswett's original experiments a number of plant pigments were separated and, because they were of colors that varied, the separated components were easily visible as colored bands in the column of absorbent. Today, chromatography is applied to almost all classes of chemical compounds, of which very few are colored, and consequently visual detection requires some chemical modification that will produce substances that will either reflect or absorb visible light. After separation on a thin layer plate, the individual substances contained in the mixture occur as invisible spots dispersed along the plate. If the chromatography was successful, the spots will be well separated from one another and if not, some spots will be merged with their neighbors.

Spot Detection in Thin Layer Chromatography

There are many chemical procedures used for rendering the spots on the plate visible; some procedures are general and they expose all the components in the mixture as visible derivatives, whereas others are very specific, and only provide colored derivatives with selected chemical types, *e.g.* amino acids. Examples of both classes of derivatization will be given but, irrespective of the type of product that is chosen, some pre-treatment will be necessary. After chromatographic development, the plate will be wet with the mobile phase, which must be removed before derivatization can take place. In many cases the solvents used in the mobile phase will be toxic and so they must be removed by heating the plate in a well-ventilated fume cupboard (fume hood) under a stream of air. If oxidation is likely to affect the derivatizing reaction, then the air can be replaced by nitrogen. The heat is usually obtained from an infrared lamp although ventilated ovens are sometime used.

Many derivatizing reagents are highly corrosive or toxic and the procedure is usually carried out using a simple spray in a spraying chamber or fume cupboard. The spraying-chamber must be made of some appropriate corrosion resistant material, usually a suitable plastic and connected directly to the laboratory fume exhaust system

An example of the type of spray used for such a procedure is shown in figure 1.

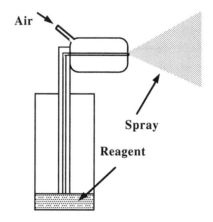

Figure 1 A Thin Layer Chromatography Reagent Spray

The spray takes the form of a simple atomizer of the type often used for the dispersion of perfume. The air is pumped over a nozzle that is connected to the reagent reservoir in the manner shown in the diagram. The materials of construction must be acid and alkali resistant and generally insoluble in most solvents.

General Spot Derivatizing Methods

There are a large number of different reagents that can be used for general derivatization and all cannot be described. However, some of the more commonly used derivatizing procedures are as follows.

Derivatizing Procedures

The Iodine Reagent

The majority of compounds become visible if exposed to iodine vapor. Although actual chemical reaction can take place with some unsaturated compounds, in the vast majority of compounds, the iodine appears to be physically bound to the material and does not chemically react with it. A spray procedure is not used in this derivatizing procedure as the iodine regent is a solid. The plate is placed in an enclosure containing iodine crystals which are preferably heated to about $50^{\circ}C$ to achieve a suitably high vapor pressure and, consequently, a reasonably fast reaction rate. Brown spots appear where the solutes are located. When the plate is removed from the enclosure, the spots rapidly disappear and so the detection procedure appears to be largely nondestructive (this would not be true for unsaturated compounds as, although they would not be destroyed by the reagent, iodine addition compounds would almost certainly be formed).

This detection method can be enhanced by exposing the plate to a very high concentration of vapor, removing the plate and allowing a few minutes for the excess iodine to evaporate. The plate is then sprayed with a 1% starch solution. The solutes then appear as blue spots on the plate. These methods will detect most organic substances and be relatively sensitive. The sensitivity of the derivatization procedure will, however, vary widely with solute type.

The Sulfuric Acid Spray

This method is destructive and the solutes cannot be recovered after detection. After drying, the plate is sprayed with a 50% v/v aqueous solution of sulfuric acid in a fume hood or spray chamber. The hood will be exposed to extremely corrosive materials and thus all fittings should be made of glass or acid resistant plastic. After spraying, and while still in the fume hood, the plate is heated to about 239°C for about 10 minutes. The solutes are partially oxidized leaving behind a charred deposit of black carbon that is easy to discern. This method will detect most involatile organic compounds.

Chromic-Sulfuric Acid Spray

This method is a modification of the sulfuric acid method but an oxidizing agent is added to facilitate the reaction and consequently, is also destructive. The plate is sprayed with a 5% w/v potassium dichromate solution in 40% v/v aqueous sulfuric acid solution. The plate is then heated for about 10 minutes at about 200°C to ensure charring is complete. This method will also detect most involatile organic compounds and it is thought to be more sensitive than the simple sulfuric acid spray.

Specific Derivatizing Methods

Specific derivatizing reagents take advantage of distinctive chemical characteristics of those solutes of interest and thus, generate colored or detectable derivatives of only those pertinent materials. Specific derivatization has two advantages: firstly it only renders visible those solutes that are of interest and secondly, the solutes do not need to be completely separated from other materials that are of different chemical type because they will not appear in the separated mixture.

Some examples of specific reagents are as follows. The N-t-butoxy protected peptides can be detected by preliminary exposure to chlorine gas followed by spraying with o-toluidine reagent or dicarboxidine reagent. Many peptides can be stained by treatment with 0.1 % solution of 4-chloro-7-nitrobenzofuran in ethanol, followed by spraying with methanol that has been adjusted to pH 11 with caustic soda. Ampicillin can be detected by spraying with a 1% starch solution/acetic acid/0.1 N iodine solution (100:8:1). There are a number of detecting sprays for lipids; a solution of 0.05% of rhodamine B in ethanol will give violet spots on a pink background, spraying with a 0.2 % solution of

2',7'-dichlorofluorescein will expose saturated and unsaturated lipids as green spots on a purple background when viewed with UV light: spraying with a 10% solution of antimony trichloride in chloroform and heating to 110°C for 1-2 minutes will produce spots of various colors on a white background. Acid base indicators have been used to detect carboxylic acids. For example, the alkaline form of bromocresol green (0.25% w/v in ethanol) when sprayed on the plate will give yellow spots on a green background for carboxylic acids. The reagent Ninhydrin (0.3 % in butanol containing 3% of acetic acid) will provide purple spots on a white background for most amino acids and many amines. Aniline phthalate gives gray-black spots for reducing sugars and many natural products can be detected by spraying the plate with diphenylboric acid b-aminoethyl ester (1% in ethanol) which can give a variety of colors for different substances. The illustrations given above are examples of the many specific and non–specific derivatizing procedures that can be used for detection in thin layer chromatography. In this book it is inappropriate to discuss further this method of detection and, for those interested, the books by Sherma and Fried [1] and Geiss [2] are recommended

Fluorescence Detection

Many organic substances have a natural fluorescence or can be derivatized to form fluorescent compounds. The production of fluorescent derivatives can be complex and there are a large number of alternative procedures available. As a consequence, the subject of synthesizing fluorescence derivatives will be discussed in detail as a separate subject in chapter 17. If the solutes of interest have a natural fluorescence, the plate can be observed in a light box illuminated with a low pressure mercury lamp or appropriate UV source. The spots will be observed as patches of usually yellow or green light against a relatively mauve or light purple background.

It should be pointed out that if a low pressure mercury lamp is employed, then the light emitted at 254 nm and below will adversely affect the eyes. Therefore, the window of the light box should be opaque to UV light and be made of soda-glass or some other appropriate UV light absorbing material.

If the materials that are separated do not naturally fluoresce, then the plate, after development and drying, must be treated with a suitable derivatizing reagent to form appropriate fluorescent products. The plate is then exposed to UV light as described previously and the solutes will again be clearly seen as fluorescent spots on a light purple background.

Fluorescent Quenching

Fluorescence quenching is the complement of fluorescent detection. In this method of detection, the spot on the plate becomes apparent as a *non–fluorescent* area surrounded by a fluorescing background. The procedure involves using a TLC plate that has been treated with a fluorescing reagent that is directly associated with the absorbing layer. It follows that when exposed to UV light, the whole plate fluoresces. It is important that the fluorescent material used is not removed by the solvent mixture employed for chromatographic development and does not interfere with the distribution of the solutes between the two phases and consequently the separation of the mixture. The plate is used to separate the solutes of interest, using normal procedures and, on completion of the development process, the plate is then exposed to UV light in a light box. The plate is seen as a bright fluorescent sheet and the solutes as dark non-fluorescent pale purple spots where the solutes have quenched the plate fluorescence. Most organic compounds and in particular aromatic and heterocyclic compounds are readily detectable by this procedure.

Scanning Densitometry

In situ scanning of TLC plates employing optical instrumentation is relatively recent but, nevertheless, is now considered essential for both the accurate location of a spot and the precise quantitative estimation of its content. The surface of the plate can be examined using either reflected light, transmitted light or fluorescent light. In addition the incident light can be either adsorbed, diffusely scattered or transmitted. It is usual to measure the light scattered, reflected or generated by fluorescence from the spot and compare it electronically with light from a part of the plate where no sample has passed (the channel between the spots). Single beam and double beam instruments are available which are diagrammatically depicted in figure 2.

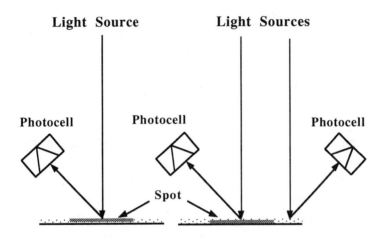

Figure 2   Single and Double Beam Densitometers

The double beam is to be preferred; one sensor monitors the sample lane and the other monitors the blank region between the lanes. The difference signal is taken as that responding solely to the sample. The incident light can be chosen to have wavelengths ranging between 200 and 700 nm. Halogen or tungsten lamps can be used to provide light at the higher wavelengths, whereas for light between 200 and 400 nm a deuterium lamp is to be preferred.

Sometimes, in order to induce fluorescence, lamps with higher energy outputs are necessary, such as the mercury lamp, which generates most of its emission at 254 nm and the xenon arc lamp which generates light over a broad wavelength, in a similar manner to the deuterium lamp, but at much higher intensities. If a greater sensitivity can be obtained by using light having a narrow band of wavelengths, a monochromator is introduced between the light source and the plate to select the required wavelength and then either the tungsten or deuterium lamp can be used as the light source.

The sensitivity of a scanning densitometer depends on a number of factors, including the basic instrument design and, in particular, the quality of the optics. The plate surface is viewed by the scanner through a slit and the major factor affecting the overall sensitivity is the slit height to spot diameter ratio. Although the slit dimensions are usually selectable, as the spots along the plate will be of different size, it is not possible to adjust the slit to an optimum size for scanning the whole of the plate.

When measuring either the adsorbed light or the fluorescent light the sensitivity is inversely related to the scan rate. It follows that the slower the scan, the greater the signal. However, carried to the extreme, this approach can extend the analysis time considerably. The relationship between the adsorbed light and the concentration of solute in the spot is not linear, and so either calibration curves must be constructed or the signal must be electronically modified in an appropriate manner to render the output linearly related to solute concentration. In any event, standard solutions must be run for calibration purposes but with manual calibration many more calibration samples are necessary. In contrast when measuring fluorescence, the output is linearly related to solute concentration; in fact, the relationship can be linear over a concentration range of about three orders of magnitude (*e.g.* 0.1 to 100 ng). Calibration is still necessary but as linear curves are obtained, linear amplifiers can be used to electronically process the fluorescence signal.

The major sources of error that arise in scanning densitometry originate largely from sample manipulation. The sample must be carefully applied to the plate in a very reproducible manner and the diameter of the spot carefully controlled. The chromatographic conditions must also be kept constant, and although this is relatively easy in GC and LC, due to the nature of chromatographic development, it is more difficult in TLC. Finally in the measuring process extreme care must be taken to ensure the spot is located in the exact center of the measuring beam or, again, errors will result. A photograph of a commercial TLC scanner manufactured by CAMAG is shown in figure 3.

Courtesy of CAMAG Inc.

Figure 3 The CAMAG Thin Layer Chromatography Scanner

The scanner is suitable for scanning both TLC plates and also electrophoretic stains and can deal with objects 200 x 200 mm in size. The spot can be monitored by either reflected light, emitted light or transmitted light. The wavelengths available range from 190 nm to 800 nm and scanning speeds as fast as 100 mm/s are available. The spatial resolution can be selected from 25 to 200 mm. A diagram of the optical layout is shown in figure 4.

There are three light sources from which to choose: a deuterium lamp providing light from 190 to 400 nm, a tungsten filament lamp providing light from 350 to 800 nm and a low pressure mercury vapor lamp, which provides high intensity line emissions at 254 nm 578 nm. Light from the lamp passes through a lens that focuses the light through a slit and onto a diffraction grating. The light from the diffraction grating is reflected by a plane mirror, through a selectable slit, and thence to another plane mirror and onto a half silvered mirror. Light is taken from the half silvered mirror to a reference photocell and the remaining light passes to the plate.

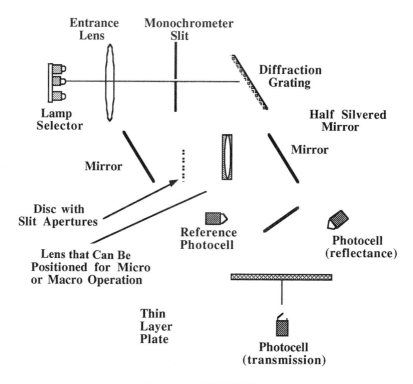

Courtesy of CAMAG Inc.

Figure 4 The Optical Layout of the CAMAG TLC Scanner

The light reflected from the plate is monitored by a photocell aligned 30° to the normal of the plate and the transmitted light is monitored by a photocell placed directly under the plate. The stage is driven by stepping motors in both the (x) and (y) directions and the reproducibility of positioning the monitor is ±50 μm in the (y) direction and ±100 μm in the (x) direction at a maximum scanning speed of 150 mm/sec. An example of the scan of a thin layer plate employing light of 220 nm is depicted in figure 5. It is seen that the scanning time is over 80 minutes and the measuring points were averages from consecutive sample pairs. Nevertheless, a very respectable chromatogram is achieved from which quantitative data can readily be obtained.

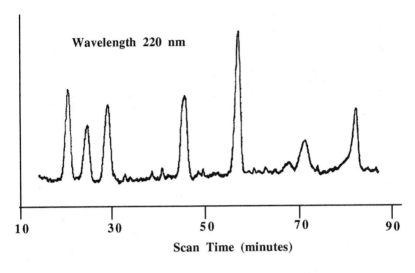

Courtesy of CAMAG Inc.

Figure 5 The Analog Curve Produced by Scanning a TLC Plate with the CAMAG Scanner

An example of the flexibility of multi-wavelength scanning is demonstrated in figure 6, where the separation of some sulfonamides and antibiotics is shown, that have been scanned over a series of different wavelengths. The wavelengths have been previously identified as optimum for certain of the components, *i.e.* each component gave a maximum response to light of the chosen wavelength. The optimum wavelengths proved to be in the UV spectrum, *viz.*, 254 nm, 365 nm, 302 nm, 313 nm and 366 nm. This presentation is similar to that produced by the diode array UV–LC detector. The three dimensional presentation provides the maximum information, with chromatographic characteristics displayed on one axis and spectroscopic characteristics displayed on the other.

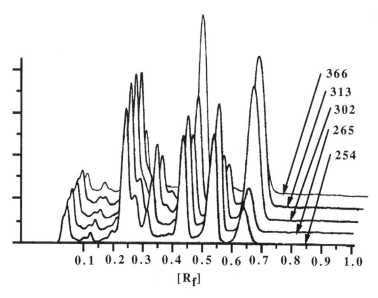

Courtesy of CAMAG Inc.

Figure 6 The Separation of Some Sulfonamides and Antibiotics from an Animal Feed Mix Monitored at Different Wavelengths

Nevertheless, the scanning procedure is time consuming and many of the advantages of TLC relative to LC are lost. Operating costs are still lower and solvent disposal problems will be significantly reduced, but the procedure will be slower and the monitoring equipment, although costing significantly less than the liquid chromatograph, is still relatively expensive.

References

1. Handbook of Thin-Layer Chromatography, (J. Sherma and B. Fried, eds.), Marcel Dekker Inc., New York (1990).
2. F. Geiss, Fundamentals of Thin Layer Chromatography, Alfred Hüthig Verlag, Heidelberg-Basel-New York (1987).

Part 4

General Detector Techniques

CHAPTER 16

SPECTROSCOPIC DETECTORS AND TANDEM SYSTEMS

Extremely complex mixtures can be separated and quantitatively analyzed by modern chromatography techniques. However, having achieved a separation, the structure of any hitherto unknown substances present in the mixture remains to be elucidated or the presumed identity of the expected components still needs to be confirmed. Generally, chromatographic data alone is insufficient evidence to substantiate the presence of a specific substance in a mixture. Supporting evidence from other analytical systems is considered essential before the identity of a substance can gain legal acceptance.

In the early days of chromatography, solutes were collected as they were eluted from the column and subsequently examined by appropriate spectroscopic techniques. Solute vapors eluted from GC columns were either condensed from the carrier gas or adsorbed on a suitable solid. The solutes were then recovered and examined applying normal spectroscopic procedures. Solutes eluted from liquid chromatography columns were collected, as a solution, in fractions of the mobile phase, the solvent was removed by evaporation or lyophilization and the residue was taken up in an appropriate spectroscopic solvent and examined. Off line spectroscopic measurement of eluted solutes proved to be tedious, protracted and difficult, particularly for multi-component mixtures. It was not long before the possibilities of combining the chromatograph directly with the spectrometer were being explored and so the first "tandem" analytical systems began to emerge. Today, as both the chromatographs and the spectrometers have become more involved, the tandem systems have likewise become very complex. Consequently the subject of tandem systems now merits a text to itself. It follows, therefore, that in a book on detectors, only the basic characteristics of tandem systems can be discussed, together with some details of the more important instruments.

Tandem systems can operate in two ways and the method selected depends on the relative speed of sample generation from the chromatograph and the speed at which the sample can be scanned in the spectrometer. If the spectroscopic data can be obtained rapidly, as with a fast scanning mass spectrometer, then the column eluent or a portion thereof, can be fed directly into the spectrometer, assuming certain precautions are taken and the appropriate interface is used. (Chromatograph/spectrometer interfaces will be discussed later.) This procedure might also be used if a Fourier transform IR were used as the identifying instrument. Conversely, if the sample to be examined is eluted from the chromatography column at a much faster rate than it can be scanned by the spectrometer, which can still happen with certain NMR instruments and other slow scanning instruments, then an interrupted elution technique can be used. As the development of the separation progresses, the column eluent is monitored by measuring its absorption at a fixed wavelength. When a peak maximum is observed to be situated in the spectrometer sensor cell, the gas supply or the mobile phase pump is stopped. The peak remains static in the sensor cell and the sample can then be scanned over a range of wavelengths in the usual way. After the spectroscopic data has been collected, the carrier gas or the mobile phase pump is started again and the development of the separation continued. In gas chromatography most of the sample remaining in the column is situated in the *stationary* phase so, during the static period, only slight peak dispersion occurs due to diffusion in the mobile phase. As a consequence, the resolution of the column is hardly affected. In liquid chromatography, the diffusivity of the solute in the mobile phase is four to five orders *less* than that in a gas and so loss of resolution on stopping the mobile phase flow is virtually insignificant. This procedure, often referred to as *interrupted elution chromatography,* can be easily automated. In fact, the automatic trapping and recovery of GC peaks for examination by both mass spectrometry and IR spectrometry was demonstrated, as long ago as 1966 [1], at the international symposium on gas chromatography held in Rome.

It should be pointed out that, although tandem systems are very effective and extremely exciting to operate, they are quite complex and also very expensive. If the analysis is not repetitive, there is adequate sample, and only one substance in the mixture being chromatographed is of consequence, then an off-line spectroscopic examination may be the more practical alternative. For single solutes it may be nearly as fast to collect the component as it is eluted and examine it spectroscopically off line, as it would be to employ a tandem system. In fact, it is probable that better quality spectra will be obtained and the solute identity confirmed with greater confidence. The off-line approach, however, is not practical for multi-component mixtures or for repetitive analyses where the results may be required for forensic purposes.

The six most common spectroscopic techniques employed for structure elucidation and to confirm substance identity are, ultraviolet spectroscopy (UV), infrared spectroscopy (IR), Raman spectroscopy, mass spectroscopy (MS), nuclear magnetic resonance spectroscopy (NMR) and atomic adsorption (AA) or atomic emission (AE) spectroscopy. There is a difference between a spectroscopic detector and a spectrometer-chromatograph combination, the latter being a tandem system. The spectroscopic detector is designed and constructed specifically as a chromatography detector, an example of which is the diode array detector. The spectrometer used in tandem instruments, although packaged with the chromatograph, is usually a standard instrument that is often fitted with a small sensor cell, appropriate for the chromatograph with which it is to be used, together with suitable low dispersion conduits and the appropriate interface. The transfer of a solute existing as a vapor in a gas stream to a spectrometer is an easier procedure than transferring the solute as a solution in the mobile phase. Thus, the initial tandem systems to be developed were associated with the gas chromatograph and these will be discussed first.

Gas Chromatography Tandem Systems

Gas Chromatography/Mass Spectrometry (GC/MS) Systems

The first association of the gas chromatograph with a mass spectrometer was successfully accomplished by Holmes and Morrell in 1957 [2] only four years after the first disclosure of GC as an effective separation technique by James and Martin in 1953. The authors connected the outlet of a GC column directly to the mass spectrometer employing a split system. The mass spectrometer was a natural choice for the first tandem system as it could easily accept samples present as a vapor in a permanent gas. In the early days of GC/MS, only packed GC columns were available and thus the major problem encountered when associating a gas chromatograph with a mass spectrometer was the relatively high flow of carrier gas from the chromatograph (ca 25 ml/min or more). This was in direct conflict with the relative low pumping rate (measured at atmospheric pressure) of the mass spectrometer. This problem was solved either by the use of a split system which only allowed a small proportion of the solute to enter the mass spectrometer or by employing a vapor concentrating device. A number of concentrating devices were developed, e.g. the jet concentrator invented by Ryhage [3] and the helium diffuser developed by Bieman [4] later known as the Bieman concentrator. A diagram of the Ryhage concentrator is shown in figure 1.

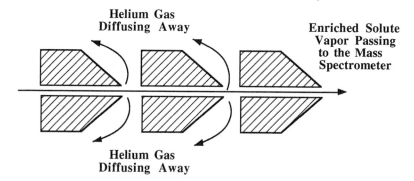

Figure 1   The Ryhage Concentrator

The concentrator consists of a succession of jets that are aligned in series but separated from each other by carefully adjusted gaps. The helium diffuses away in the gap between the jets and is removed by appropriate vacuum pumps. In contrast, the solute vapor, having greater momentum, continues into the next jet and finally into the mass spectrometer. The concentration factor is a little greater than an order of magnitude depending on the jet arrangement. The sample recovery can be in excess of 25%.

A diagram of the Bieman concentrator is shown in figure 2 and functions on an entirely different principle. It consists of a heated glass jacket surrounding a sintered glass tube. The eluent from the chromatograph passes directly through the sintered glass tube and the helium diffuses radially through the porous walls and is continuously pumped away. The helium stream enriched with solute vapor passes into the mass spectrometer.

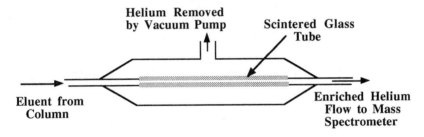

Figure 2   The Bieman Concentrator

Solute concentration and sample recovery is similar to the Ryhage device but the apparatus, though more bulky, is somewhat easier to operate. An alternative system was devised that employed a length of porous polytetrafluorethylene (PTFE) tube, as opposed to one of sintered glass, but otherwise functioned in the same manner.

The development of the open tubular columns eliminated the need for concentrating devices, as the mass spectrometer pumping system could dispense with the entire carrier gas flow from such columns. Consequently, the column flow was passed directly into the mass spectrometer and the total sample entered the ionization source. One of the first capillary column-mass spectrometer systems was described by Banner *et al.* (5) and, for historical interest, a diagram of their original apparatus is shown in figure 3.

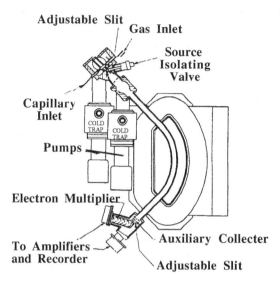

Figure 3 The First Tandem Capillary Column Mass Spectrometer System

The mass spectrometer used by Banner and colleagues was a rapid scanning magnetic sector instrument that easily provided a resolution of one mass unit. Nowadays, mass spectrometers (giving vastly improved resolution) are mostly used with capillary columns and operated in a very similar manner, with the column eluent passing directly into the ionization source of the spectrometer. The most advanced system involves the triple quadrupole mass spectrometer, which gives extremely impressive in-line sensitivity, selectivity and resolution.

A diagram showing the basic operating principle of a triple quadrupole mass spectrometer is shown in figure 4. The sample enters the ion source from the column, via a suitable interface if necessary, and, in GC, is usually fragmented by either an electron impact or chemical ionization process. In the first analyzer the various charged fragments are separated in the usual way and they then pass into the second quadrupole section, sometimes called the collision cell.

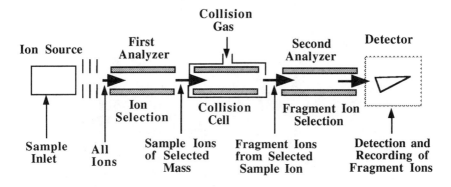

Figure 4　The Triple Quadrupole Mass Spectrometer

In the center quadrupole section a particular ion mass is selected for further fragmentation. The new fragments, formed from the ionization of the selected molecular or fragment ion, then pass into the third quadrupole which functions as second analyzer. The second analyzer segregates the new fragments into their individual masses which are detected by the sensor, producing the mass spectrum. In this way, the exclusive mass spectrum of a particular molecular or fragment ion can be obtained from the myriad of ions that may be produced from the sample in the first analyzer. It is seen that this is an extremely powerful analytical system that can handle exceedingly complex mixtures and very involved molecular structures. The system has a high resolving power and is ideal for structure elucidation. It is clear that the triple quadrupole mass spectrometer, if used in conjunction with a suitable interface and ionizing procedure, would be equally effective when used in tandem with the liquid chromatograph. The device is not inexpensive; nevertheless, it is although a complex instrument, it is not difficult to use and can be easily operated by a trained analyst. It is likely that, ultimately, this type of mass spectrometer will become the most popular tandem instrument to be used with both the gas chromatograph and the liquid chromatograph.

A photograph of the triple quadrupole mass spectrometer manufactured by GG Organic Inc. is shown in figure 5.

Courtesy of VG Organic Inc.

Figure 5 The Triple Quadrupole Mass Spectrometer Manufactured by VG Instruments

The combination of the gas chromatograph or the liquid chromatograph with the triple quadrupole mass spectrometer, or the single quadrupole mass spectrometer, are the most commonly used tandem systems. They are used extensively in forensic chemistry, in pollution monitoring and control and in metabolism studies. The quadrupole mass spectrometers exhibit both very high sensitivity and good mass spectrometric resolution. As a consequence they can be used with gas chromatography open tubular columns on the one hand, and microbore LC columns on the other. An sample of the use of the single quadrupole monitoring a separation from an open tubular column is shown in figure 6. The column was 30 m long, 0.25 mm I.D. and carried a 0.5 µm film of stationary phase. A 1 µl sample was used and the column was programmed from 50° to 300°C at 10°/min.

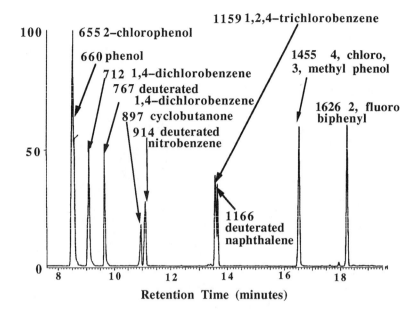

Courtesy of VG Organic Inc.

Figure 6 A Separation from an Open Tubular Column Monitored by a Single Quadrupole Mass Spectrometer

The Ion Trap Detector

The ion trap detector is a spectrometric system designed specifically for chromatographic detection and, consequently, is a true spectroscopic detector. The electrode orientation of the quadrupole ion trap mass spectrometer is shown in figure 7. It was shown in figure 4 that the quadrupole spectrometer contains four rod electrodes, whereas the ion trap consists of three cylindrically symmetrical electrodes which are made up of two end caps and a ring. The device is small, the opposite internal electrode faces being only 2 cm apart. Each electrode has accurately machined hyperbolic internal faces. An r.f. voltage and an additional d.c. voltage is applied to the ring and the end caps are grounded. The r.f. voltage causes rapid reversals of field direction so any ions are alternately accelerated and decelerated in the axial direction and *vice versa* in the radial direction. At a given voltage, ions of a specific mass range are held oscillating in the trap. Initially, the electron beam is used to produce ions and after a given time the beam is turned off and all ions except those selected by the magnitude of the applied r.f. voltage are lost to the walls of the trap. The remainder continue oscillating in the trap.

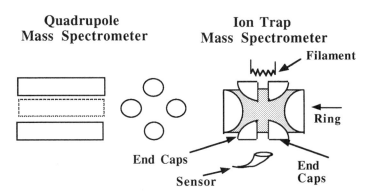

Figure 7 Pole Arrangement for the Quadrupole and Ion Trap Mass Spectrometers

The potential of the applied r.f. voltage is then increased and the ions sequentially assume unstable trajectories and leave the trap via the aperture to the sensor in order of their increasing m/z values. The original models were not very efficient but it was found that the introduction traces of helium in the ion trap significantly improved the quality of the spectra. The improvement resulted from ion–helium collisions reducing the energy of the ions and allowing them to concentrate in the center of the trap. As well as detection, the ion trap will provide satisfactory spectra for solute identification by comparison with reference spectra. However, the spectrum produced for a given substance will probably differ considerably from that produced by the quadrupole mass spectrometer. The separation of a number of commonly assayed drugs monitored by the ion trap detector is shown in figure 8. Each component is present at a level of 18 ng. It seen that the ion trap is fairly sensitive and the identity of each component can be confirmed from its spectra providing they are compared with reference spectra run under the same conditions.

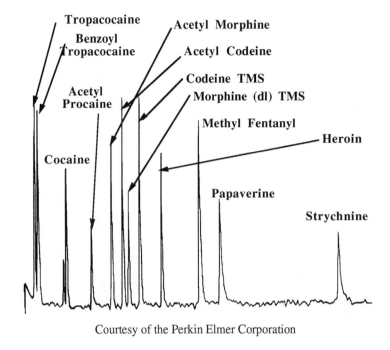

Courtesy of the Perkin Elmer Corporation

Figure 8 The Separation of Some Narcotic Drugs Monitored by an Ion Trap Detector

The Time of Flight Mass Spectrometer

The time of flight mass spectrometer was invented many years ago but due to the factors controlling resolution not being clearly recognized and also to certain design defects in the first models, it exhibited limited performance and was eclipsed by other rapidly developing techniques. However, with improved design, and the introduction of Fourier transform techniques, the performance has been vastly improved. As a result, there has been a resurgence of interest in this particular form of mass spectroscopy. A diagram of the time of flight mass spectrometer is shown in figure 9. The sample is volatilized into the space between the first and second electrodes and a burst of electrons over a 1 μs period produces ions. The extraction voltage E is then applied for a short period and this focuses the ions because those further form the second electrode will experience a greater force than those closer to the second electrode.

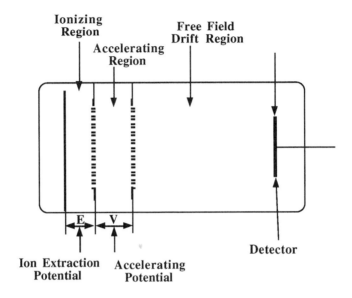

Figure 9 The Time of Flight Mass Spectrometer

After focusing the accelerating potential (V) is applied for a much shorter period than that used for ion production (*ca* 100 nsec) so that all the ions in the source are accelerated almost simultaneously. The ions then pass through the third electrode into the drift zone and are then collected by the sensor electrode. The velocity of the ions after acceleration will be inversely proportional to the square root of the ion mass. With modern ion optics and Fourier transform techniques Erickson *et al.* (6) could sum twenty spectra per second for subsequent Fourier transform analysis. The advantage of the time of flight mass spectrometer lies in the fact that it is directly and simply compatible with direct desorption from a surface, and thus can be employed with laser desorption and plasma desorption techniques.

An excellent discussion on general organic mass spectrometry is given in Chapman [7].

Gas Chromatography/Infrared Spectrometry Systems

IR spectral data is not nearly so informative as that obtained from the mass spectrometer. An IR spectrum can certainly be used with confidence for solute confirmation, but has somewhat limited use in structure elucidation, when the compound is completely unknown. The mass spectrum can usually provide the molecular weight of the solute and, if the spectrometer has sufficient resolution, also an empirical formula. The mass spectrum, particularly if produced by electron impact ionization, can also provide a considerable number of fragments from which the actual structure of the compound can often be educed. In contrast, the IR spectrum will only provide confirmation of the presence of specific chemical groups in the solute molecule. This feature, however, is often useful as support to a mass spectrum obtained by electron impact ionization in the elucidation of a chemical structure. As already mentioned, GC samples for IR examination were originally obtained by trapping the solute vapor at low temperatures and then examining them off-line. One of the first automatic GC/IR systems was developed by Scott *et al.* [1], who developed a device that trapped the solute vapor after it had passed through a heated IR sample cell. The column flow was then automatically stopped and the solute regenerated back into the IR adsorption cell and the spectrum taken. The apparatus included a complicated valve system to ensure that the resolution of the chromatograph was not impaired by the stop/start process and included a very primitive MS 10 low resolution mass spectrometer that simultaneously produced mass spectra. A diagram of the IR cell and adsorption trap is shown in figure 9. The technique of collecting the sample in a cooled trap, often containing a stationary phase, has been used for many years, a more recent example being that of Hawthorn and Miller [8]. It should be noted that the cell employed by Scott *et al.* was gold plated internally to avoid loss of stray light, thus increasing the efficiency of the system. The use of "light pipes," as such devices are called, was developed very early in the GC/IR development and one of the first references is that of Wilks and Brown in 1964 [9]. Most modern GC/IR systems also employ light pipes to increase the sensitivity of the IR cell but in addition they are used in conjunction with Fourier Transform IR spectrometers.

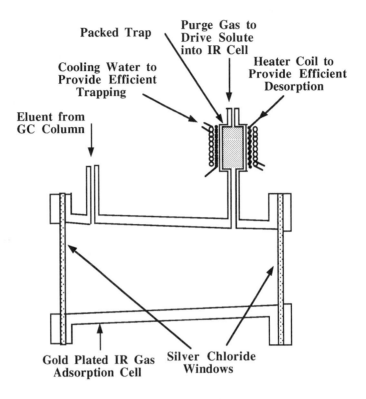

Figure 9 The IR Cell and Adsorption Trap of an Early GC/IR System

An example of a modern GC/IR instrument is that produced by the Perkin Elmer Corporation, and a diagram of the optical arrangement of the interface is shown in figure 10. The interface is appropriate for the macrobore capillary column, which is led into the interface through a heated tube right up to the light pipe. Concentric to the column, and through the same heated tube, is fed a stream of scavenging gas that carries the solute through the IR light tube and thus maintains the integrity of the separation at the expense of some solute dilution and consequent loss of sensitivity. If the solute bands were not swept out by the scavenging gas, the solute peaks from the column would accumulate in the IR light pipe and as a consequence, several solutes would be detected and measured simultaneously and resolution would be lost.

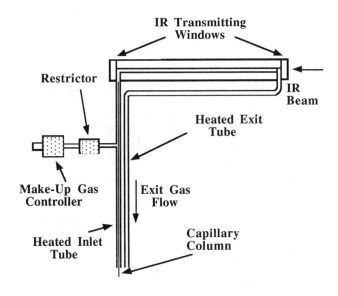

Courtesy of the Perkin Elmer Corporation

Figure 10 The Optical Arrangement of the Perkin Elmer GC/IR Instrument

An interesting application of the instrument to the analysis of the essential oils of *basil* is shown in figure 11. The oils were extracted by supercritical fluid techniques. In the example given the herb basil was extracted with liquid carbon dioxide at 60°C and 250 atmospheres pressure. The extract was usually decompressed through a length of silica capillary into an appropriate solvent or trapped on a suitable adsorbent and thermally desorbed. The column, a macrobore open tubular column, is 50 m long, 0.32 mm in diameter and carries a 5 micron film of a methyl silicone. The chromatogram was obtained by plotting the integrals of each adsorption curve against time. A spectrum taken at the peak maximum of linalool (peak 4) is shown in figure 12. It is seen that a clean spectrum is obtained and the resolution from the *macrobore capillary column* does not appear to be denigrated.

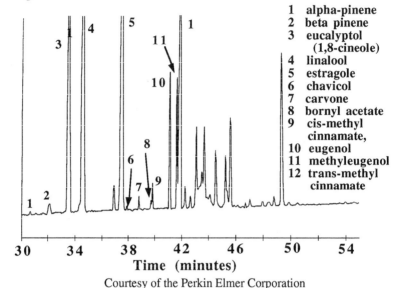

Courtesy of the Perkin Elmer Corporation
Figure 11 The Separation of a Sample of Basil Oil

Spectra, with adequate resolution for solute identification, have been obtained from as little as 10 ng of material, but the sample size will depend strongly on the extinction coefficient of the solute at the critical adsorption wavelengths of the respective compounds.

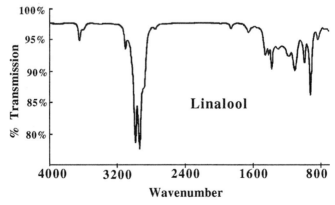

Courtesy of the Perkin Elmer Corporation
Figure 12 Spectrum of Linalool from a Light Pipe Interface

The Atomic Emission Detector

The atomic emission spectroscopic detector is extremely versatile with very high sensitivity and selectivity. Atomic emission is achieved by means of a helium plasma spectrometer and consequently, it can be rather complex and costly. It does, however, have some unique and valuable characteristics that make it essential for certain types of analyses, *e.g.* in forensic chemistry. A diagram of the helium plasma atomic emission spectrometer is shown in figure 13. The plasma is microwave induced into a helium stream employing a water cooled transducer. Carrier gas from the column mixes with the pure helium make-up gas, enters the plasma and the elements present in the solute emit light, the wavelength of which is characteristic for each element. The eluent subsequently passes to waste. The light emitted passes through a quartz window and is then focused by a quartz lens and spherical mirror onto a diffraction grating.

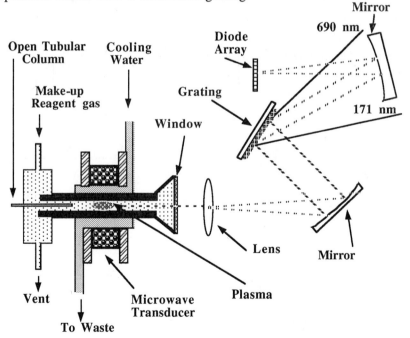

Courtesy of the Hewlett-Packard Corporation

Figure 13 The Helium Plasma Atomic Emission Spectrometer

The dispersed light is then focused by a second mirror on to a diode array. The diode array can have any number of diodes but the detector manufactured by Hewlett-Packard has 211. The total wavelength range covered is from 171 nm to 800 nm. The diode array is scanned continuously during the development of the chromatogram and the data from each diode is stored in a computer. In this way, by selecting the wavelength pertinent to a particular element, a chromatogram can be constructed that monitors that specific element. An example of the multi-element monitoring of a 15 component mixture is shown in figure 14.

The components were present in amounts ranging from 20 to 30 ng. The column was made of fused silica, 23 m x 0.32 mm with a film thickness of 0.17 µm. It is seen that the column resolution is well maintained and the overall sensitivity appears to be good although actual sensitivity values in terms of minimum detectable concentration were not given.

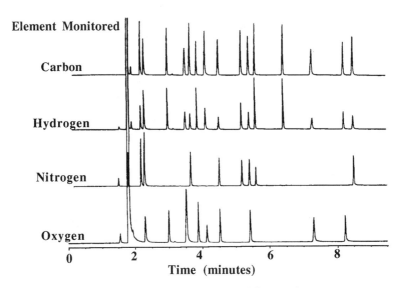

Courtesy of the Hewlett-Packard Corporation

Figure 14 A Separation Monitored by an Atomic Emission Spectrometer for Four Different Elements

It is also seen that single element monitoring is possible which is similar, in principle, to single ion monitoring in GC/MS or LC/MS. It follows that the presence of all the compounds in the mixture that contain a specific element can be selectively and exclusively displayed. This procedure also reduces the need for such high resolution from the column as any interfering substances that do not contain the specific element of interest are not displayed.

Liquid Chromatography Tandem Systems

Liquid chromatography tandem systems evolved much later than their gas chromatography counterparts. This was partly due to the late development of liquid chromatography itself and partly due to the difficulties involved in connecting a liquid chromatograph to a tandem instrument. It was found that removal of the mobile phase from the column eluent, or choosing conditions where the mobile phase did not influence the performance of the associated spectrometer was a challenge in itself. In the early days, off-line procedures were the most popular, collecting each fraction and presenting it as an individual sample to the appropriate spectrometer. These procedures, however, became very cumbersome as the resolution possible from LC columns increased and more complex mixtures were being separated. These difficulties emphasized the work put into the development of liquid chromatography tandem techniques and, as a result. today there are at least as many LC tandem systems as there are GC tandem systems commercially available.

Liquid Chromatography/Mass Spectrometry (LC/MS) Systems

In GC/MS ionization is usually either by electron impact or chemical ionization. Chemical ionization takes place at higher pressures than electron impact ionization (about 1 torr) and results from the collision of the sample molecules with the reagent ions. The reagent ions are themselves produced by electron impact ionization and so chemical ionization is often a two stage process. Chemical ionization produces sample ions close in molecular weight to that of the uncharged molecule (often M+H or M+CH$_3$ if methane is being used as the reagent gas) and usually very few smaller fragments. As chemical ionization can take place at significantly higher pressures than electron impact ionization, it is a convenient ionization process to use with GC tandem systems, as the demands made on the pumping system to remove the carrier gas or mobile phase vapor are not so stringent.

The ionization of solutes eluted from LC columns is far more difficult. Solutes separated by LC are often very polar or have very high molecular weights and thus are highly involatile. As a result quite different ionization techniques are frequently required. In 1976 Benninghoven *et al.* [10] demonstrated that high energy primary ions at low current densities could be used to produce intact molecular ions from organic compounds adsorbed on surfaces and this procedure was given the term *secondary ion mass spectrometry* (SIMS). A similar technique, *fast atom bombardment (FAB),* uses a neutral primary atom beam aimed at a liquid sample matrix containing the solute of interest to produce ionization [11,12]. *Laser desorption ionization* (LDI) [13], another ionization method, was improved by Kistemaker [14], who demonstrated that sub-microsecond laser pulses could be used to generate pseudo-molecular ions from solid surfaces. Unfortunately, however, these ionization techniques require a fairly complicated interface for their successful use and their employment in LC/MS is confined to relatively high molecular weight solutes.

In liquid chromatography the mobile phase must be eliminated, either before entering or from inside the mass spectrometer so that the production of ions is not adversely affected. The volatilization of the mobile phase generates a considerable volume of solvent vapor. The problem can be reduced by using a splitting system in much the same way as the early GC/MS tandem systems operated or by employing the solvent vapor as a chemical ionization agent. In fact the first LC/MS system developed was that by McLafferty [15] who used the direct sampling technique. A small portion of the column eluent was allowed to pass directly into a high resolution mass spectrometer together with the eluting solutes and, as result, produced chemical ionization spectra of the eluents.

Transport Interfaces: The Wire Transport Detector

In 1974 Scott *et al.* [16] developed a transport system for transferring the solute from the LC column to the mass spectrometer. This interface allowed electron impact spectra to be obtained and so all the expected ion fragments were obtained to facilitate interpretation of the spectra. A diagram of the transport LC/MS interface is shown in figure 15.

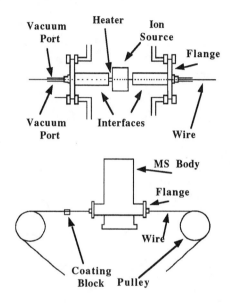

Figure 15 The Transport Interface for LC/MS

The basic principle was similar to that of the moving wire LC transport detector. The eluent from the column passed over a moving stainless steel wire, coating it with a film of column eluent. The wire passed through two orifices into two chambers connected in series. Each chamber was connected to a vacuum pump that reduced the pressure in the first to a few microns and in the second to about 10^{-5} mm of mercury. During this process the solvent evaporated from the wire leaving a coating of the solute on the wire surface. A current of a few milliamperes was passed continuously through the wire, which only became hot when it entered the high vacuum of the ion source where it could no longer lose heat to its surrounding. The temperature of the wire rose rapidly to about 250°C in the ion source, vaporizing the sample directly into the electron beam. The wire left the ion source in the same manner as it entered, via a second pair of differentially pumped chambers. A diagram of the interfaces is shown in figure 16.

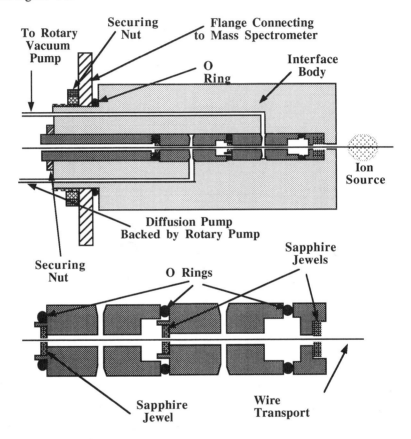

Figure 16 The Wire Transport Interfaces

The main body of the interface is constructed of stainless steel and is fitted to side flanges of a Finnigan quadrupole mass spectrometer, such that the interfaces are re-entrant to the ion source and terminate a few millimeters from the electron beam. The two chambers are separated and terminated by ruby jewels 0.1 in I.D. and 0.018 in thick. The jewels in the left-hand interface, where the sample is introduced have apertures 0.010 in I.D. The jewels in the right hand interface, where the wire transport leaves the mass spectrometer to the winding spool have apertures 0.007 in. I.D. The larger diameter apertures on the feed side are employed to reduce "scuffing" of the wire and possible solute loss. Ruby jewels were necessary to prevent frictional erosion of the apertures by the stainless steel wire. The first chamber of each interface was connected to a 150 l/min. rotary pump which reduced the pressure in the first chamber to about 0.1 mm of mercury. The second chamber of each interface was connected to an oil diffusion pump backed by a 150 l/min rotary pump. The pressure in the second chambers was reduced to about 5-10 µm of mercury. The entrance and the exit of each interface were fitted with a helium purge that passed over the aperture through which the wire was entering or leaving and ensured that only helium was drawn into the interfaces. In this way background signals from air contaminants were greatly reduced. The purge also allowed the use of methane as a chemical ionization agent if it were required.

Figure 17 The Original Moving Wire LC/MS Interface

The pressure in the source was maintained at about 1×10^{-6} mm of mercury. A photograph of the original transport interface is shown in figure 17.

The sensitivity of the device to diazepam, monitoring the eluent by total ion current, was found to be about 4×10^{-6} g/ml, which in the original chromatographic system was equivalent to about 7×10^{-10} g per spectrum. A total ion current chromatogram of a sample of the mother liquor from some vitamin A acetate crystallization is shown in figure 18. The total ion current is taken as proportional to the sum of all the mass peaks in a specific spectrum. The separation in figure 18 was carried out employing incremental gradient elution using 12 different solvents. It is seen that a good separation is obtained and the integrity of the separation is maintained after passing through the interface.

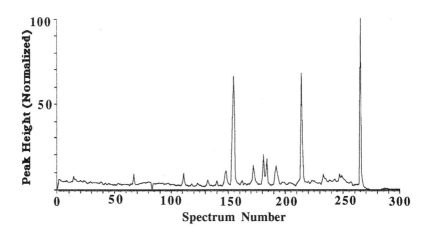

Figure 18 The Total Ion Current Chromatogram of a Sample of Mother Liquor from a Vitamin A Acetate Crystallization

It is also seen that the system is entirely independent of the solvent used in the separation, which ranged in polarity from the very dispersive *n* paraffins to highly polar aliphatic alcohols and included chlorinated hydrocarbons, nitroparaffins, esters and ketones.

The Belt Transport Detector

The wire transport interface was later modified by McFadden *et al.* [17], the wire being replaced by a continuous belt made of high temperature plastic which could be thermally cleaned after passing through the ion source and prior to re-entering the coating block. The belt is actuated by a motor driven pulley. A diagram of the McFadden interface is shown in figure 19. The column eluent is taken up on a stainless steel or high temperature plastic ribbon (about 3.2 mm wide and 0.05 mm thick) and transported to the vacuum locks by the moving belt. On the way to the vacuum locks, the coating can be heated by an infrared heater to facilitate evaporation of the solvent. The remaining solvent is removed in the vacuum-locks and about 10^{-7} g/sec enters the mass spectrometer.

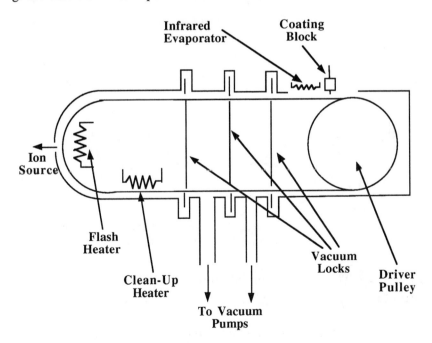

Figure 19 The Belt GC/MS Interface

The first pump removes air at a rate of 500 l/min and the second at about 300 l/min. The pressure in the first vacuum-lock is maintained at about 1 to 20 torr and that in the second about 0.1 to 0.5 torr. As a consequence the mass spectrometer source can be easily operated at about 10^{-6} torr. Flash vaporization of the solute occurs by radiant heating in a small chamber that butts directly onto the solid probe entrance to the ionization chamber and the vapor passes through a small hole directly into the ion source. The flash heater is either a nichrom coil or a quartz heater tube. The slots in the vacuum-locks are made of sapphire strips. An example of the use of the belt interface to monitor the separation of a pesticide mixture is shown in figure 20.

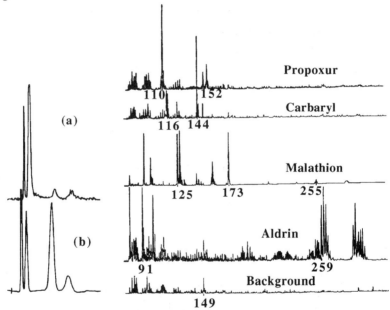

(a) Separation Monitored by Total Ion Current
(b) Separation Monitored by UV Detector

Figure 20 The Separation of Some Pesticides Using the Belt Interface

It is seen that there is little loss of resolution in the interface and very little cross contamination occurs between the peaks due to incomplete removal of each solute from the belt prior to the next coating. The sensitivity of the system to carbaryl was claimed to be about 1 ng but the actual sensitivity in g/ml was not possible to calculate from the data available.

Other Transport Systems

Alcock *et al.* [18] has reviewed the various types of transport interfaces with particular reference to their use with small bore columns. Games *et al.* [19] examined the peak dispersion that could take place in the transport interface itself together with the electronics associated with the mass spectrometer.

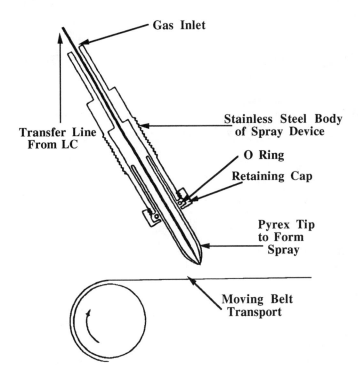

Figure 21 The Spray Deposition Device for Belt Transport Interfaces

They concluded that the LC/MS system incorporating the transport interface behaved as a low dispersion LC detector and consequently, could be used very effectively with small bore columns. Hayes *et al.* [20] conducted a systematic study of the effect of the method of solvent deposition on the transport medium on the overall performance of the LC/MS system. They designed a nebulizer to spray the column eluent onto the moving belt, a diagram of which is shown in figure 21. A stainless steel tube, 0.0625 in O.D. and 0.007 in I.D., is placed concentrically inside a Pyrex tube that carries the nebulizing gas. The Pyrex tube is held inside an outer steel tube by means of a screw cap and an O ring. The effect of nebulizer temperature and gas flow rate on the integrity of the elution curve produced by the interface were investigated. From the results of their studies the authors claimed that small bore columns could be used satisfactorily with the interface without degrading the resolution of the column and, at the same time, provide good quality mass spectra.

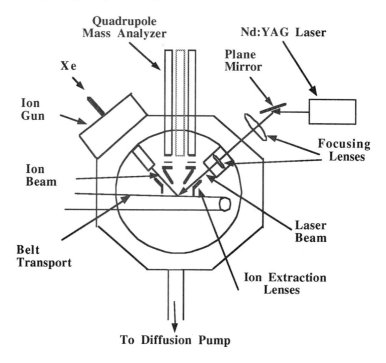

Figure 22 The Dual Ionization Transport Interface

They also claimed detection limits of 40 pg and linear dynamic ranges extending over four orders of magnitude. Fan *et al.* [21] developed a belt interface system for LC/MS that could provide both secondary ion mass spectrometry and laser desorption mass spectrometry. A diagram of the dual ionization source is shown in figure 22. It consisted of a quadrupole mass spectrometer with a moving belt and a Finnigan/INCOS data system. Samples were deposited on the belt by a type of thermospray process. Both the ion beam and the laser light were aligned to strike the belt at 45° and normal to each other. Xenon gas, ionized by electron impact, was used as the source of primary ions, which are focused by an Einzel lens and steering plates through an aperture onto the belt forming a spot covering and area of about 5 mm^2. The beam strikes the belt with an energy of about 3 keV. The laser light can be focused on the same spot by a suitable set of mirrors and lenses. Ultimately, the authors employed as the transport medium a carbon steel belt that had a blackened surface and consequently absorbed more energy from the laser beam. Providing laser energies were kept reasonably low there was no degradation of the surface. Comparing the results from the two methods of ionization, laser desorption had the advantage that it provided more reliable molecular weight information. Furthermore, by varying the power of the laser, different degrees of fragmentation could be achieved and consequently more structural information could be obtained. However, as it was difficult to control the laser energy absorbed, the mass spectra were less reproducible. In contrast, continuous ionization by primary ion bombardment provided more characteristic fragment peaks, without losing molecular weight information and both the mass spectra and the chromatograms were less noisy and more reproducible.

The Thermospray Interface

The interfaces that effectively replaced the transport system were the thermospray and electrospray sample introduction systems. The thermospray interface, a diagram of which is shown in figure 23, is a development from the direct inlet system of McLafferty. The successful use of the thermospray interface was first reported by

Covey and Henion [22]. The device consisted of a central stainless steel tube (0.004 in. I.D.) that passed through a heated copper block containing a cartridge heater. In the block was embedded a thermocouple that monitored the temperature and provided an output for the temperature controller. At the end of the tube was a pinhole through which the heated solute was vaporized. As the pumping system of the mass spectrometer was limited, either the eluent from a normal packed column was split, or a microbore column was used. The use of the thermospray system was further developed by Voyksner *et al.* [23]. They noted that the LC/MS thermal spray system frequently produced molecular weight information (parent ions) and exhibited lower detection limits than the other forms of LC/MS interfaces. They found that when they used a thermal spray system with a 0.1 M ammonia acetate buffer with a mobile phase that contained a high proportion of water, high sensitivities were achieved. The optimum interface temperature varied with solvent composition and could be determined by maximizing the solvent buffer ion intensities. Thermospray ionization resembles chemical ionization using ammonia as the chemical ionization reagent.

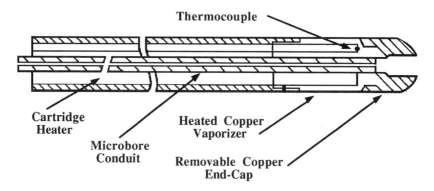

Figure 23   The Thermospray LC/MS Interface

The system produces protenation, ammonium addition and proton-bound solvent molecular clusters. The ionization procedure with this system was reported to be very soft resulting in very few molecular fragment ions being formed. The system was used very successfully in the analysis of triazine herbicides and organo-phosphorus pesticides, excellent specificity and sensitivity was achieved.

Blakely and Vestal [24] employed the thermospray system with the quadrupole mass spectrometer and demonstrated that it could provide stable vaporization and ionization at flow rates up to 2 ml/min with an aqueous mobile phase. If the mobile phase contained a significant amount of ions in solution (*ca.* 10^{-4} to 1.0 M) no extra thermal ionization source is required to achieve detection of many non-volatile solutes at the sub-microgram level. They found that with weakly ionized mobile phases, a conventional electron beam needs to be used to provide gas-phase reagent ions for the chemical ionization of the solute. The thermospray system has been effectively used by Voyksner *et al.* [25] in the analysis of cancer drugs.

The Electrospray Interface

Another ion producing inlet system that is now commonly used in LC/MS is *electrospray ionization,* reported by Whitehouse *et al.* [26]. The explanation given for the formation of ions is as follows. The column eluent, usually containing ions of some type, is sprayed into a dry gas such as nitrogen. This process produces a cloud of droplets that rapidly evaporate and, as a consequence, become smaller. The accompanying increase in charge density and the reduced radius of curvature of the droplets result in strong electric fields being formed that are sufficient to cause the droplets to explode producing ions many of which can contain multiple charges. A diagram of the Hewlett-Packard electrospray ionization LC/MS interface is shown in figure 24. The column eluent is mixed with a nebulizing gas and the spray jet directed onto a disc target. In the center of the target is a pinhole entry into the interface and the jet is directed to the side of this aperture. The fine droplets at the periphery of the spray are drawn into a chamber held at a reduced pressure through the pinhole aperture. Inside the chamber, the droplets are entrained in a stream of hot nitrogen gas that rapidly evaporates the solvent producing ions in the manner described above.

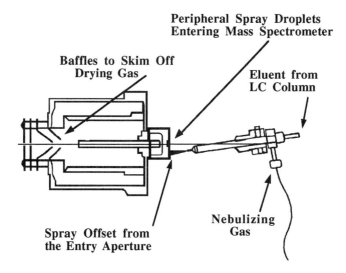

Courtesy of the Hewlett-Packard Company.

Figure 24   The Hewlett-Packard Electrospray Ionization LC/MS Interface

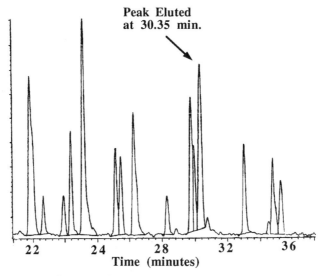

Courtesy of the Hewlett-Packard Company.

Figure 25 The Total Ion Current Chromatogram of a Sample from the Tryptic Digest of Lysozyme w/R&A

The core of the jet is skimmed by conical screens to remove the drying gas and the ions then pass directly into the mass analyzer. As ions with multiple charges can be produced, each charge being provided by an associated proton, this in effect increases the mass range of the spectrometer. For example, an ion of molecular weight 1000, carrying three charges, will appear on the spectrum at about a mass 333. The total ion current chromatogram of a sample from the tryptic digest of Lysozyme w/R&A is shown in figure 25.

A Vydac C-18 reversed phase column 250 mm long and 2.1 mm I.D was used which was thermostatted at 50°C and operated at a flow rate of 200 µl per min. The mobile phase composition was programmed from pure solvent A (0.1% trifluoro-acetic acid (TFA) in water) to 60% solvent B (0.1% TFA in acetonitrile) over a period of 60 minutes.

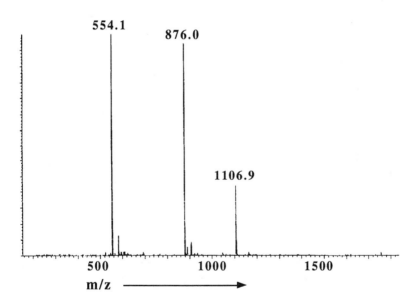

Courtesy of the Hewlett-Packard Company.

Figure 26 The Mass Spectrum of the Peak Eluted at 30.35 Minutes in the Tryptic Digest Chromatogram of Lysozyme w/R&A

The total column eluent was passed into the interface but, as seen above, due to the nature of the ionization process, only a portion of the solute actually enters the mass analyzer. The mass spectrometer employed was the Hewlett-Packard MS Engine Quadrupole Mass Spectrometer. The mass spectrum of the peak eluted at 30.35 minutes is shown in figure 26. The figure shows three peaks at m/z values of 554.1, 876.0 and 1106.8. It is seen from figure 25 that the peak was not completely resolved from its neighbors and thus it was necessary to decide whether all three peaks originated from the same solute and whether any pair of the peaks resulted from multiple charges on the same molecule. A program is included in the Hewlett-Packard data handling software that tests the interrelationship between the individual mass peaks of the mass spectrum. The software determines whether the heights of any pair or group of peaks are linearly related.

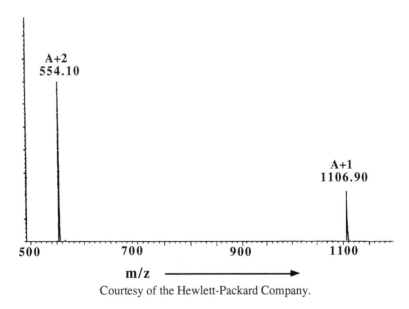

Courtesy of the Hewlett-Packard Company.

Figure 27 Mass Spectrum Showing Multiple Charged Peaks of a Parent Molecule

If the respective peaks increase or decrease proportionally with one another, then they must originate from the same parent molecule and probably represent multiple charges on the same molecule. The peaks in figure 10 were tested in this manner and the results are shown in figure 27. The data processing demonstrated that the peak at 876.0 m/z was not related to the other two and, furthermore, the peaks at 554.1 and 1106.9 were doubly and singly charged species of the same molecule.

e.g. $\dfrac{1106.9 - 1(H+)}{2} + 2(H+) = 554.95$ c.f. 554.1 (from spectrum)

The Atmospheric Ionization Interface

Another popular and efficient inlet system for the LC/MS combination is the *atmospheric pressure chemical ionization* process. This system has some similarity to the electrospray interface and can also cope with flow rates of up to 2 ml/min. and thus the total column eluent can be utilized without splitting the flow.

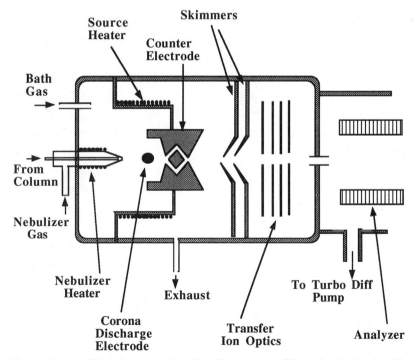

Figure 28   The Atmospheric Pressure Chemical Ionization Interface

However, only a portion of the eluent actually provides ions that enter the mass spectrometer. A diagram of the atmospheric pressure chemical ionization interface is shown in figure 28.

The sample is first vaporized in the center of a heated coaxial tube into a nitrogen stream that is carried in the outer tube. The sample then drifts through a chamber containing a corona discharge set up by a potential difference of about 2,000 volts applied between a simple electrode arrangement. The reactant ions that are formed in the corona discharge collide with the sample molecules and largely give *sample molecule plus a proton* (hydrogen positive ions), *i.e.* $[M+H]^+$. This is a process very similar to chemical ionization. The ions so formed then pass through two differentially pumped chambers to reduce the pressure sufficiently to allow the ionized sample to enter the mass spectrometer. In the first chamber the ions are electrically focused through two apertures and the excess unionized material skimmed off to waste. In the second chamber, more gas is removed, the pressure is reduced further and the ions are electrically focused onto the entrance aperture of a quadrupole mass spectrometer. The sample then enters the electron beam of the electron impact ionization source of the mass spectrometer and fragmentation takes place. As the sample entering the mass spectrometer is virtually a parent ion already, the fragmentation pattern is very similar to that obtained in MS/MS. In this way, the system differs fundamentally from the electrospray interface. The ionization process is soft, very sensitive and gives good characteristic spectra that can be used for both sample identification and structural elucidation.

A recent modification of the atmospheric pressure ionization technique involving a special low dead volume interface was described by Thomson *et al.* [27]. They employed packed microbore columns (170 µ, 320 µ, and 500 µ I. D. with lengths ranging from 5 to 15 cm) in conjunction with a low-volume, wall-coated capillary column as an interface. The total ion current chromatogram of the tryptic digest sample of about 1 picomole of human growth hormone is shown in figure 29. The column was packed with an octadecyl bonded phase have a mean pore size of 300 Å and a particle diameter of 7 μ. A gradient from 20% solvent (A), (0.1% TFA in water) to 80% solvent (B) (75% 0.1% TFA and 25% acetonitrile) was employed over a period of about 1 hour. Flow rates of about 80 to 100 μl per minute were used with about 3 μl passing to the capillary column and entering the interface.

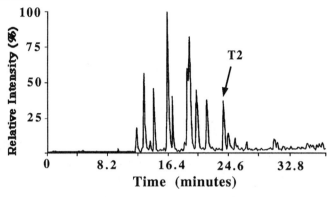

Courtesy of the Perkin Elmer SCIEX Corporation

Figure 29 Total Ion Current Chromatogram of a Tryptic Digest Sample of Human Growth Hormone

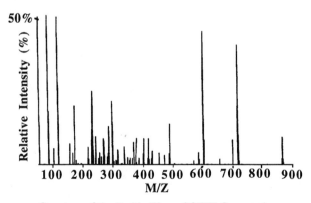

Courtesy of the Perkin Elmer SCIEX Corporation

Figure 30 Spectrum of a Product from the Tryptic Digest of Human Growth Hormone Obtained from a Low Dead Volume Atmospheric Ionization Interface

It is seen that an excellent separation is obtained and apparently little resolution is lost in the capillary interface. The mass spectrum of the peak marked T2 in the chromatogram is shown in figure 30. It is clear that good quality spectra can be obtained up to ion masses of at least 900. Such a combination of techniques can be invaluable for the structure elucidation of compounds generated in biochemical research.

Liquid Chromatography/Infrared (LC/IR) Systems

The problems involved in the association of an infrared spectrometer with a liquid chromatograph are twofold. Firstly the majority of solvents absorb light in the infrared range, often over those ranges that would be most informative in structure elucidation and identity confirmation. This means that the solvent needs to be removed, or an infrared transparent solvent must be employed, which may not be compatible with the mobile phase requirements of the liquid chromatograph. Secondly, the sensitivity of the infrared spectrometer is several orders of magnitude less than that of the UV spectrometer or the mass spectrometer and thus larger samples are required, and this may also be unacceptable to the chromatographic system. For these reasons prior to about 1975, LC/IR was carried out off line, the peak being collected, the solvent removed by appropriate procedures, the solute redissolved in a suitable IR solvent and the spectrum obtained in the usual way.

The situation has not change dramatically since that time although with Fourier transform IR now generally available the problem of sensitivity has been partly solved. Furthermore, as the spectra are stored, the background spectra of the solvent can be subtracted leaving the actual spectra of the solute. Unfortunately, this involves subtracting two very large signals and thus the signal to noise of the difference is rather low and such spectra are often very noisy causing uncertainty in the spectra interpretation and identity confirmation. In 1975 Kizer *et al.* [28] demonstrated that this technique would work but, as would be expected, it provided very poor sensitivity. Kuehl and Griffiths [29] approached the problem in another way employing a rather crude but, nevertheless, effective transport system. They tried moving ribbon devices with pre-concentrating techniques rather unsuccessfully and finally used a carousel of cups containing potassium chloride. A diagram of their carousel is shown in figure 31. The carousel resembles a fraction collector and consequently the device is more like an off-line system than an in-line system. The carousel has 32 cups fitted with a fine mesh screen each containing potassium chloride powder.

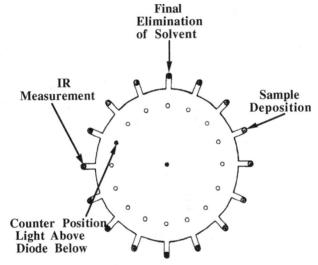

Figure 31 Carousel Transport for On-Line IR Monitoring of LC Column Eluents

The position of the carousel is controlled automatically and only three positions are actively used. In the first position the eluent is deposited on the potassium chloride, in the second position a stream of air is drawn through the potassium chloride to remove the solvent and in the third position the spectrum is taken. The use of the carousel containing potassium chloride powder certainly increased the sensitivity of the LC/IR combination, but the finite intervals of sample collection made the system unsuitable for modern high efficiency columns. Jino and Fujimoto [30,31] employed a potassium bromide plate as a transport system. The eluent from a small bore column (flow rate 5 µl/ min) was allowed to fall on the plate and evaporate, the plate being moved as each peak was eluted. The eluent was also monitored by a UV detector to identify when the plate should be moved. After the chromatogram had been developed, the plate was scanned by an IR spectrometer and the spectrum of each solute obtained. Good sensitivity was reported but not precisely defined in chromatographic terms. However the procedure (and for that matter, the procedure of Kuehl and Griffiths [29]) were little more than novel methods of fraction collecting and, in fact, were off-line procedures. In 1983, Brown and Taylor [32] introduced a micro IR cell, 3.2 µl in volume, that they employed with a small bore column and claimed an overall increase in mass sensitivity of about 20 orders of magnitude relative to the standard 4.6 mm I.D. column. They also employed an FT/IR spectrometer but the actual improvement was obscured by the fact that the column length of the small bore column was significantly different from that of the standard column.

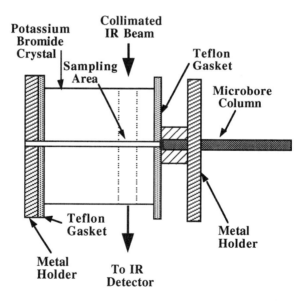

Figure 32   Zero Dead Volume Micro IR Cell

An interface for use with microbore columns in conjunction with FTIR was described by Johnson and Taylor [33] and with their system reduced the detection limit to 50 ng. A diagram of their flow cell is shown in figure 32. It consisted of a crystal block of either calcium fluoride or potassium bromide, 10 x 10 x 6 mm with a 0.75 mm hole drilled through it. The IR beam passes through the block normal to the flow of mobile phase. Since the focal diameter of the FT/IR spectrometer was 3 mm and the hole in the cell only 0.75 mm, a beam condenser was used to reduce the focal diameter to that of the hole. The authors observed that although a maximum IR signal could be obtained when the peak maximum was in the light path, they found that the maximum signal-to-noise was obtained by summing the scans across the peak as it passed through the cell between ± 1.53 σ.

Conroy and Griffiths [34] developed a solvent extraction device that could be employed with a LC/FTIR combination. The device involved an extraction procedure that took the column eluent and continuously extracted the dissolved solute into dichloromethane. The dichloromethane was then concentrated and finally dispersed onto a plug of potassium chloride powder. This device appears a little clumsy and, fact, is really and off-line fraction collecting procedure.

Sabo *et al.* [35] developed an LC/FTIR interface for both normal and reversed phase chromatography using an attenuated total reflectance cell. The cylindrical internal reflectance flow cell contained in a ZnSe crystal, had a nominal volume of 24 µl and an equivalent transmission cell path length of 4-22 µm over the usable range. On-the-fly spectra of the components from a 100 µl sample of a solution containing 2% of acetophenone and ethyl benzoate and 1% of nitrobenzene gave clearly identifiable spectra. Nevertheless, relative to other LC/FTIR systems this was not a very sensitive device.

Johnson *et al.* [36] developed a rather unique extraction cell for an LC/IR system. They used the segmented flow of an aqueous effluent from a reversed phase system and separated the extraction solvent by means of a "hydrophobic" (dispersive) membrane. A diagram of their apparatus is shown in figure 32. There are two pumps, one provides the solvent for the chromatographic development, and the other the extraction solvent which can be either chloroform or carbon tetrachloride. After passage through the column the two streams are mixed at a T junction and form a segmented flow as both solvents are virtually immiscible. The segmented flow passes through an extraction coil to a separator, a diagram of which is shown in the lower portion of figure 33.

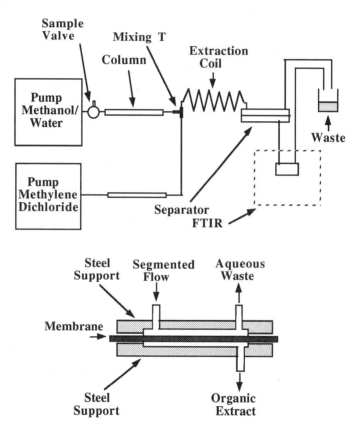

Figure 33 An Extraction Interface for LC/IR

The separator was made of stainless steel and the volume on either side of the membrane was about 16 µl, the membrane itself having pores about 0.2 µm in diameter. The membrane was sealed against the steel supports by compression. The amount of solvent that was passed through the membrane was controlled by adjusting the differential pressure across the membrane. The device would obviously cause serious peak dispersion and would be unsuitable for use with high efficiency or small bore columns. Spectra were obtained from samples containing 300 µg of material indicating relatively poor sensitivity. A photograph of a modern, commercial LC/IR system fabricated by Lab Connections Inc. is shown in figure 34.

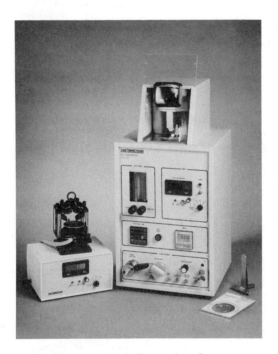

Courtesy of Lab Connections Inc.

Figure 34 The Lab Connections, Inc. LC/IR Interface System

The device is unique in the sense that it is a transport type of interface but uses a germanium disc as the carrier which is transparent to IR light over the range of wavelengths that provide structural information. A section of the germanium disc is shown in figure 35. The disc is about 2 mm thick and the lower surface is coated with aluminum to reflect any IR light that passes through it. The interface consists of two modules, the coating module and the optical module. Diagrams of the two modules are shown in figure 36.

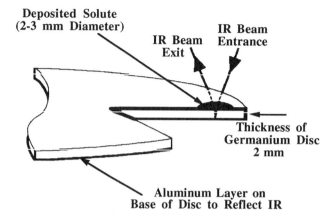

Courtesy of Lab Connections Inc.

Figure 35 Section of the Germanium Disc Transport System

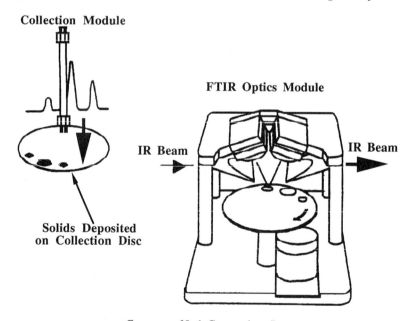

Courtesy of Lab Connections Inc.

Figure 36 The Coating and Optical Modules of the Interface

In fact, this type of interface is not really an in-line device, but a fraction collector that is constructed in a form that can also be used to spectroscopically examine the collected samples. The LC column is connected to the chromatography module and eluent from the column passes into an ultrasonic nebulizer. The nebulizer sprays the solvent in a tightly focused jet onto the surface of a monocrystalline germanium sample-collection disc. The temperatures of both the ultrasonic nebulizer and the collection disc are carefully controlled. The disc rests on a heated stage in an evacuated compartment that rotates slowly during the elution of the sample. All the solvents evaporate leaving a deposit of solvent-free solutes on the disc surface. After the separation is complete, the disc is removed and placed on the stage of the optical module, which is located in the sample compartment of the FTIR bench. The compartment is purged with carbon dioxide free air. The stage rotates and the in-coming IR beam passes though the sample, through the disc, and is reflected back though the sample by the layer of aluminum on the underside of the disc and back to the spectrometer. This procedure produces a two-pass transmission spectrum of the sample. The FTIR spectra obtained from a calibration sample of benzyl benzoate and testosterone cypionate are shown in figure 37.

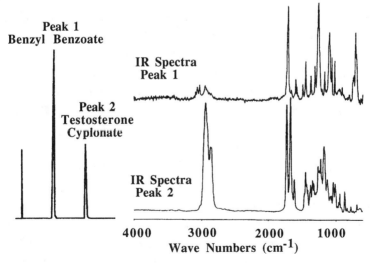

Figure 37 IR Spectra of Benzyl Benzoate and Testosterone Cypionate Obtained from the LC/IR Interface

It is claimed that satisfactory spectra can be obtained for identification purposes from 100 ng of sample. This level of sensitivity is quite satisfactory and an advance in the technique for drug process control. However, it might not be so suitable for biological samples such as blood and urine.

In general, tandem systems involving the combination of the liquid chromatograph in-line with the infrared spectrometer have not been very successful and most IR spectra of LC eluents are obtained by what are, in effect, off-line procedures, as in the example given above. The FTIR spectrometer, in its present form, demands too large a sample size and is too insensitive for successful in-line association with modern high-efficiency microbore LC columns. Fortunately, the demand for in-line production of IR spectra from LC eluents is not great and, in most cases, the off-line methods are quite satisfactory for the majority of LC/IR applications.

Liquid Chromatography Nuclear Magnetic Resonance (LC/NMR) Systems

NMR spectroscopy is the only spectroscopic technique that can, under many circumstances, unambiguously identify the structure of an hitherto unknown compound. It follows that the association of the liquid chromatograph with the NMR spectrometer would be a very powerful analytical tool for the separations and identification of unknown substances. There are, however, some serious difficulties in the association of the two techniques. The four main problems that must be solved were outlined, many years ago by Bayer *et al.* [37], who were one of the first research groups to combine the liquid chromatograph on-line with the NMR spectrometer. Firstly, the intensity of the NMR signal depends on the flow rate and as the flow rate increases, the signal decreases. However, the reduction in signal can be restricted to a reasonable level at flow rates between 0.5 and 2 ml/min. Secondly, in order to realize high NMR resolution, the magnetic field throughout the sample must be very homogeneous and to achieve this, the sample tube is usually spun at fairly high speeds. So far, this has proved impossible in flow-through cells and consequently, in the past, considerable resolution has been lost in these types of cell. Thirdly, the solvent itself usually contains protons or ^{13}C nuclei and thus can interfere with the spectrum of the solute. In the case of proton spectroscopy, solvents may be chosen that do not contain protons such as carbon tetrachloride or deuterated solvents but the former restricts chromatographic performance and the latter can become very expensive. The use of small bore columns would significantly reduce the solvent consumption and render the use of special solvents more economically viable but such columns would demand very small cell volumes and minimum extra-column dispersion. Finally the sensitivity of the NMR spectrometer is not nearly as great as that of the UV spectrometer or the mass spectrometer and consequently much larger charges must be placed on the column. Nevertheless, over the years, with improved techniques, such as higher resolution NMR spectrometers operating at 500 and 800 MHz using superconducting magnets coupled with ^{13}C spectroscopy, practical LC/NMR systems have been successfully developed and are now commercially available.

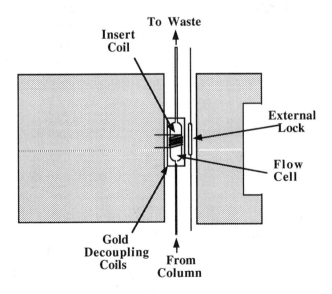

Figure 38 The Basic NMR Spectrometer Flow-Through-Sample Cell

A diagram of the original flow-through cell of Bayer [37], which incorporates the basic principles of the earlier LC/NMR systems that operated with electromagnets, is shown in figure 38. The volume of the cell had to be kept as small as possible to minimize band dispersion, but, at the same time, the geometry of the flow cell had to provide optimum synchronization with the NMR sensing coil. Consequently, in the older electromagnetic instruments, the wall of the cell had to be straight and parallel with the axis of the coil. Such flow-through cells had volumes in excess of 400 µl (too large for modern LC columns) and thus at a flow rate I ml/min the average residence time of the sample was about 25 sec.

Modern high sensitivity, high resolution NMR instruments have superconducting magnets and thus have entirely different sensing coils and cells. The coils are no longer wound concentrically on the sample tube but are in the Helmholtz configuration and, furthermore, because of the geometry of the superconducting coils and cryostatic environment, the column eluent can not pass vertically through the field entering at the top of the instrument and exiting at the base. The element from the column therefore normally passed through a U shaped conduit, one limb of which will be the actual sample cell. Modern flow-through cells have volumes ranging from 20 to 50 µl but recently, Sweedler *et al.* [38] have claimed the successful use of sample tubes with capacities that are measured in nanoliters with improved signal to noise ratio of about two orders of magnitude. The sensing coil is only 1 mm long, 0.5 mm O.D. and the capillary running through the coil has a sample capacity of 5 nl. The secret appears to be in surrounding the coil/capillary assembly in a "commercial perfluorinated organic liquid" that has the *same magnetic susceptibility* as the copper microcoil. This is claimed to provide a more uniform magnetic field in the sample region leading to improved resolution and peak shape. This discovery could quickly lead to vastly improved LC/NMR systems. A diagram of a modern LC/NMR tandem system is shown in figure 39. For the most part, modern LC/NMR systems are not in-line devices but actually function as automatic fraction collectors that pass the sample to the NMR spectrometer for examination by a normal static procedure. They are unique, in that the cell is designed for a flow through function and to have a small sample volume.

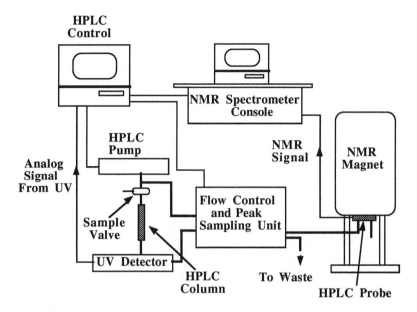

Figure 39 The Layout of Modern LC/NMR Tandem System

They also have special valving arrangements to direct the column and solvent flows appropriately. Nevertheless, the spectra are generally run on a stop flow principle; the sample being held in the cell while the spectra are obtained. This is a direct result of the sample being generated by the chromatograph faster than the spectrometer can acquire the necessary data from it. The sample cell must be small enough not to destroy the chromatographic resolution but be large enough to contain sufficient sample to allow the spectrometer to acquire valid data in the residence time available. To date, these two criteria have not been efficiently achieved. As seen in figure 39 the tandem system is basically a liquid chromatograph and a NMR spectrometer joined by a valving system that allows any given peak to be passed to the spectrometer or, if the spectrometer is acquiring data, to be stored in sample loops until the spectrometer is free. A diagram of the flow control and sampling unit is shown in figure 40.

Basically, the flow control and sampling unit allows three alternative methods of operation. Firstly the eluent from the column can flow directly from the UV detector to the NMR sample tube and the spectra can be continuously monitored during the development of the separation. The success of this procedure will depend on the volume of the cell, the sample size, the column flow rate, the resolution of the NMR spectrometer and the rate of data acquisition by the computer. In general, unless the new micro-cell facilities mentioned above are exploited, this procedure will rarely be successful, particularly if microbore columns are used and multi-component mixtures are being examined.

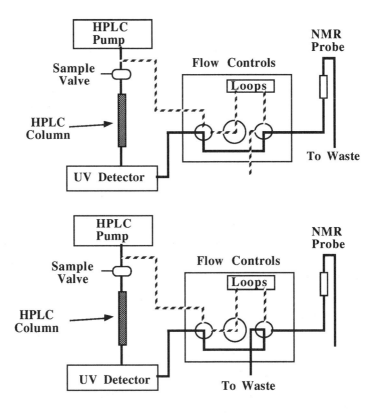

Figure 40 The Flow Control and Peak Sampling Unit for a LC/NMR System

The second alternative is to direct specific samples to the NMR that are of particular interest. The sample can then be trapped in the cell and data acquired from an adequate number of pulses to provide the required resolution. Subsequently, the sample can be expelled from the cell using solvent supplied directly from the chromatography pump. The third alternative is to direct the eluent from the column to a sample loop where it can be stored until the spectrometer is available to take data. If necessary, a number of solutes can be stored in different loops and they can be examined when convenient. When the data has been acquired from one sample, the solute stored in the next loop can then be displaced into the NMR cell. Samples that have been examined can either be displaced to waste or collected for further examination. A photograph of the Varian flow control device for the LC/NMR system is shown in figure 41.

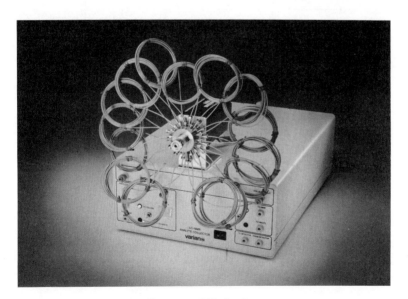

Courtesy of Varian Inc.

Figure 41 The Flow Control Device for the LC/NMR System

An interesting example of the use of the Varian LC/NMR system in handling two unresolved peaks is shown in figure 42. The spectra were obtained using their new LC-NMR Microflow probe, which was developed to maintain good chromatographic resolution and also provide the necessary high NMR sensitivity. The cell had a sensor volume of 60 µl, which is a little large for microbore columns but can be used with wider bore columns. The spectra were obtained by stopping the column flow when the peaks were in the sensor cell and running the spectra in the normal way. Using this technique, it is clear that a multiple peak can be scanned. It is seen from the insert chromatogram obtained from the UV detector that there are two unresolved solutes present in the peak and the NMR spectra taken across the peak clearly differ as shown in figure 42.

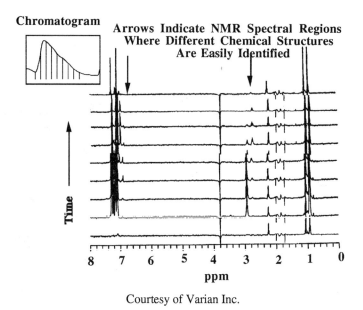

Courtesy of Varian Inc.

Figure 42 NMR Spectra Taken Across an Unresolved Peak Using the Varian Interface System, the LC/NMR System

Even though the components are not resolved, a shoulder on the main component peak yields an identifiable spectrum that is quite different from that taken from the front portion of the peak. The areas of the

NMR spectra that demonstrate the change that results during the final elution of the first peak and the start of the elution of the second peak are depicted by arrows. This type of tandem system can provide unambiguous confirmation of solute identity that is ideal for forensic purposes.

Liquid Chromatography Atomic Absorption Spectroscopy (LC/AAS) Systems

The association of a spectrometer with a liquid chromatograph is usually to aid in structure elucidation or the confirmation of substance identity. The association of an *atomic absorption spectrometer* with the liquid chromatograph, however, is usually to detect specific metal and semi-metallic compounds at high sensitivity. The AAS is highly element-specific, more so than the electrochemical detector; however, a flame atomic absorption spectrometer is not as sensitive. If an atomic emission spectrometer or an atomic fluorescence spectrometer is employed, then multi-element detection is possible as already discussed. Such devices, used as a LC detector, are normally very expensive. It follows that most LC/AAS combinations involve the use of a flame atomic absorption spectrometer or an atomic spectrometer fitted with a graphite furnace. In addition in most applications, the spectrometer is set to monitor one element only, throughout the total chromatographic separation.

One of the main applications of LC/AAS is to help determine metal speciation in samples, not merely to identify the presence of a particular element. It is not sufficient to detect the presence of lead, mercury or chromium, one must also to be able to identify the form in which they are present. Depending on the chemical form of a mercury compound, it may or may not be highly toxic. Similarly, if chromium is present in the tertiary form it is not particularly dangerous; conversely, in its sixth valency state it is strongly carcinogenic. It follows that the liquid chromatograph can be employed to separate the different species and the atomic absorption spectrometer can identify and confirm the presence of a specific element eluted at a particular retention time.

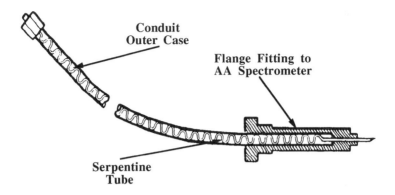

Figure 43 The LC/AAS Serpentine Tubing Interface

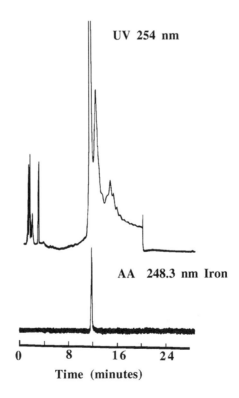

Figure 44 Chromatograms of a Blood Sample Monitored by the UV Detector and the Atomic Absorption Spectrometer with the Serpentine Tube Interface

In order to successfully utilize the LC/AAS combination, both the chromatograph and the spectrometer must be optimized, which has been discussed in some detail in a number of publications [39-41]. It has been claimed [39] that the poor sensitivity that has been obtained from the LC/AAS system, relative to that from the atomic spectrometer alone, was due to the dispersion that takes place in the column.

Although substantially true, this misunderstanding arises from the fact that the spectroscopist views the chromatograph as just another sampling device and not as a separation system. The point of interfacing a liquid chromatograph with an absorption spectrometer is to achieve a separation before detection. Consequently, the important dispersion characteristics are *not* those that take place in the column, but those that occurs in the interface between the chromatograph and the spectrometer and in the spectrometer itself.

These two sources of dispersion can not only reduce element sensitivity but also destroy the separation originally achieved in the column. The magnitude of the extracolumn dispersion is particularly important if high speed columns, packed with very small particles, are being used since such columns produce very narrow peaks a few microliters in volume. High-speed columns seem ideally suited for use with AAS instruments as they can be operated at the very high flow rates necessary for efficient solvent aspiration into the spectrometer nebulizer. Unfortunately, due to the basic design of most AAS instruments a significant length of tubing is necessary for the interface, if normal operation of the spectrometer is not to be impeded.

Katz and Scott [42] solved this problem by the use of low dispersion serpentine tubing as the interface between the exit from the UV detector of the liquid chromatograph and the spectrometer. A diagram of their interface is shown in figure 43. The principle of low dispersion tubing has already been discussed and it is sufficient to say that the outer interface tube was 49 cm long, 0.25 cm I.D. and merely protected the serpentine tube contained inside. The inner serpentine tube had a peak-to-peak amplitude of 1 mm. An example of the chromatograms obtained from a blood sample monitored by both a UV detector and the AAS detector employing the serpentine interface is shown in figure 44. The AAS unambiguously picks out the peak containing iron and the width of the iron peak clearly substantiates the efficacy of the serpentine interface.

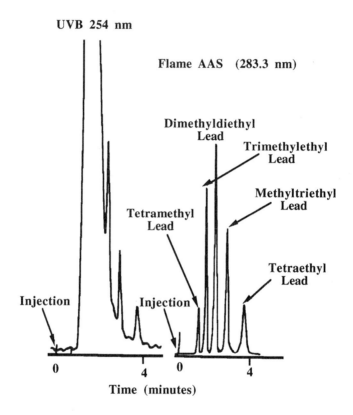

Figure 45 An Example of a Separation of Lead Compounds Demonstrating the Selectivity of the LC/AAS System

The LC/AAS has been employed for many years and Holak [43] used it to monitor the separation of a number of mercury containing drugs, mersalyl, thimerosal and phenyl mercuric borate. Suzuli et al. [44] used the technique to identify the heavy metals bound to isoproteins extracted from liver tissue. Robinson and Boothe [45] used the selectivity of the LC/AA system to monitor the alkyl lead compounds in sea water and Messman and Rains [46] separated four alkyl leads, tetramethyl lead (TML), trimethylethyl lead (TMEL), dimethyldiethyl lead (DMDEL), methyltriethyl lead (MTEL) and tetraethyl lead (TEL) in gasoline. An example of their separation is shown in figure 45. It is seen that excellent selectivity can be obtained, eliminating the need for high resolving power columns, as the multitude of interfering hydrocarbons present in gasoline are not detected.

Thin Layer Chromatography Tandem Systems

Linking thin layer chromatography with a tandem instrument is a more difficult procedure than combining GC or LC with an appropriate spectrometer. Furthermore, due to the nature of the development process in TLC, the combination must inevitably be an off-line procedure and therefore not a truly tandem system. Scanning the thin layer plate for UV absorption or fluorescence emission, after development and drying, has already been discussed but the combination of TLC with IR spectroscopy and mass spectroscopy is more difficult. The number of examples appearing in the literature involving TLC tandem systems are relatively few compared with those involving GC or LC. If a reflectance IR spectrometric system can be used, then the plate can be scanned after development and subsequent drying, to provide IR reflectance spectra of each spot but, again, this is an off-line procedure. Furthermore, suitable mechanical scanning apparatus must be available to accurately position the plate in the spectrometer. However, a more interesting tandem system, recently reported [47], is the combination of the time-of-flight mass spectroscopy with thin layer chromatography using matrix assisted laser desorption. This technique takes advantage of the unique possibilities that time-of-flight mass spectroscopy offers for this type of sample ionization which was emphasized when the time-of-flight mass spectrometer was discussed.

Thin Layer Chromatography/Mass Spectroscopy

This combination depends firstly on the development of the TLC plate followed by mobile phase removal by drying in an appropriate manner. Secondly, the matrix ionization aid is formed separately as a thin film on a stainless steel plate. The reagents, dissolved in an appropriate solvent, are sprayed onto the steel plate and dried. The thin layer plate is then sprayed with an appropriate solvent and while wet, the stainless steel plate is pressed onto the thin layer plate (the coating being in contact with the thin layer plate). The solvent on the TLC plate serves two purposes. Firstly it extracts the solutes into solution so that they can come in contact with the matrix film.

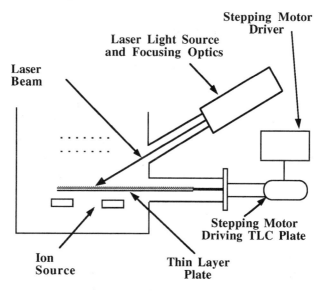

Figure 46 The TLC Plate Sampling Device for the Time of Flight Mass Spectrometer

Secondly, the matrix film takes up the solvent and thus incorporates the solutes into the matrix layer. The steel plate is then removed, leaving the matrix layer containing the solutes on the surface of the thin layer plate. The thin layer plate is then placed in the laser desorption apparatus shown in figure 46. The desorption device consists of a TLC plate holder, driven by a stepping motor, that can be traversed both in the (x) and (y) axis so that the plate can be precisely scanned in two dimensions. The plate is situated in the ion source of the time of flight mass spectrometer and by pulsing the laser beam the spectra can be obtained and, if necessary, accumulated in the usual manner. An example of the spectra obtained for rhodamine–B is shown in figure 47.

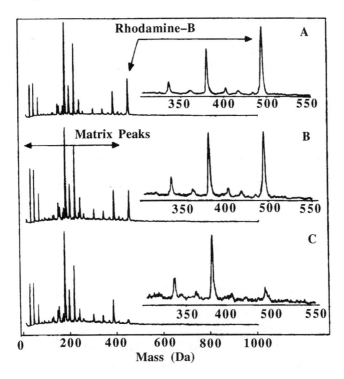

Figure 47 The Mass Spectra of Rhodamine B Separated on Aluminum Backed TLC Plates with an Organic Binder

The lower spectrum is a blank, showing no Rhodamine–B present. The first and second spectrum were irradiated with laser energies 40% and 100% above threshold respectively. Matrix peaks occur at masses ranging from 250 to 400 Da and thus interfere with the spectrum. Consequently, a limit is set on the mass measurement of low molecular weight materials. The sensitivity quoted was 40 pg but the size of the spot was not given. Nevertheless, it is seen that very useful mass spectra can be obtained by directly scanning the thin layer plate and taking advantage of the unique properties of the time of flight mass spectrometer.

* * *

A final comment on tandem techniques. For routine analytical work with forensic overtones, tandem systems can be essential. For the *ad hoc* occasion when solute confirmation is necessary, fraction collection followed by the usual spectroscopic examination will often give results that will be at least as reliable and probably more precise. In addition, the equipment will certainly be less expensive to purchase and maintain.

References

1. R. P. W. Scott, I. A. Fowlis, D. Welti and T. Wilkins, *"Gas Chromatography 1966"* (Ed. A. B. Littlewood), The Institute of Petroleum (1967)318.
2. J. C. Holmes and F. A. Morrell, *Appl. Spec.*, **11**(1957)86.
3. R. Ryhage and E. von Sydow, *Acta Chem. Scand.*, **17**(1963)2025.
4. J. T. Watson and K. Bieman, *Anal. Chem.* **37**(1965)844.
5. A. E. Banner, R. M. Elliott and W. Kelly, "Gas Chromatography 1964", (Ed. A. Goldup), J. Inst Pet., London (1964)180.
6. D. J. Erickson, C. E. Enke, J. F. Holland and J.T.Watson, *Anal. Chem.*, **62**(1990)1079.
7. Practical Organic Mass Spectrometry, (Ed. J. R. Chapman) John Wiley and Sons, Chichester and New York (1994)
8. S. B. Hawthorne and D. J. Miller, *J. Chromatogr. Sci.*, **24**(1986)258.
9. P. A. Wilks and R. A. Brown, *Anal. Chem.*, **36**((1964)1896.
10. A. Benninghoven,D. Jaspers, and W. Lidtermeann, *Appl. Phys. Lett.* **11**(1976)35.
11. K. L. Rinehart, Jr., Science, **218**(1982)254.
12. M. Barber, R. S. Bordoli, R. D. Sedgwich and A. N. Tyler, *J. Chem Soc., Chem. Comm.*, **7**(1981)325.
13. F. J. Vastola and A. J. Pirone, *Adv. Mass Spectrom.*,(1968)167.
14. M. A. Pothumus, P. G. Kistemaker and H. L. C. Menzelaar, *Anal. Chem.*, **50**(1978)985.
15. M. A. Baldwin, and F. W. McLafferty, *Biomed. Mass Spectrom.*, **1**(1974)80.
16. R. P. W. Scott, C. G. Scott, M. Munroe and J. Hess. Jr., *"The Poisoned Patient: The Role of the Laboratory"*, Elsevier, New York (1974) 395.
17. W. M. McFadden, H. L. Schwartz and S. Evens, *J. Chromatogr.*, **122**(1976)389.
18. N. J. Alcock, C. Eckers, D. E. James, M. P. L. Games, M. S. Lant, M. A. McDowall, M. Rossiter, R. W. Smith, S. A. Westwood and H.-Y. Wong, *J. Chromatogr.*, **251**(1982)165.

19. D. E. Games, M. J. Hewlins, S. A. Westwood and D. J. Morgan, *J. Chromatogr.* **250**(1982)62.
20. M. J. Hayes, E. P. Lanksmeyer, P. Vouroo and B. L. Karger, *Anal. Chem.* **55**(1983)1745.
21. T. P. Fan, A. E. Schoem, R. G. Cooks and P. H. Hemberger, *J. Am. Chem. Soc.* **103**(1981)1295.
22. T. Covey and J. Henion, *Anal. Chem.,* **55**(1983)2275.
23. R.D.Voyksner, J. T. Bussey and J. W. Hines, *J. Chromatogr.*, **323**(1985)383.
24. C. R. Blakely and M. L. Vestal, *Anal. Chem.*, **55**(19830750.
25. R.D. Voyksner,J. T. Bussey and J. W. Hines, *J. Chromatogr.*, **323**(1985)383.
26. C. M. Whitehouse, R. N. Dreger, M. Yamashita and J. B. Fenn, *Anal. Chem.* **573**(1985)675.
27. B. Thomson, Tom Covey, B. Shushanm M. Allen, and Takeo Sakuma, Perkin Elmer Corporation, Private Communication.
28. K. L. Kizer, A. W. Mantz and L. C. Bonar, Am. Lab. **May** (1975)
29. D. Kuehl and P. R. Griffiths, *J. Chromatogr. Sci.* **17**(1979)471.
30. K. Jino and C. Fujimoto, *J. High Resolut. Chromatogr.*, (1981)10277.
31. K. Jino, C. Fujimoto, and Y. Hirata, *Appl. Spectosc.*, **36,1**(1982)67.
32. R. S. Brown and L. T. Taylor, Anal. Chem., **55**(1983)1492.
33. C. C. Johnson and L. T. Taylor, *Anal. Chem.*, **56**(1984)2642.
34. C. M . Conroy and P. R. Griffiths, *Anal. Chem.*, **56**(198402636.
35. M. Sabo, J. Gross, J. Wang and I. E. Rosenberg, *Anal. Chem.*, **57**(1985)1822.
36. C. C. Johnson, J. W. Hellgeth and L. T. Taylor, *Anal. Chem.*,**57**(1985)610.
37. E. Bayer, K. Albert, M. Nieder, E. Grom and T. Keller, *J. Chromatogr.,* **186**(1979)497.
38. V. Sweedler, Science, 270(1995)1967.
39. D. R. Jones, II. C. Tung and S. E. Manchen, *Anal. Chem.*, **48**(1976)7
40. D. R. Jones and S. E. Manchen, *Anal. Chem.*, **48**(1976)1897.
41. J. A. Koropchak and G. N. Coleman, *Anal. Chem.*, **52**(1980)1252.
42. E. D. Katz and R. P. W. Scott, *Analyst*, (**March**)(1985)253.
43. W. Holak, *J. Liq. Chromatogr.*, **8(3)**(1985)563.
44. K. T. Suzuki, H. Sunaga and T. Yajma, *J. Chromatogr.*, **303**(1984)131.
45. J. W. Robinson and E. D. Boothe, *Spectrosc. Lett.* **17(11)**(1984)689.
46. J. D. Messman and T. C. Rains, *Anal. Chem.,* **53**(1981)1632.
47. A. I. Gusev, O. J. Vasseur, A. Proctor, A. G. Sharkey and D. M. Hercules, *Anal Chem.*, **67**(1995)4565.

CHAPTER 17

PRACTICAL DETECTOR TECHNIQUES

One of the problems that frequently faces the analyst or chromatographer is the choice of a detector for a given application. The technical capabilities, ease of operation and cost all play a part in the decision. Some of the properties of the detector that have been previously discussed will be again very briefly summarized and presented in a way that may help in a purchasing decision.

All detectors that are supplied by reputable instrument companies will include a manual giving detailed working instructions for their product. Wherever possible, the analyst should adhere to the operating conditions that are recommended by the manufacturer. Modern LC detector sensors and their associated electronic equipment can be very complex and expensive to repair. Consequently, suggested maintenance procedures should always be carried out exactly when recommended and only those servicing operations such as cleaning, consumable-part replacement, etc., that are advised by the manufacturer should be undertaken by the user. All other servicing or repair exigencies should be carried out by a qualified service engineer who is familiar with the product and who has been trained to service the particular instrument concerned.

It is apparent from the discussions that have taken place in this book that a detector can be versatile, sensitive and relatively expensive, or it can have a restricted field of application, be relatively insensitive but also economic to purchase and maintain. There is a place for both types of detector in the analytical laboratory. For method development, a very versatile detecting system is essential and probably a range of different detectors should be available in order to identify the most pertinent for a given analysis. In contrast, however, once the analytical method has been developed, and the analysis is to be carried out routinely on a dedicated instrument in a control laboratory, then the detector should be chosen to possess only those specifications that are necessary for the analysis concerned and nothing more. As a consequence, the required analytical accuracy and precision will be realized but, in addition, the instrument will be easier to operate, less expensive to buy and, furthermore, less expensive to maintain. It is a little surprising that, for some reason, many analysts find it difficult to purchase instrumentation that only just meets the specifications that are needed. It appears that popular opinion dictates that over-specifying the instrumentation gives some insurance against unexpected future demands. This thinking is not rational and not in the best interests of technical or business efficiency; it helps to bear in mind that, *what if...* rarely ever happens.

General Detector Operation

One of the most important decisions that is left to the analyst when operating a liquid chromatograph is the choice of detector sensitivity. In some instruments the output from the sensor is monitored continuously over its entire dynamic range and so sensitivity is not an optional experimental parameter. Nevertheless, in this case, the sample size determines the concentration range over which the eluted solutes are monitored and thus an optimum sample size must be chosen. The detector should never be operated at its maximum sensitivity unless such conditions are enjoined by limited sample size or column geometry. Provided that there is adequate sample available, and the sample concentration when eluted is within the linear dynamic range of the detector, the *maximum* sample size that the column can tolerate should be used. This ensures that the detector noise is always minimal with respect to the magnitude of the signal and thus provides the best quantitative accuracy and precision. Although higher sensitivities may be readily available, there is no intrinsic advantage in using them. Conversely, there is a distinct *disadvantage* in employing high sensitivities when they are not needed as, at high sensitivities, the noise can become commensurate with the magnitude of the signal causing degeneration in the quality of the analysis. Fortunately, those detectors that exhibit the highest sensitivity are often very specific in response, *e.g.* the electron capture detector in GC or the fluorescence detector in LC. Because their response is specific, the noise level is always relatively low, as many of the mobile phase contaminants that would produce noise in detectors with a universal response are not sensed by the specific detector.

Detector Selection

The choice of detector is normally determined by the nature of the sample that is presented for analysis but where the detector is to be used to analyze a wide range of different materials, it is usually chosen to have a catholic response. Ideally the detector should have a high sensitivity with a wide linear dynamic range and be relatively insensitive to ambient conditions. Sometimes cost is also an important factor that influences the choice of the detector. It should be emphasized that although a given detector may be available from different vendors at very different prices, the performance of a given detector is not necessarily reflected in its cost. Although the Cadillac is, indeed, a very fine car, the Chevrolet will take you most places, safely, with reasonable comfort, and for a fraction of the cost. There are a number of instrument companies that manufacture dependable, low cost chromatography instruments, including detectors and, where cost is important, most of these products will be found to be reliable and accurate. The low cost detectors of Gow-Mac are well designed, linear and sensitive and, although possibly a little devoid of glamour, from a technical point of view, they will be found to meet most of the requirements of the majority of chromatography users. Economic considerations with respect to analytical instrumentation becomes more and more important as mandated tests become more stringent and extensive.

Detectors for Gas Chromatography

There are a large number of different GC detectors available and each has a specific area of application for which it is most suited. However, the most commonly used GC detectors are probably the flame ionization detector, the nitrogen phosphorus detector, the electron capture detector and the katherometer detector. Of these four, the most popular and, without doubt, the most useful GC detector is the flame ionization detector (FID). The FID has the widest linear dynamic range (about five orders of magnitude), a sensitivity of about 10^{-11} g/sec and is relatively immune to changes in ambient conditions. Its main disadvantage is the need for three separate flow-controlled gas supplies. Because its response is proportional to the mass entering it per unit time, it is ideal for use with open tubular columns and small diameter packed columns. A scavenger gas flow can be used to reduce the effect of conduit and sensor volume on peak dispersion. The sensor is simple in construction and the associated electronics consist of basic high-impedance linear amplifiers. The ion optics are not critical and so it can be manufactured very inexpensively.

The next most popular detector is probably the nitrogen phosphorus detector (NPD) which is a modification of the FID. It is sensitive but specific in response, and in design and cost it resembles the FID in many ways. It has comparable sensitivity to the FID but a smaller linear dynamic range. It also suffers from the same disadvantage as the FID in that it requires three gas supplies. Unfortunately, as discussed in an earlier chapter, the alkali bead that provides its specific response has a limited life and must be replaced regularly. The ancillary electronics are also similar in form to those of the FID. As many compounds that are included in legislative regulations contain nitrogen or phosphorus, the NPD detector is frequently used for forensic purposes.

The electron capture detector (ECD) is another specific detector with an exceedingly high sensitivity and is thus in great demand for trace analysis. This detector has the advantage that it can function with a single gas supply but the older models utilize a radioactive source. More recently the device has been shown to work well using a helium plasma. However, this type of ECD has only relatively recently been made available commercially and although, so far, it has proved very successful, its long term stability and performance remain to be established. Nevertheless, the helium plasma ECD does appear to have an improved performance over that employing a radioactive source. The detector is not strongly affected by changes in ambient conditions, has a very high sensitivity and a linear dynamic range that, depending on the electrode geometry, can extend over three orders of magnitude.

The katherometer detector is used extensively in gas analysis and for this purpose is the most popular detector. It has only moderate sensitivity, which, however, is quite adequate for most gas analysis applications. It has a linear dynamic range that covers about three orders of magnitude and has the distinct advantage of requiring only one flow-controlled gas supply. Unfortunately, it is very sensitive to changes in flow rate and ambient temperature and thus must be well thermostatted. It is compact, rugged, requires very simple ancillary electronic equipment and thus can be relatively inexpensive. Besides its popular use for gas analysis it is also popular as a teaching instrument.

Specific Problems Associated with GC Detectors

Gas chromatography detectors are relatively easy devices to operate and maintain. Some have special foibles. The FID flame is ignited by a small heater filament situated close to the flame. Difficulty in flame ignition is frequently experienced and it is almost always because the flow rates of the three gases have not been properly adjusted to the values recommended for ignition by the manufacturer. The electrodes of the FID will become corroded with time. If there has been extensive use of silicone stationary phases, or silyl derivatives have been regularly chromatographed, silica deposits will form around the electrodes and jet. In due course these contaminants will produce excessive noise. Under such circumstances the sensor will need cleaning and possibly the jet and electrodes replaced. Inevitably there will be periods when excessive noise appears on the detector output, usually as long term noise or drift. If the sensor electrodes are clean, long term noise and drift is often a result of leaks in the column or detector gas lines. In the author's experience 95% of all cases of excessive long term noise or drift have been caused by either leaks somewhere in the pneumatic system or column bleed. All unions should be examined for leaks using soap film tests. Drift and noise from column bleed can be identified as it increases with column temperature. Unless guard columns are used, column bleed will eventually occur from accumulated sample residues and cause noise and drift. Noise and drift can also happen if the stationary phase is heated above its recommended operating temperature. The conduit between the column and the detector, and the detector oven itself, must always be operated at a temperature 20°C above the maximum temperature the column is set to reach; otherwise sample condensation will distort peaks and also produce noise and drift. The temperature differential of 20°C between column and sensor oven temperatures should always be maintained irrespective of the GC detector being used.

The NPD detector has maintenance requirements very similar to those of the FID. After extended use, the response of the NPD will seriously deteriorate and this will be due to exhaustion of the alkali bead, which must be replaced. The bead replacement should preferably be carried out by the maintenance engineer and not by the analyst. The NPD detector will not suffer to the same extent from column bleed and gas line leaks because the detector has a specific response to nitrogen and phosphorus. Nevertheless, the same precautions should be taken as those advised for the FID. If used with silicone stationary phases, or for the separation of silyl derivatives, the electrodes and jet will also become covered with silica and must be cleaned. In general, problems with detector electronics are very rare indeed. Contemporary detector electronics are almost exclusively solid state and consequently, if the electronic system functions well for 28 days it is likely to continue functioning for 28 years. This is because virtually all solid state failures occur in the first four weeks of operation. Detector malfunctions are almost always due to the use of incorrect operating conditions or due to sensor problems.

The electron capture detector has an extremely high sensitivity and, as a result, is very sensitive to electrode contamination. In general, the electron capture sensor will require cleaning twice as frequently as other types of detectors. If the sensor contains a radioactive source, which most that are active in the field at this time do, the sensor should be cleaned by a service engineer. In fact, the service engineer will usually replace the sensor, and subsequently clean the contaminated sensor under the appropriate conditions at the service station. Modern radioactive sources are well protected and, provided the operating conditions recommended by the manufacturer are adhered to, there is no danger associated with their use. Due to the extremely high sensitivity of the ECD, column contamination can be a particular problem and can produce baseline instability, significant noise and drift. Stringent chromatographic hygiene must be practiced when using the ECD. Sample containers and syringes and sample loops must be carefully washed between samples and a guard column is usually essential if routine samples are being analyzed on a continuous basis. Sample carryover can very easily occur producing spurious peaks or impaired quantitative accuracy. The usual temperature differential between column and sensor oven must also be maintained.

The katherometer detector is a rugged, very forgiving detector that can be quite seriously abused in operation and still provide accurate quantitative results. This, of course, is a direct result of its relatively low sensitivity. The main problem with the katherometer is instability due to poor control of operating and ambient conditions. The katherometer is very sensitive to temperature changes and changes in columns flow rate (the reasons have already been discussed). Consequently, the flow controller and the sensor oven controls must be well maintained to ensure a precise column flow rate and a precise sensor oven temperature. The katherometer is most commonly used for gas analysis and so sensor contamination is not very common and column contamination is even less likely. The linear dynamic range of the katherometer is best assumed to be little more than two orders of magnitude to ensure accurate quantitative results. When the katherometer is used for gas analysis, care must be taken to avoid leaks in the gas supply system; otherwise there will be significant drift and spurious peaks may be observed.

Detectors for Liquid Chromatography

The four most commonly used LC detectors are the UV detector, the fluorescence detector, the electrical conductivity detector and the refractive index detector. Despite there being a wide range of other detectors to choose from, these detectors appear to cover the needs of 95% of all LC applications. This is because the major use of LC as an analytical technique occurs in research *service* laboratories and industrial *control* laboratories where analytical methods have been deliberately developed to utilize the more straight forward and well established detectors that are easy and economic to operate. LC detectors are more compact than their GC counterparts and need much less ancillary support. Most operate solely on the mobile phase and need no other fluid supplies for their effective use. All LC detectors are 3-5 orders of magnitude less sensitive than their GC counterparts and thus sensor contamination is not so severe, and generally less maintenance is required.

The workhorse detector in LC is undoubtedly the UV detector, usually in the form of the diode array detector. In fact the diode array detector is more commonly used as a multi-wavelength detector than a device for producing UV spectra of column eluents. The choice of wavelength makes the response of the UV absorption detector close to that to that of a universal detector, because a specific wavelength(s) can usually be selected that will sense all the solutes of interest in a given mixture. The low cost alternative is the fixed wavelength UV detector. This detector sacrifices some versatility because the light source being largely monochromatic and, as a result, it costs less. On the other hand, the fixed wavelength detector is fundamentally more sensitive than the multi-wavelength alternative by nearly an order of magnitude. Relative to other LC detectors, the UV absorption detector has a good sensitivity and a linear dynamic range that extends over three orders of magnitude.

The fluorescence detector is the most commonly used high sensitivity detector. It can be simple to operate, sensitive and reliable, with a linear dynamic range of over two orders of magnitude. The scope of the fluorescence detector is extended by choosing an appropriate excitation wavelength or by making fluorescent derivatives. Thus the fluorescence detector that contains a monochromator to select the wavelength of the excitation light is to be recommended and is probably the most cost effective. The fluorescence detector is often used for trace analysis in environmental monitoring, pharmaceutical control analysis and forensic chemistry. The sensor cell is not easily contaminated and if a background fluorescence from contamination is observed it can usually be removed with appropriate solvents. Column contamination with fluorescent materials is far more likely and, due to the high sensitivity of the detector, can cause serious problems and produce a high background signal and drift. The use of guard columns can reduce this problem and regular column cleaning with appropriate solvents will also help reduce the buildup of fluorescent materials.

The electrical conductivity detector is exclusively used for ion detection in ion chromatography. The device is sensitive and has a linear dynamic range of over three orders of magnitude. Both the sensor and the associated electronics are very simple and the device should be very inexpensive. This detector is particularly rugged and surprisingly free from operating problems. Spurious peaks, high back-ground signal, noise and drift are almost always due to poor operating conditions, column contamination or the aging of suppressor columns.

The refractive index detector has poor sensitivity, very limited linear dynamic range and can not be used with flow programming or gradient elution. The sensor is very sensitive to temperature changes and, thus, must be carefully thermostatted. Despite these disadvantages, it is still commonly used as a detector for those substance that do not absorb UV light, are not conducting and do not fluoresce. It is widely used in the analysis of sugars, polysaccharides, aliphatic alcohols and other similar types of material that other detectors can not sense. Basically, the refractive index detector is a universal detector, as it will detect all substances that have a refractive index that differs from that of the mobile phase. It is this property that, in spite of its disadvantages, has allowed it to survive for over 50 years.

Specific Problems Associated with LC Detectors

Many sensors that measure light absorption, fluorescence or refractive index are provided with both a sample cell and a reference cell. The two cells must provide similar signals to the two photocells and amplifier and thus both should be filled with mobile phase. If the column is changed, and an alternative mobile phase is used, then the solvent in the reference cell must also be changed. Sometimes when using non-absorbing mobile phases such as n-heptane in a UV detector, the reference cell is operated empty and the output from the two cells balanced electrically. This somewhat defeats the objective of the reference cell but the arrangement can function satisfactorily. If balance is difficult to obtain, even when both cells are filled with the same solvent mixture, then it is likely that the cell windows are contaminated and need to be cleaned. Cell windows can usually be cleaned by passing acetone or chloroform through the cell. If solvents fail to clean the cell, then a 10% aqueous solution of nitric acid can be used. However, after cleaning, the acid must be rapidly washed from the cell with water to prevent it from corroding the stainless steel fittings and other materials from which the cell is fabricated. If the cell still remains contaminated even after cleaning with acid, it must be disassembled and the windows cleaned as directed by the manufacturer. It is extremely important to follow the manufacturers instructions closely when disassembling and assembling the cell; otherwise the cell may be seriously damaged. In particular, when reassembling the cell after cleaning, care should be taken not to tighten the cell locking screws too tightly; otherwise, the cell windows or lenses may be fractured. Nevertheless, the cells must be tightened sufficiently to ensure there are no leaks.

Bubbles can be a particular problem in LC flowthrough cells. The symptoms are violent off-scale signals that may be either positive or negative. Bubbles can form from dissolved air in the mobile phase particularly when aqueous mixtures of methanol or acetonitrile are used. It is good practice to continuously sparge all solvents that are to be used as mobile phase components with helium. To aid in the elimination of bubbles, should they form, the column should be connected to the lower cell connection of the sensor and exit through the upper tube (that is, assuming the inlet and outlet tubes are oriented in this way). The upper exit tube should be bent over in the form of a U so that solvent leaving the cell applies a negative pressure to the cell equivalent to the height of liquid in the side limb of the U. Consequently, any bubbles of gas entering the cell at the base will rise through the cell, and be aspirated out of the cell by the negative pressure applied by the U exit tube. If bubbles still remain in the cell, then the mobile phase flow rate should be increased and slight back pressure applied to the exit of the cell by restricting the flow. This back pressure reduces the size of the bubble and often allows it to pass out of the cell. If the bubble is particularly recalcitrant, then the union between the column and the detector should be loosened and a flow of polar solvent such as acetone (or some other solvent miscible with the mobile phase) forced back through the cell causing the bubble to be removed *via* the loosened union. It should be emphasized that when back pressure is being applied to the cell to remove bubbles, the pressure should not be allowed to rise sufficiently to cause leaks between the windows and the cell gaskets.

Spurious Peaks

During the development of a chromatogram, unexpected and foreign peaks are sometimes observed. There can be a number of reasons for these peaks.

Elution of sample solvent. In some cases a sample has been dissolved in a solvent that does not have the same composition as that of the mobile phase. Under such circumstances, the solvent components of the sample may be eluted as peaks in the same chromatogram or as broad peaks in some subsequent chromatogram. It must be remembered that any substance that is not a component of the mobile phase will appear as a sample solute, even though it may be merely a foreign component in the sample solvent. The only satisfactory solution to this problem is to dissolve the sample in an aliquot of the mobile phase. In most cases spurious peaks are not caused by detector malfunction except for one instance which will be discussed below.

Contaminated syringe. If the syringe used to handle the sample, or the sample valve, is not adequately cleaned between samples, then small contaminating peaks will be eluted interfering with the chromatogram of the new sample. The syringe and sample valve should be well washed with mobile phase between samples and the syringe and valve washed at least once with every new sample before it is placed on the column. Again this problem is not a detector problem.

Air dissolved in the sample. Air dissolved in the sample will usually be eluted close to the dead volume and will modify the refractive index of the mobile phase. The change in refractive index will produce a detector disturbance when the refractive index detector or any light absorption detector is being used. The problem can be easily eliminated by degassing the sample with helium, provided the sample does not contain any volatile components.

Detector displacement effects. Detector displacement effects are about the only way a detector can produce spurious peaks. If the materials of construction of the sensor cell, or any part of the sensor that comes in contact with the mobile phase, can adsorb materials contained in the mobile phase, then these can be a source of spurious peaks. If solutes are eluted from a sample that are adsorbed more strongly on the sensor parts than the solvent components, then the mobile phase components will be displaced into the mobile phase and appear as a spurious peak. This problem is only solved by selecting materials that are non-adsorbent for fabricating the sensor cell.

Baseline Instability

Detector noise in its various forms has already been discussed but, in practice, measures can often be taken to reduce some of them.

Short term noise. Short term noise appears as "grass" on the baseline and it is this type of noise that can often be eliminated. If a potentiometric recorder is being used, and the short term noise persists at low sensitivities, then it can often be eliminated by reducing the recorder *gain* or increasing the degree of *damping*. If short term noise is only present at high sensitivities, then the source of the noise usually lies in the detector electronics. The noise can be reduced by interposing an appropriate passive or active filter between the detector output and the recorder or data acquisition system. Another source of noise arises from pump oscillation and can usually be identified as being in phase with the pump strokes. Pump noise can be reduced by incorporating a pulse dampening device between the pump and the column. The noise can also be reduced by improving the temperature control of the detector sensor.

Long term noise. Long term noise has already been defined as having a frequency similar to that of the eluted peaks and thus is far more difficult to reduce. Long term noise can increase progressively over the life time of a column as it becomes contaminated by trace materials from samples that have been analyzed. These trace materials accumulate in the stationary phase and eventually elute irregularly. A silica column can often be cleaned by eluting with a series of solvents. Six column dead volumes each of n-heptane, dichloromethane, ethyl acetate, acetone, ethanol and water are passed sequentially through the column. This procedure completely deactivates the column and in the process will wash out the contaminating materials. The column has then to be reactivated by repeating the process using the same solvents in reversed order. If this procedure does not eliminate the long term noise, then the column must be replaced. Reversed phase columns are cleaned in a similar manner but by using different solvents. The column is washed sequentially with acetonitrile, acetone, methylene dichloride and n-heptane. The same solvents are then passed through the column in reversed order to reactivate it. This procedure usually cleans reversed phase columns satisfactorily but again, if it does not, then the column will need to be replaced.

Drift. There are two main sources of drift, both due to non equilibrium conditions in the column and the detector. If the detector, column and mobile phase are not in thermal equilibrium, then serious drift will occur. This can be eliminated by careful temperature control of column and detector. Another and more common source of drift arises when the stationary phase and mobile phase have not been given sufficient time to come into equilibrium. This type of drift often occurs when changing the mobile phase composition and mobile phase should be pumped through the chromatographic system until a stable baseline is achieved. Trace impurities in the mobile phase can cause prolonged drift and long-term noise and so very pure solvents must be used for the mobile phase. Distilled in glass solvents may not necessarily be sufficiently pure to ensure drift-free detector operation.

Differential and Integral Detector Operation

Chromatography detectors are designed to provide an output that is linearly related to the concentration of solute in the mobile phase passing through it. Consequently, the curve produced by the recorder or computer print-out is a true representation of the Gaussian profile of the eluted peak. In the vast majority of LC separations this type of output is desirable but there are a number of applications where an alternative form of presentation can be useful. All detectors that employ reference cells to compensate for changes in the mobile phase can be connected to a column system in such a way as to provide alternative outputs. Differential and integral outputs can be obtained by very simple modifications to a standard chromatographic system. It should be emphasized that with computer data acquisition and handling, differential and integral responses can be obtained very easily by software only. However, when a computer is not available

453

and where it is desirable to use alternative data presentation to provide improved chromatographic performance, such procedures may be of interest.

The Differential Detector

The differential form of the Gaussian function has already been discussed and is sigmoid in shape with a positive maximum at the first point of inflexion of the Gaussian curve and a minimum at the second point of inflexion. If the peaks are completely resolved in the normal chromatogram, then they can be clearly and unambiguously identifiable in their differential form. If, however, the peaks are not completely resolved, then the differential curve of the unresolved peaks are confused and extremely difficult to interpret and for this reason the differential form of the Gaussian function is rarely used. Nonetheless, if the elution profile of the solutes are not Gaussian in form, the differential detector can be extremely useful.

The concentration of a solute (X_V) eluted after v volumes of mobile phase has passed through the column will be some function of (v), (f(v)),

thus, $\quad X_V = f(v)$

and $\quad \dfrac{dX_v}{dv} = \dfrac{d(f_{(v)})}{dv} = f'(v)$

For finite values of (dv), i.e. (Δv), where (Δv) is small,

$$\dfrac{\Delta X_v}{\Delta v} = f'(v)$$

and

$$\dfrac{\Delta X_{v+\Delta v} - \Delta X_v}{\Delta v} = f'(v)$$

or $\quad \Delta X_{v+\Delta v} - \Delta X_v = \Delta v f'(v) \quad\quad (1)$

Further, if (Δv) is a constant, then $(\Delta X_{v+\Delta v} - \Delta X_v)$ will be directly proportional to $f'(v)$, the differential function of the eluted peak.

The value of $(\Delta X_{v+\Delta v} - \Delta X_v)$ can easily be measured in practice when using a katherometer of refractive index detector as both the GC detector and the LC detector normally have reference cells. The column is connected to the detector in the manner shown in figure 1.

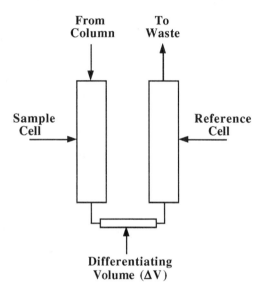

Figure 1 Cells Connected to Provide a Differential Response

This technique was first employed in GC by Boeke [1] and was later introduced to LC by Essigmann and Catsimpoolas [2]. The column eluent passes directly to the sample cell and thence to the reference cell by way of a short length of tubing, the volume of which is (Δv). As the signal from the reference cell acts electrically in opposition to the signal from the sample cell, the output from the detector is $(\Delta X_{v+\Delta v} - \Delta X_v)$, as given in equation (1). Consequently, the output will provide the differential of the function that describes the concentration of the solute in the mobile phase. An application of the differential mode of detection is shown in the chromatogram at the top of figure 2.

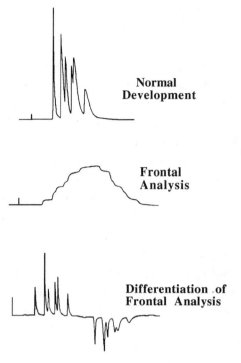

Figure 2 The Separation of a Mixture of Tetracyclines Monitored by Differential Detection

The chromatogram shows the separation of a mixture of different tetracyclines and it is seen that by normal monitoring there is serious peak tailing. This tailing results from the poor desorption characteristics of the solute from the silica gel stationary phase. The quantitative results from such a chromatogram would not have a high precision or accuracy. It is noted that the front of each peak in the normal chromatogram is fairly sharp and, if the tails could be made equally sharp, the quantitative accuracy would be improved. The center chromatogram shows the mixture separated by frontal analysis still using the normal mode of detector response. The frontal analysis chromatogram first shows a series of ascending steps as each solute is eluted and then a series of descending steps as the frontal sample leaves the column. Each step corresponds to the elution of a component superimposed on the top of those previously eluted. It should be noted that the rising steps are fairly sharp, whereas the descending steps are rather diffuse. In the lower chromatogram the effect of differentiating the frontal analysis curves is shown. It is seen that greatly improved separation is observed from the differentiation of the ascending steps and any quantitative results obtained would also be improved. In contrast differentiation of the descending steps produces a chromatogram that is significantly worse than that obtained from normal elution. It is seen that under certain conditions the preferable kinetics of the chromatographic process can be advantageously selected resulting in improved analytical results. Theoretically, there should be some examples where the desorption rate of the solute from the stationary phase is faster than the adsorption rates, in which case differentiation of the descending steps of the frontal analysis curve would provide more accurate results. In the author's experience these conditions have never been observed either in GC or LC. Another example of the advantages of the differential mode of detector operation is shown in figure 3.

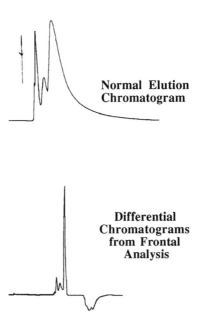

Figure 3 The Separation of a Fermentation Extract Monitored by Differential Detection

In the upper chromatogram is shown the separation of an extract of a fermentation broth. It is seen that a very poor separation is obtained due to the tailing peaks. In the lower chromatogram is shown the result of differentiating a frontal analysis of the same sample. It is seen that the differentiation of the ascending steps of the frontal analysis provides a much improved separation from which reasonable quantitative analysis could be obtained. It is interesting to note that virtually all the resolution is lost in the differential of the descending steps of the frontal analysis curve. Again this curve illustrates the poor desorption kinetics of the chromatographic distribution system.

The use of the differential mode of detector operation can be extremely useful in cases where the normal chromatographic development gives very poor separations due to poor distribution kinetics between the two phases. However, the technique does require significantly more sample for frontal analysis than for normal elution development so that sufficient sample must be available. Furthermore, the response of the detector operated in the differential mode is nearly two orders of magnitude less than that when used in the normal mode and so adequate detector sensitivity must be available.

Integral Detection

Any detector can be arranged to give an integral response if so desired and, for this purpose, no special use of a reference cell is required. In practice, the need for a detector having an integral response is very rare as electronic or digital integrators can carry out the same function more efficiently and with greater precision. Nevertheless, for those who, for some reason or another, can not utilize the more conventional methods of integration, the alternative use of a detector as an integrator will be described. If (n) solutes are completely eluted from a column in volume (v), then,

$$\int X_1 dv + \int X_2 dv + \int X_3 dv + \cdots \int X_n dv = M_1 + M_2 + M_3 + \cdots M_n$$

where (X_1), (X_2)...(X_n) are the concentrations of solutes (1), (2),...(n) after a passage of mobile phase volume (v) and (M_1), (M_2),...(M_n) are the masses of the individual solutes in the mixture. Thus on the elution of the solutes (1), (2),...(n) the resulting integral curve will be in the form of a series of steps, the height of the step for each solute being proportional to the mass of that solute. It should be pointed out that analysis by normalization of each step can only be carried out if the detector response to each solute is the same. Alternatively, the product of the step height and the respective response factor for each solute can be normalized to provide a quantitative analysis.

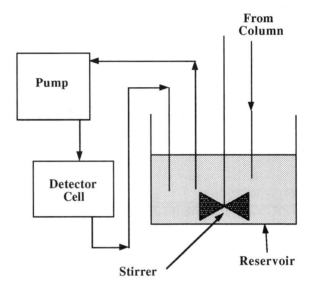

Figure 4 Apparatus to Provide Chromatographic Integration

The experimental arrangement used to provide an integral of the output from a detector is shown in figure 4. Mobile phase from a reservoir is pumped continuously through the sample cell of the detector and back again to the reservoir. The column eluent is allowed to flow continuously into the reservoir and ideally the volume of mobile phase in the reservoir should be maintained constant. The volume of mobile phase in the reservoir should, therefore, be at least 50 times the volume of mobile phase required to elute all the solutes. If all the solutes are eluted in 10 ml of mobile phase, it follows that the volume of mobile phase in the reservoir should be at least 500 ml.

The reservoir should be continuously stirred with a magnetic stirrer and the volume of mobile phase circulated through the detector and back to the reservoir should be maintained at about 1% of the total volume of mobile phase. An example of an integral chromatogram obtained by this procedure is shown in figure 5.

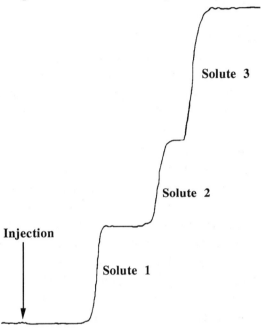

Figure 5   An Elution Curve Obtained by Integral Detection

It is seen that a typical integral chromatogram is obtained that is quite suitable for quantitative analysis. As the total mass of each solute eluted from the column is, in effect, diluted to 500 ml by the mobile phase in the reservoir, high detector sensitivities are required.

The reservoir can also act as a post-column reactor. The appropriate reagent is dissolved in the mobile phase and will react immediately with any solute eluted from the column to form a derivative that can be sensed by the detector. Using the integral detector as a post-column reactor, however, can only be successful if the derivatives are not labile and do not decompose or change before all the solutes in the mixture are eluted. For example, Fluoropa derivatives of amino acids would not maintain their fluorescent properties sufficiently well for use with this technique.

Vacancy Chromatography

The term "vacancy chromatography" was introduced by Zhukhovitski and Turkel'taub [3] and is used to describe the chromatographic technique in which the mobile phase consists of a solvent, or solvent mixture, containing a group of components which are maintained at a constant concentration. Upon injection of a "sample" the resulting chromatogram may show negative or positive peaks depending on whether the components detected were present in the sample at a lesser or greater concentration than they were in the mobile phase. Under these circumstances the solutes are continuously passing through the detector at a constant concentration which can increase or decrease as a "peak" is eluted from the column. If the mobile phase (containing a solute at concentration (X_0)) is fed continuously onto a chromatographic column and equilibrium is allowed to become established, the eluent leaving the column will also contain solute at concentration (X_0). If a sample of the same solute dissolved in the mobile phase at a concentration (X_1) is now injected onto the column where ($X_1 \neq X_0$), then this will produce a perturbation on the concentration (X_0) as it leaves the column. From the Plate Theory the equation for this perturbation will be given by

$$X_N = (X_1 - X_0) \frac{e^{-\frac{w^2}{2N}}}{\sqrt{2\pi N}} \quad (2)$$

where (X_N) is the concentration of the solute in the (N th) plate (*i.e.* the concentration sensed by the detector),
(v) is the volume flow of mobile phase measured in plate volumes,
(N) is the number of theoretical plates in the column,
and (w) is defined as $w = v - N$.

It is clear from equation (2) that if $(X_1 > X_0)$ a positive peak will be produced and if $(X_1 < X_0)$ then a negative peak will be produced.

In normal chromatographic procedure the sample is injected into the column containing mobile phase only and, thus, the eluted peaks are always positive, representing an increase in the solute concentration in the detector. However, the chromatographic process will operate in exactly the same way if the mobile phase contains a given concentration of the solutes, and the concentration of solutes in the sample is *less* than that in the mobile phase. In this case, on injection negative pulses of solute concentration will be propagated down the column resulting in negative peaks being eluted at the same retention time as the positive peak would be eluted. An example of such a chromatogram is shown in figure 6.

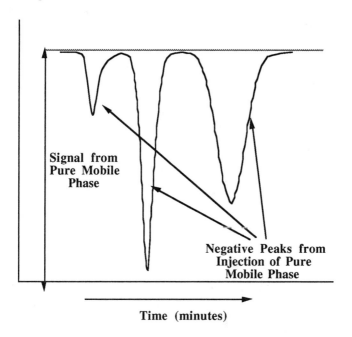

Figure 6   Negative Peaks from Vacancy Chromatography

It is seen that as the peaks are eluted, the concentration falls below that normally in the mobile phase and negative peaks are produced. This system can be useful in monitoring a mixture in process control. A sample of the mixture of substances is chosen that has the composition that is required to be maintained in the product and is dissolved in a known concentration in the mobile phase. The mobile phase is then pumped continuously through the chromatographic system until equilibrium is reached. The detector output will then be equivalent to the sum of all the components in the product at the required concentration. A sample of the product is then taken periodically and diluted in the same ratio as the mobile phase and injected onto the column. If the composition is correct, there will be no signal from the detector as there will be no concentration perturbation effective in the column. If, however, the composition differs from that of the desired product, then peaks will be produced that will represent the difference between the value in the product at the time of sampling and the standard product. An example of the type of chromatogram that may be produced is shown in figure 7.

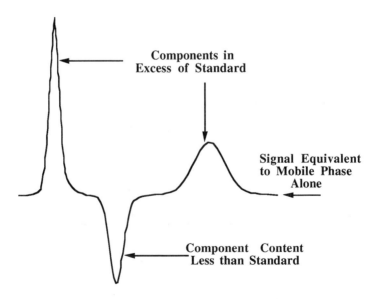

Figure 7 Monitoring the Composition of a Product by Vacancy Chromatography

It is seen that the first and third components are present in excess of the desired composition, whereas the second component is present at a concentration below that of the standard.

Scott et al. [4] applied the technique to the separation of nucleic acid bases and an example of their results is shown in figure 8. A solution of uracil, hypoxanthine, guanine and cytosine was used as the mobile phase and the upper chromatogram shows a peak for cytosine that was present 0.4 mg/ml in excess of the concentration in the mobile phase. As the charge was 600 μl the peak represents a total excess mass of 0.24 μg. No other peaks are shown, as the other solutes are present in the sample in the same concentration as the mobile phase.

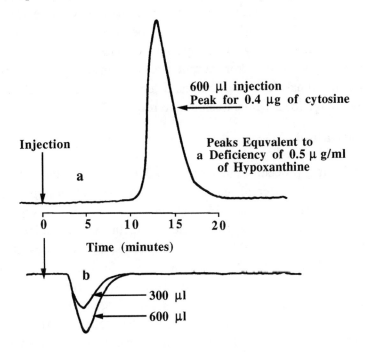

Figure 8 Chromatograms Obtained from Vacancy Development of a Mixture of Uracil, Hypoxanthine, Cytosine and Guanine

The lower chromatogram show peaks from 300 μl and 600 μl samples which contain all the solutes at the same concentration as that in the mobile phase except there is no hypoxanthine present. Hypoxanthine is present in the mobile phase at 0.5 µg/ml; thus the negative peaks represent -0.15 and -0.3 µg samples. The value of this system is that small concentration differences in hypoxanthine content can easily be determined without interference from uracil, which, under normal elution conditions, is completely unresolved.

Vacancy chromatography functions under conditions where all the solutes are in equilibrium with the two phases. Consequently, adsorption/desorption processes are far more symmetrical and peak deformation is very rare. Many substances that could not be separated under normal development conditions can be resolved and cleanly separated by vacancy chromatography, as all the highly active sites are saturated and the desorbtion isotherm becomes linear. However, vacancy chromatography makes special demands on the detector, as small changes in detector output are measured against a high background signal. It follows that the detector must have a wide linear dynamic range for accurate quantitative results.

Derivatization

Derivatization procedures are employed in both GC and LC, but usually for quite different reasons. In GC, derivatization is generally carried out to render the solutes volatile so that they can be separated by development in a gaseous mobile phase. In LC, derivatization is largely carried out to add UV chromaphores, fluorophores etc. to solute molecules so that they can be sensed by UV or fluorescent detectors. In the early days, derivatization was often carried out merely to apply gas chromatography instead of liquid chromatography to the analysis of involatile and strongly polar materials. At the time, GC was more accurate and reliable, so GC would be preferred, even when LC was, perhaps, the more appropriate technique. Then as LC became as accurate and dependable as GC, the reverse occurred. Volatile substances that were quite suitable for separation by GC were derivatized to make them detectable by LC because LC was the simpler and more popular technique. It is almost unbelievable, but the trend is now reversing once again. LC produces copious amounts of solvents, the disposal of which often involves difficult and expensive waste chemical processes. In contrast GC produces little or no problems of waste disposal and thus has a clear economic advantage. It follows that, once again, derivatization is being used to make solutes volatile and amenable to analysis by GC, whereas previously they have been separated by LC. It is clear that there are a number of reasons, both technical and economic, why sample derivatization can be important and consequently the procedures that are available warrant some discussion.

Derivatization Procedures Used in GC Analysis

Involatility can arise from two causes; the sample is highly polar, *e.g.*, aliphatic acids, carbohydrates or the higher molecular weight alcohols, or the sample can have a very high molecular weight, *e.g.*, waxes, synthetic polymers, biopolymers, etc. Substances that are involatile as a result of strong polarity can often be successfully derivatized and made relatively volatile. Conversely, substances that are involatile as a result of their high molecular weight can rarely be rendered volatile merely by derivatization. It follows that whereas it is easy to produce a volatile derivative of a low molecular weight aliphatic acid, it is virtually impossible to make a volatile derivative of a biopolymer.

GC derivatizing agents can be classified on the basis of the chemical nature of the materials they render volatile. For example, different reagents are used to make volatile derivatives of acids, alcohols, amino acids, etc., In fact, the reagents will be classified on the basis of the chemical properties of the solutes rather than by the chemical nature of the reagents themselves. The various reagents will be described under this classification.

Esterification. Esterification is used largely to render organic acids volatile. There are a number of derivatization processes available; one of the most common and useful is esterification with an alcohol in the presence of hydrochloric acid. A few milligrams of the material is heated with 100-200 µl of the alcohol containing 3 M HCl for about

30 minutes at 70°C. The alcohol is removed by evaporation in a stream of nitrogen (assuming the derivative is considerable less volatile than the alcohol). Boron trifluoride and boron trichloride are also useful esterifying reagents. The reagent is commercially available as 14%s solution in methanol specifically for making methyl esters. BF_3 catalyzed reactions are fairly rapid and after heating in a boiling water bath are complete in a few minutes. A few milligrams of the sample are added to 1 ml of the reagent and heated in boiling water for about 2 minutes and the esters extracted into *n*-heptane. One of the best reagents for producing methyl esters is diazomethane, but, unfortunately, this reagent is extremely carcinogenic and must be *handled with great care*. *N*-methyl-*N*-nitro-*N*-nitrosoguanidine is often used as the starting material for generating diazomethane, but it too, is exceedingly carcinogenic and the De Boer and Backer reagent (Diazald) (*N*-methyl-*N*-nitroso-*p*-toluenesulphonamide) [5] is preferable, Diazomethane is a yellow gas, sometimes employed as a solution in diethyl ether in the presence of methanol. It is often prepared, as required, using a simple generating apparatus described by Schlenk and Gellerman [6]. A few milligrams of the acid are dissolved in 2 ml of ether containing 10% of methanol and the gas generated from 2 mmol of Diazald is bubbled into the solution in a stream of nitrogen. When the yellow color persists the ether/methanol is evaporated off in a stream of nitrogen providing a residue of the ester. The trimethyl silyl reagents, such as trimethylsilyl chloride, readily produce trimethyl silyl esters although there are a number of other silicone compounds that can be used to produce silyl esters. A few milligrams of the acid are dissolved in 600 μl of pyridine and 200 μl of di(chloromethyl)tetramethyldisilazane are added together with 100 μl of chloromethyldimethylsilyl chloride. The mixture is then left for about 30 minutes at room temperature to react. Another reagent that works well and is frequently used for esterification using a similar procedure is hexamethyldisilazane.

<u>Acylation.</u> Acylation can greatly reduce the polarity of amino, hydroxy, and thiol groups, which can significantly improve their chromatographic behavior reducing tailing in GC and streaking in

TLC. There are two main types of reagent for acetylation and they are the acid anhydrides and the acid chlorides. Between 1 and 5 milligrams of the sample is dissolved in chloroform (about 5 ml) and warmed with 0.5 ml of acetic anhydride and 1 ml of acetic acid for 2–16 hours at 50°C. Excess of reagent is removed by evaporation in vacuum (assuming the product is relatively involatile) and taken up in chloroform for subsequent GC analysis. Sodium acetate is often used as a basic catalyst for acetylation, particularly for carbohydrates extracted from urine. In the treatment of urine, the dried residue is first oximated to derivatize the carbonyl groups and then the hydroxyl groups are acetylated with acetic anhydride in the presence of sodium acetate at 100°C. Acetyl chloride is not so widely used for acetylation because hydrochloric acid is evolved in the reaction, so the presence of an appropriate base is considered necessary to scavenge the hydrochloric acid as it is produced. The N-acetyl methyl ester derivative of hydroxyanthranilic acid, which is a metabolic product of tryptophan, is often used for the GC analysis of the material. The sample is first esterified with diazomethane to give the methyl ester. The product is then evaporated to dryness in a stream of nitrogen and acetylated with a 50 % benzene/acetyl chloride mixture [7]. The reaction mixture is dried in a stream of nitrogen and taken up in methanol for GC analysis. Only a very small number of the possible reagents that can be used for rendering substances volatile for GC analysis have been mentioned. For further details the reader is directed to Blau and Halket [8].

Derivatization Procedures Used in LC Analysis

In LC, derivatization is almost solely used to improve the detection limits of a specific substances, *e.g.*, peptides, proteins, etc. Occasionally, derivatives are made to improve the selectivity of a particular stationary phase for certain materials, but the use of such reagents in this manner is not germane to the subject of this book. There are various classes of reagents that link specific chemical groups to the solute molecule, *e.g.*, UV chromophores or fluorophores, and thus would be termed fluorescence reagents, UV chromaphore reagents, etc. Each reagent class can be divided into groups similar to the GC reagents based on the chemical nature of the solutes with which they react. For example, there are fluorescent reagents that are used to specifically derivatize different solutes, *e.g.*, acids, alcohols, amino acids, etc. The LC derivatizing reagents will also be described under this classification.

<u>UV absorbing reagents</u>. Two of the most common reagents used to derivatize amino groups are benzoyl chloride and *m*-toluoyl chloride. Both reagents add a benzene ring to the molecule that contributes a strong UV absorbing chromaphore. The *p* nitrobenzoyl chloride reagent imparts an even stronger UV chromaphore than the simple aromatic ring and is often preferred for this reason. The reaction is fairly rapid and takes place quite satisfactorily at room temperature. Pyridine is usually employed either as, or together with, a solvent to scavenge the hydrochloric acid that is released. Similar reagents are the toluenesulphonyl chloride and the benzene sulphonyl chloride, which both produce sulphonamides with the amines. As an example, gentamicin, a polyfunctional amino compound, can be analyzed by reverse phase chromatography after labeling with benzene sulphonyl chloride. The derivatization is completed in about 10 minutes at 75°C. 1-Fluoro-2,4-dinitrobenzene is also commonly used for derivatizing amino compounds, for example, the amino glycosides such as neomycin, fortimicin, amikacin, tobramicin, gentamicin and sissomicin.

Carboxylic acids are a large group of naturally occurring compounds such as fatty acids, prostaglandins, bile acids and other organic acids all of which have relatively weak UV and visible absorption. Common reagents used to introduce a UV chromaphore into such compounds are phenacyl bromide and naphthacyl bromide. Benzoyl chloride, *m*-toluoyl chloride and *p*-nitrobenzoyl chloride are also used to derivatize compounds with hydroxy groups and to introduce a really strong UV chromaphore 3,5-dinitrobenzoyl chloride can be used. The two most common reagents for derivatizing substances that contain a carbonyl group are 2,4-dinitrophenylhydrazine and *p*-nitrobenzyl-hydoxylamine. For example, these reagents are commonly used to derivatize aldehydes, ketones, ketosteroids and sugars. *p*-Nitrobenzylhydoxylamine has been used successfully to insert UV chromaphores into prostaglandins extracted from biological samples.

Fluorescent reagents. One of the most popular fluorescent reagents is 5-dimethyl aminonaphthalene-1-sulphonyl chloride (dansyl chloride, DNS-chloride or DNS-Cl). Dansyl chloride reacts with phenols and primary and secondary amines under slightly basic conditions to form a fluorescent suphonate ester or suphonamide. The quantum efficiency of dansyl derivatives is high; whereas the reagent itself does not fluoresce. Unfortunately, the hydrolysis product, dansylic acid, is strongly fluorescent and causes interference with water-soluble derivatives. The derivatives, however, are often removed by the subsequent chromatographic process. The detection limits of the dansyl derivatives are often in the low nanogram range (ca 1 x 10^{-9} g/ml) and the excitation and emission maxima can vary between 350-370 nm for excitation and 490-540 nm for emission. This reagent has been used successfully in the analysis of amino acids, alkaloids, barbiturates and pesticides.

4-Chloro-7-nitrobenz-2,1,3-oxadiazole (NBD chloride) reacts with aliphatic primary and secondary amines to form highly fluorescent derivatives. Aromatic amines, phenols and thiols yield weakly or non-fluorescent derivatives; consequently, the reagent is specific for aliphatic amines. The reaction is carried out under basic conditions and the products are extractable from aqueous mixtures by solvents such as benzene or ethyl acetate. The fluorescence can be significantly reduced by the presence of water and so the solution should be dry and the reagent can obviously not be used to form derivatives for reversed phase chromatography. Detection limits are in the fraction of a nanogram range (2–5 x 10^{-10} g/ml). The advantage of this reagent over dansyl chloride is that both the reagent and its hydrolysis products are non fluorescent. The excitation and emission wavelengths are also higher (480 nm excitation and 530 nm emission). NBD chloride derivatives have been used for the analysis of amino acids, amphetamines, alkaloids and nitrosamines.

Fluorescamine (4-phenylspiro(furan-2-(3H),1'-phthalan)3,3'-dione) is also a commonly used fluorescence reagent. It reacts almost instantly and selectively with primary amines, while the excess of the reagent is hydrolyzed to a non-fluorescent product. The reagent itself is non-fluorescent. The reaction is carried out in aqueous acetone at a pH of about 8–9 and the derivatives can be chromatographed directly. The excitation and emission wavelengths are 390 nm and 475 nm respectively. Two disadvantages of the reagent are its cost and the fact the products are less stable, cannot be stored and should be injected onto the column immediately after formation. Fluorescamine has been employed in the analysis of polyamines, catecholamines and amino acids.

A less costly alternative to fluorescamine is o-phthaldehyde (OPT), the derivatives of which are more stable and consequently can be stored overnight if necessary. It is used in a similar manner to fluorescamine the detection limits being about 0.1 ng (*ca.* 4 x 10^{-10} g/ml). OPT has been used in the analysis of dopamine, catecholamines and histamines. Other fluorescence reagents that are sometimes used include 4-bromoethyl-7-methoxycoumarin, diphenylindene, sulphonyl chloride, dansyl-hydrazine and a number of fluorescent isocyanates.

For further information on derivatizing reagents the reader is strongly recommended to refer to the *Handbook of Derivatives for Chromatography* edited by Blau and Halket [8].

Post-Column Derivatization

Post-column derivatization does not merely require the selection of the most appropriate reagent to react with the solute to render it detectable, but also involves the modification of the chromatographic system to allow the reaction to take place prior to entering the detector. This necessitates the insertion of a post column reactor between the column and the detector. Such a reactor can easily interfere with the resolution obtained from the column and consequently the reactor system must be designed with some care to minimize extra-column dispersion. The post-column reactor is required to fulfill the following functions:

1/ Provide a source of reagent and a means of mixing it efficiently with the column eluent.

2/ Ensure the reaction is complete before the derivatized product enters the detector.

3/ Minimize the dispersion that takes place in the reactor so that the integrity of the separation achieved by the column is maintained.

A diagram of a post-column reactor system is shown in figure 9.

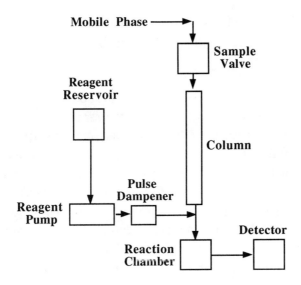

Figure 9 Chromatograph with Post-Column Reactor

The post-column reactor consists essentially of four parts, a reagent reservoir, a pump, mixing T and a reaction chamber. The pump should provide a pulse-free flow of reagent as there is little resistance downstream to the detector and most detectors are flow sensitive to a greater or lesser extent. Preferably, either a syringe pump should be employed or if a reciprocating pump is used, it should be fitted with an efficient pulse dampener. The former is preferable, but syringe pumps can be clumsy to refill.

The design of the mixing T can be critical and different kinds of mixing systems have been studied by Scholten et al. [9]. The mixing T must ensure that the column eluent is intimately mixed so that there are no localized volumes of mobile phase that do not contain sufficient reagent to allow complete derivatization. Furthermore, the mixing T must not cause any peak dispersion and impair the separation achieved by the column. The reaction chamber provides a hold-up volume to give time for the slower derivatizing reactions to proceed to completion. This volume must also be carefully designed to reduce any peak dispersion to a minimum. Low dispersion tubing such as serpentine tubing can be used, or if a relatively large reaction time is necessary then a short length of column packed with glass beads might be suitable.

The reagents used in post-column derivatization will be specific for the solute of interest and will often be the same, or similar to those used in pre-column derivatization. Reagents that produce derivatives that absorb in the visible light range are popular, as relatively simple detectors can be employed. In general, derivatization, particularly post-column derivatization, is only used as a last resort to provide sensitivity or selectivity, which is difficult or impossible to obtain by any other means. Post-column reactors complicate an already complex instrument, render it more difficult to operate and make it more expensive. In addition, however well the post-column reactor is designed it will inevitably impair, to some extent, the separation achieved by the column.

References

1. J. Boeke, *Gas Chromatography 1960* (ed. R. P. W. Scott), Butterworths, London (1960)88.
2. J. M. Essigmann and N. Catsimpoolas, *J. Chromatogr.*, **103**(1975)7.
3. A. A. Zhukhovitski and N. M. Turkel'taub, *Dokl. Acad. Nauk. USSR*, **143**(1961)646.

4. R. P. W. Scott, C. G. Scott and P. Kucera, *Anal. Chem.*, **44**(1972)100.
5. T. J. de Boer and H. J. Backer, *Recl. Trav. Pays. Bas.* **73**(1954)229.
6. H. Schklenk and J. L. Gellerman, *Anal. Chem.*, **32**(1991)8.
7. D. P. Rose and P. A. Toseland, *Clin. Chim. Acta.*, **17**(1967)235.
8. *Handbook of Derivatives for Chromatography* (Ed. Karl Blau and John Halket), John Wiley and Sons, New York (1993).
9. A. M. M. T. Scholten, U. A. Brinkman and R. W. Frei, *J. Chromatogr.*, **218**(1981)3.

CHAPTER 18

QUANTITATIVE ANALYSIS

Accurate quantitative analysis can only be achieved if three provisions are met. Firstly, the components of the mixture must be adequately separated by the chromatographic system. Secondly, the detector must respond in a constant and predictable manner to changes in the concentration of each solute in the mobile phase as it passes through the sensor. Thirdly, the output from the detector must be either manually or electronically processed in the correct manner to provide an accurate analysis.

The first *proviso* stipulates that the components of the mixture must be separated sufficiently to allow the unique area or height of each peak to be accurately measured. It follows that the degree of resolution that is necessary to achieve adequate quantitative accuracy will need to be considered.

Chromatographic Resolution

The resolution of a pair of closely eluted solutes is usually defined as the distance between the peaks, measured in units of the average standard deviation of the two peaks. However, it is necessary to decide the degree of separation that constitutes adequate resolution for accurate quantitative analysis. In figure 1, five pairs of peaks are shown, separated by values ranging from 2σ to 6σ, the area of the smaller peak being half that of the larger peak. It is clear, from figure 1, that a separation of 6σ would be ideal for accurate quantitative results. Unfortunately, such resolution will often demand very high efficiencies from the column system, which may also entail very long analysis times.

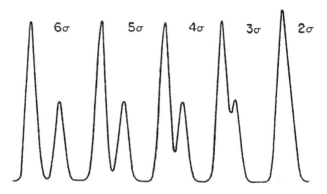

Figure 1 Peaks Showing Different Degrees of Resolution

In addition, even if manual measurements are employed, a separation of 6σ is far more than *necessary* for accurate quantitative analysis. In fact, accurate quantitative results can be obtained with a separation of only 4σ and it will be shown later that if peak heights are employed for quantitative measurement, then even less resolution may be tolerated. Duplicate measurements of peak area or peak height on peaks separated by 4σ should not differ by more than 2%. If the chromatographic data is acquired and processed by a computer, then with modern software, even a separation of 4σ could be more than necessary. Poor resolution can often be accommodated by deconvoluting the peak and measuring the peak heights or areas of the deconvoluted peaks.

Peak Deconvolution

Many data acquisition and processing systems include software that mathematically analyzes convoluted (unresolved) peaks, identifies the individual peaks that make up the composite envelope and then determines the area of the individual peaks.

The algorithms in the software must embody certain tentative assumptions in order to analyze the peak envelope, such as, the peaks are Gaussian in form or are described by some other specific distribution function and are symmetric. Furthermore, it must also assume that *all* the peaks can be described by the *same* function (*i.e.*, the efficiency of all the peaks is the same), which, as has already been discussed, is also not always true. Nevertheless, providing the composite peak is not too complex, deconvolution can be reasonably successful. If resolution is partial and the components are present in equal quantities, then the deconvolution approach can be very successful. An example of the convolution of two partially resolved peaks of equal size is shown in figure 2.

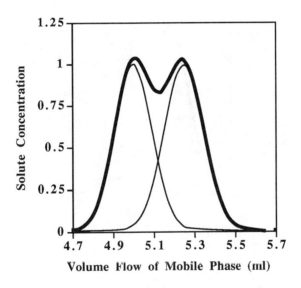

Figure 2 The Deconvolution of Two Partially Resolved Peaks Representing Solutes Present in Equal Quantities

It is seen that two Gaussian shaped peaks can be easily extracted from the composite envelope and the software could also supply values for either the heights of the deconvoluted peaks or their area. It should be noted, however, that the two peaks are clearly discernible and the deconvoluting software can easily identify the approximate positions of the peak maxima and assess the peak widths. Such information allows the software to arrive at a valid analysis quickly and with reasonable accuracy. Nevertheless, most quantitative software provided by commercial data processing suppliers would construct a perpendicular from the valley to the baseline, thus bisecting the combined peak envelope. The area of each half would then be taken as the respective area of each peak. If the peaks are of significantly different sizes the problem becomes more difficult but nevertheless, is solvable. An example of the analysis of a composite envelope for two peaks of different size is shown in figure 3.

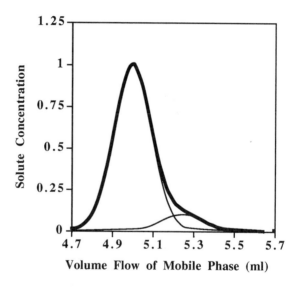

Figure 3 The Deconvolution of Two Partially Resolved Peaks Representing Solutes Present in Unequal Quantities

It is clear that the resolution will not permit peak skimming (a technique that will be discussed in detail later) with any hope of a reasonable degree of accuracy. It is also seen that the position of the peak maximum, and the peak width, of the major component is easily identifiable. The software can thus accurately determine the Gaussian function (or any other appropriate distribution expression that may be used) of the major component, and the reconstructed profile of the major component would then be subtracted from the total composite peak leaving the small peak as difference value.

However, it should be pointed out that in the examples given the solutes were at least partly resolved. Unfortunately, as the resolution becomes lower and lower and the need for an accurate deconvolution technique becomes even greater, the value of the software presently available appears to become minimal. A typical example, taken at the other extreme, where a deconvolution technique would be useless is given in figure 4.

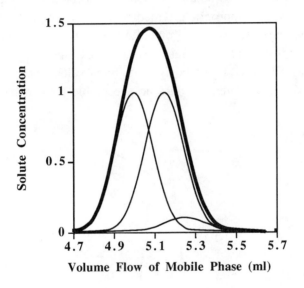

Figure 4 The Deconvolution of Three Completely Unresolved Peaks

It is seen that however sophisticated the software might be, it would be virtually impossible to deconvolute the peak into the three components. The peaks shown in the diagram are only discernible because the peaks themselves were assumed and the composite envelope calculated. The envelope, however, would provide little basic information on which the computer could start the deconvolution; there is no hint of an approximate position for any peak maximum and absolutely no indication of the peak width of any of the components or even of how many peaks are present.

Deconvoluting software has limited application and must be used with considerable caution. Every effort should be made to separate all the components of a mixture by chromatography and peak deconvolution should be employed only as a last resort.

The Detector Response

The detector response index has been featured as a means of defining linearity, but it can also be used to take into account any non-linearity that is present and appropriately modify the peak height or peak area calculations and thus improve quantitative accuracy. An example of two peaks constructed from Gaussian functions using response factors of 0.95 and 1.05 are shown in figure 5. Such values were considered to be outside those which would be acceptable for a detector to be defined as linear.

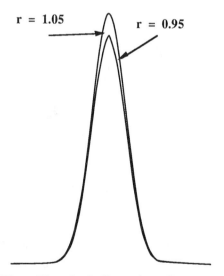

Figure 5 Two Identical Gaussian Curves Drawn with Response Indices of 0.95 and 1.05

It is seen that there is a clear difference between the peaks, and the error involved is quite considerable, as discussed in chapter 2 where detector linearity is discussed in detail. It follows that in order to obtain a value that truly reflects the solute concentration being measured, the values of detector output must not only be corrected for the linearity constant but also the numerical value of the response factor. This is not a difficult calculation, as will be shown shortly, but should be automatically incorporated into any data processing software that is not dealing with the *ideal* detector

From the theory given on detector response,

$$V = Ac^r$$

where (V) is the voltage output from the detector,
(A) is the linearity constant,
and (r) is the numerical value of the response index.

Consequently the output that would truly represent the solute concentration would be

$$c = \sqrt[r]{\frac{V}{A}} \tag{1}$$

If (r) is unity, then the expression reduces to the well-known simple relationship that is normally used in quantitative analysis,

$$c = \frac{V}{A} \tag{2}$$

Unfortunately, (r) is not usually unity and therefore equation (2) should be used in all accurate quantitative analyses.

The Quantitative Evaluation of the Chromatogram

A chromatogram can be evaluated for quantitative analysis in one of two ways, by the measurement of either peak areas or peak heights. Virtually all modern chromatographs are fitted with data acquisition hardware and processing software and consequently appropriate algorithms are built into the processing software for the accurate measurement of the chosen parameters. However, a considerable number of chromatographic instruments are still being used where manual measurement of peak areas and peak heights is still carried out. Furthermore it is advisable for the analyst to understand the advantages and disadvantages or the different methods of area and height measurement to be able to choose the most appropriate data processing method. Consequently, the two methods of chromatogram evaluation will be considered in some detail.

Peak Area Measurements

The area of a peak is the integration of the mass per unit volume (concentration) of solute eluted from the column with respect to time. If the flow rate is constant, and the detector is a concentration sensitive device, the integration will also be with respect to *volume flow* of mobile phase through the column. It follows that the total area of the peak is proportional to the total mass of solute contained in the peak. Measurement of peak area accommodates peak asymmetry and even peak tailing without compromising the simple relationship between peak area and mass. Consequently, peak area measurements give more accurate results under conditions where the chromatography is not perfect and the peak profiles are not truly Gaussian or Poisson or described by some other specific function.

Unfortunately, whether a chart recorder is used, and measurements are made directly on the chart, or computer data acquisition and processing is available, both systems are based on time as the variable and not volume of mobile phase flowing through the column. It follows that when using peak area measurements a high quality flow controller (in the case of GC) or a high quality high pressure pump (in the case of LC) must be used to ensure that all *time* measurements are *linearly related* to *volume flow of mobile phase*. It is interesting to note that the flame ionization detector (FID) (the most commonly used detector in GC) responds to *mass* of solute entering it per unit time and thus the peak area is directly related to solute mass. This highly desirable property makes the accuracy of the quantitative analysis relatively indifferent to small changes in flow rate. Unfortunately, there is no equivalent detector in LC or, at least, not one that is commercially available.

Peak areas can be measured manually in a number of ways, the simplest being the product of the peak height and the peak width at 0.6065 of the peak height (2σ). This does not give the true peak area but providing the peak is Poisson, Gaussian or close to Gaussian it will always give accurately the same proportion of the peak area. One of the older methods of measuring peak area was to take the product of the peak height and the peak width at the base (4σ). As already stated, the peak width at the base is the distance between the points of intersection of the tangents drawn to the sides of the peak with the baseline produced beneath the peak. This technique has the disadvantage that the tangents to the elution curve must also be constructed manually, which introduces another source of error into the measurement. Anothee method involves the use of a planimeter (an instrument that provides a numerical value for the area contained within a perimeter traced out by a stylus), which is a very tedious method of measuring peak area and (partly for the same reason) is also not very accurate. The most accurate manual method of measuring peak area (which, unfortunately is also a little tedious) is to cut the peak out and weigh it. A copy of the chromatogram should be taken and the peaks cut out of the copy and weighed. This last procedure can accommodate any type of peak malformation and still provide an accurate measurement of the peak area. It is particularly effective for skewed or malformed peaks where other methods of manual peak area measurement (with the exception of the planimeter) fail dismally and give very inaccurate results. The recommended method is to use the product of the peak height and the peak width at 0.6065 of the peak height but this does require adequate resolution of the components of the mixture. There are a number of techniques employed in the computer calculation of peak areas that should be mentioned. These techniques are used for partially unresolved peaks and are not deconvoluting routines but approximation procedures. In figure 6 the technique used for unresolved peaks of approximately equal heights is illustrated. In practice, the area of the first peak is obtained by integrating from the extreme of the envelope to the minimum between the peaks. The area of the second peak is obtained in a similar manner by integrating between the minimum between the peaks to the end of the envelope. Manual measurements are taken in the same manner by dropping a perpendicular from the minimum between the peaks to the baseline.

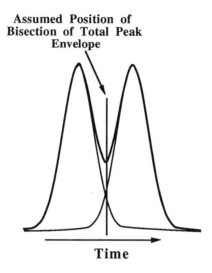

Figure 6 Area Assessment of Two Unresolved Peaks of Equal Height

The area on either side of the minimum is taken as that for each peak. It is seen that the procedure would give a fairly accurate value for each peak area. In fact the method often gives area measurements that are more accurate than employing the *skimming* procedure.

Peak *skimming* is carried out when resolution is incomplete and it is necessary to estimate the boundary between two peaks or a sloping baseline. A simple example of the comparative use of peak cutting and peak skimming is given in figure 7. The ratio of the area of the small peak to that of the larger is five. In the first instance the area of each peak is taken by integrating between the start of the envelope to the minimum between the peaks and from the minimum to the end of the envelope. The same can be achieved manually by dropping a perpendicular from the minimum to the baseline as shown and the area of each peak taken as that on either side of the bisector. It is seen that the division is fairly accurate, the first peak including a small portion of the area of the second peak and *vice versa*. The diagram also illustrates the process of peak skimming. Depending on the software available, the contour of the larger peak is projected under the smaller peak and the area of the smaller peak is taken as the difference between the area of the total envelope and the area of the first peak taken beneath the projected contour. The accuracy of this procedure depends entirely on the method used for projecting the baseline.

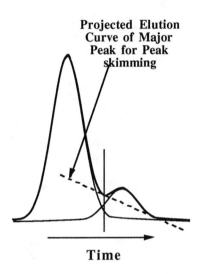

Figure 7 Measurement of the Area of Two Unresolved Peaks by Peak Cutting and Peak Skimming

The method used in the diagram utilizes a linear extrapolation under the second peak and it can be seen that this leads to extensive error. Employing a second order polynomial or an exponential extrapolation would probably provide more accurate results. The weakness of this procedure lies in the fact that each unresolved pair of peaks provides a unique contour and thus any standard function that is employed for extrapolation purposes can only be an approximation.

Another problem that often arises is the area measurement of a peak superimposed on a sloping baseline or on the tail of a large overloaded

in figure 8. It is seen that the shape of the small peak is extensively distorted and that the use of a linear function to project the curve of the major peak below that of the smaller is virtually useless.

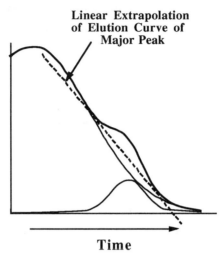

Figure 8 The Elution of a Small Peak on the Tail of a Large Peak

Again a second or third order polynomial or exponential function would certainly provide improved results. However, in this case, accurate values for the peak areas would probably only be obtained by peak deconvolution. The first peak would be reconstructed using parameters obtained from the ascending portion of the peak that is free from contamination from the smaller peak. The area of the first peak would be obtained by direct integration and that of the smaller peak by the difference between the area of the total envelope and that of the first peak. A skimming procedure can often be used when measuring the area of peaks on a sloping baseline and this is probably the most worthwhile utilization of the skimming technique. An example is depicted in figure 9. It is seen that if the slope is small, a linear function might be used to extrapolate the baseline beneath the peak. For baselines with greater slopes (*e.g.* that shown in the diagram) a second order polynomial will usually provide an accurate contour of the baseline below the peak and allow an accurate assessment of the peak area. In the case of manual measurement, the baseline beneath the peak can usually be quite accurately constructed with the aid of a set of French curves.

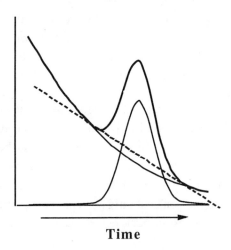

Time

Figure 9 The Measurement of Peak Areas on a Sloping Baseline

It should be pointed out that these subterfuges are almost always taken because the quality of the chromatography is inadequate. Although many and varied correction procedures can be devised to help improve the quantitative accuracy that is obtained from poorly shaped or poorly resolved peaks, this is not the best solution. It is advised that the effort should be put into improving the chromatographic conditions to achieve the necessary resolution and, in doing so, provide not only more accurate data but also more reliable results.

Peak Height Measurements

As the peak height is inversely related to the peak width, then, if peak heights are to be used for analytical purposes, all parameters that can affect the peak width must be held constant. This means that the capacity factor of the solute (k') must remain constant and, consequently, the solvent composition is held stable. The temperature must also be held steady and a highly repeatable method of sample injection must be used. If computer data acquisition and processing are employed, then a direct printout of the peak heights is obtained and, with most systems, the analysis is calculated and also presented. If the peak heights are to be measured manually, which even today is the procedure used in many chromatography analyses, then the baseline is produced beneath the peak and the height between the extended baseline and the peak maximum measured. In general, measurements should be made by estimating to the nearest 0.1 mm.

If peak heights are being measured, then a minimum resolution must be achieved and mathematical adjustments are not an alternative to chromatographic resolution. In theory there should be no part of an adjacent peak situated under the maximum of the peak height to be measured. However, this is impossible due to the nature of the Gaussian function. In practice the peaks should be separated sufficiently to allow a maximum error of 1.0% in peak height measurement.

It follows that the extent of overlap will affect the relative magnitude of the height of a small peak to a greater extent than that of a larger. If the peaks are of equal height, then the criteria of 1% error is met by a separation of about three standard deviations.

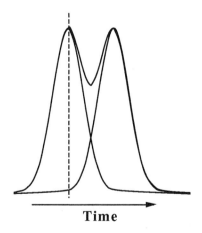

Figure 10 Two Peaks Separated by Three Standard Deviations

This situation is depicted in figure 10. It is seen that the overlap of the two peaks at the point of the peak maximum of the first peak is very slight. In fact, if the height of the first peak is taken as unity, the height of the envelope of the two peaks at the point of the maximum of the first peak is still only 1.011. It is interesting to determine the resolution required to achieve a minimum error of 1% in the peak height measurement, for a range of mixtures of differing composition. In figure 11 the minimum separation in terms of the standard deviation of the first peak is plotted against the percentage error for different binary mixtures.

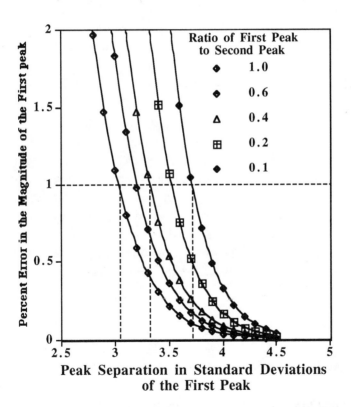

Figure 11 Resolution Required to Maintain a Maximum Error of 1% Employing Peak Height Measurements

It is seen that, as one might expect, the smaller the minor peak, the greater must be the resolution to maintain a maximum error of 1%. When the two peaks are equal in height, they need only be separated by 3.1σ; however when the smaller peak is only about 60% of the second larger peak, the necessary separation increases to about 3.35σ. At the extreme where the minor component is only about 10% of the larger peak the resolution must be about 3.7σ . Nevertheless the degree of resolution would appear to be significantly less than that required for the accurate measurement of peak areas. This is particularly true if the sizes of the two peaks are significantly different. It follows that peak heights are still frequently employed in quantitative analysis, particularly where a chromatograph is being used that does not include data acquisition and processing equipment.

Quantitative Analytical Methods for GC and LC

Essentially, there are two procedures used in GC and LC for quantitative analysis: the first employs a reference standard to which the peak areas or peak heights of the other solutes in the sample are compared; the second is a normalization procedure where the area (height) of any one peak is expressed as a percentage of the total area (heights) of all the peaks. The most common method used in both GC and LC is that employing reference standards. However, there are certain circumstances, *e.g.* where the detector employed has the same response to all the solutes in the mixture, where the normalization method may be advantageous.

Quantitative Analysis Using Reference Standards

There are two procedures for using reference standards: in the first method a weighed amount of the standard can be added directly to the sample and the area or heights of the peaks of interest compared with that of the standard. This procedure is called the *internal standard method*. In the second procedure a weighed amount of the standard can be made up in a known volume of solvent and chromatographed under exactly the same conditions as the unknown sample, but as a separate chromatographic run. In this case the peak area or height of the standard solute in the reference chromatogram is compared to the peak areas or heights of the solutes of interest in the sample chromatogram. This procedure is called the *external standard method*. The details of the two methods will now be described separately.

The Internal Standard Method

The most accurate quantitative results in GC and LC are probably obtained by the use of an internal standard. However, it can be difficult to employ, as the procedure depends upon finding an appropriate substance that will elute in a position on the chromatogram where it will not interfere or merge with any of the natural components of the mixture. Unfortunately, for a multi-component sample, this can become difficult to the extent of being virtually impossible, under which circumstances the external standard method must be used. Having identified an appropriate reference standard, the response factors for each component of interest in the mixture to be analyzed must be determined. It should be noted that this usually does not include all the components. In many instances only certain components of the mixture need to be assayed. A synthetic mixture is then made up containing known concentrations of each of the components of interest together with the standard.

Now, if there are (n) components, and the (p) component is present at concentration (c_p), the standard at a concentration (c_{st}) and the response index of the detector can be considered to be unity.

Then, $$\frac{c_p}{c_{st}} = \frac{a_p}{a_{st}} \alpha_p \quad (3)$$

where (a_p) is the area of the peak for component (p),
(a_{st}) is the area of the peak for the standard,
and (α_p) is the response factor for component (p).

Consequently, the response factor (α_p) for the component (p) can be calculated as follows:

$$\alpha_p = \frac{c_p a_{st}}{c_{st} a_p} \tag{4}$$

If peak heights are used instead of peak areas, then by simple replacement,

$$\frac{c_p}{c_{st}} = \frac{h_p}{h_{st}} \alpha_p \tag{5}$$

where (h_p) is the height of the peak for component (p),
(h_{st}) is the height of the peak for the standard,
and (α_p) is the response factor for component (p).

Thus, by rearrangement, the response factor (α_p) for the component (p) is given by

$$\alpha_p = \frac{c_p h_{st}}{c_{st} h_p} \tag{6}$$

A weighed amount of standard is now added to the sample and the sample is chromatographed. Let the concentration of the standard be (C_{st}), the concentration of any component (p) be (C_p) and the peak areas of the component (p) and that of the standard (A_p), (A_{st}) respectively; then

$$C_p = \frac{A_p C_{st}}{A_{St}} \alpha_p \tag{7}$$

If peak heights are employed instead of peak areas,

$$C_p = \frac{H_p C_{st}}{H_{st}} \alpha_p \tag{8}$$

where (H_p) is the peak height of component (p) in the sample,
and (H_{st}) is the peak height of the standard in the sample.

Now if the response index (r) is not unity, which will be the case for virtually all practical detectors, then the value of (r) should be taken into account in the expressions used to obtain quantitative results. Consider firstly the computer integration of the area. The area is calculated by summing all the signals from the detector over the period of the peak. Now, if $(r \neq 1)$ then, from equation (2), the area that will incorporate the non-linearity will be

$$a_{(r)} = \sum \sqrt[r]{V}$$

where $(a_{(r)})$ is the corrected area of the peak for a non-linear detector with a response index of (r).

It should be noted that the areas, as supplied by normal data acquisition and processing software, will not produce this corrected area without program modification.

Thus, equation (3) must be modified to

$$\frac{c_p}{c_{st}} = \frac{a_{p(r)}}{a_{st(r)}} \alpha_p \qquad (9)$$

equation (4) must be modified to

$$\alpha_p = \frac{c_p a_{st(r)}}{c_{st} a_{p(r)}} \qquad (10)$$

Now equation (5) employs peak heights and so the values for the peak height can be taken as those provided by the computer but must be taken to the (r) th root; thus,

$$\frac{c_p}{c_{st}} = \frac{\sqrt[r]{h_p}}{\sqrt[r]{h_{St}}} \alpha_p \qquad (11)$$

and equation (6) becomes

$$\alpha_p = \frac{c_p \sqrt[r]{h_{St}}}{c_{St} \sqrt[r]{h_p}} \qquad (12)$$

Again, a weighed amount of standard is now added to the sample and the sample is chromatographed. Let the concentration of the standard be (C_{st}), the concentration of any component (p) be (C_p) and the peak areas *corrected for the response index* (r) of the component (p) and that of the standard ($A_{p(r)}$), ($A_{st(r)}$) respectively; then

$$C_p = \frac{A_{p(r)} C_{st}}{A_{St(r)}} \alpha_p \qquad (13)$$

If peak heights are employed instead of peak areas,

$$C_p = \frac{\sqrt[r]{H_p} C_{st}}{\sqrt[r]{H_{St}}} \alpha_p \qquad (14)$$

Thus, the concentration of any (or all) of the components present in the mixture can be determined, providing they are adequately separated from one another. It is interesting to note that if the maximum accuracy and precision is required, and the data is to be corrected for a response index that is other than unity, either peak heights must be used or the chromatogram must be processed manually. For repeat analyses of the same type of mixture, the operating conditions can be maintained constant and, as there is no extreme change in sample composition, the response factors will usually need to be determined only once a day.

The External Standard Method

When using the external standard method, the solute chosen as the reference is chromatographed separately from the sample. However, as the results from two chromatograms, albeit run consecutively, are to be compared, the chromatographic conditions must be maintained extremely constant. In fact, to reduce the effect of slight changes in the operating conditions of the chromatograph, the sample and reference solutions are often chromatographed alternately. The data from the reference chromatograms run before *and* after the sample are then used for calculating the results of the assay.

The external standard method has a particular advantage in that the reference standard (or standards) that are chosen can be identical to the solute (or solutes) of interest in the sample. This also means that the relative *response factors* between the reference solute and those of the components of interest are no longer required to be determined. In addition, using the external standard method, the reference mixture can be made to have standard concentrations that are closely similar to the components of the sample. As a consequence, errors that might arise from any slight non-linearity of the detector response are significantly reduced.

Assuming the first instance that the response index of the detector is unity, if the (p)th solute in the mixture is at a concentration of ($c_{p(s)}$) in the sample and ($c_{p(st)}$) in the standard solution, then

$$c_{p(s)} = \frac{a_{p(s)}}{a_{p(st)}} c_{p(st)} \qquad (15)$$

where ($a_{p(s)}$) is the area of the peak for solute (p) in the sample chromatogram,
($a_{p(st)}$) is the area of the peak for solute (p) in the reference chromatogram,
and ($c_{p(st)}$) is the concentration of the standard in the reference solution.

If peak heights are employed, then

$$c_{p(s)} = \frac{h_{p(s)}}{h_{p(st)}} c_{p(st)} \qquad (16)$$

where ($h_{p(s)}$) is the area of the peak for solute (p) in the sample chromatogram,
($h_{p(st)}$) is the area of the peak for solute (p) in the reference chromatogram,
and ($c_{p(st)}$) is the concentration of the standard in the reference solution.

If the value or the response index, (r), is not unity, then the corrected area and height must be used. Thus again assuming

$$a_{(r)} = \sum \sqrt[r]{V}$$

then equation (15) becomes

$$c_{p(s)} = \frac{a_{p(s)(r)}}{a_{p(st)(r)}} c_{p(st)} \qquad (17)$$

and equation (16) becomes

$$c_{p(s)} = \frac{\sqrt[r]{h_{p(s)}}}{\sqrt[r]{h_{p(st)}}} c_{p(st)} \qquad (18)$$

Theoretically, providing the chromatographic conditions are kept constant, the reference chromatogram need only be run once a day. However, in practice, it is advisable to run the reference chromatogram at least every two hours and, as suggested above, many analysts run a reference chromatogram immediately before and after each sample. It is a common task in many control laboratories to analyze a large number of repeat samples of a very similar nature for long periods of time, often with automatic sampling. Under such circumstances, reference samples and assay samples are often run alternately throughout the whole batch of analyses. Not only will this give more accurate results, but it will also alert the operator to any change in operating conditions, which can then be immediately rectified.

The Normalization Method

The normalization method is the easiest and most straightforward to use and requires no reference standards or calibration solutions to be prepared. Unfortunately, to be applicable, the detector must have the same response to all the components of the sample. Thus, it will depend on the detector that is employed and the sample being analyzed. Unhappily, it is the least likely method to be appropriate for most LC analyses. In GC the response of the flame ionization detector (FID) depends largely on the carbon content of the solute. Thus the technique can be used in GC when employing the FID for compounds of similar types such as a mixture of high molecular weight paraffins. An exceptional example in LC, where the normalization procedure is frequently used, is in the analysis of polymers by exclusion chromatography using the *refractive index detector*. The refractive index of a specific polymer is a constant for all polymers of that type having more than 6 monomer units. Under these conditions normalization is the obvious quantitative method to use. The method of calculation is also very simple but again will depend on whether the value of (r) is assumed to be unity or not. Assuming, firstly, that (r) is unity, the percentage x(p)% of any specific polymer (p) in a given polymer mixture can be expressed by

$$x_p = \frac{a_p}{a_1 + a_2 + a_3 + \cdots + a_n} 100 \qquad (19)$$

where (a_p) is the peak area of polymer (p)

or
$$x_p = \frac{a_p}{\sum_{p=1}^{p=n} a_p} 100 \qquad (20)$$

If peak heights are used, the percentage x(p)% of any specific polymer (p) in a given polymer mixture can be expressed by similar equations,

$$x_p = \frac{h_p}{h_1 + h_2 + h_3 + \cdots + h_n} 100 \qquad (21)$$

where (h_p) is the peak area of polymer (p)

or
$$x_p = \frac{h_p}{\sum_{p=1}^{p=n} h_p} 100 \qquad (22)$$

In most computer data acquisition systems there is an iterative program that allows the user to identify the reference and sample chromatograms and the pertinent peaks for processing. The software will then carry out the necessary calculations and provide the output in the form of a printed analytical report.

However, assuming that $(r \neq 1)$ then equation (19) becomes,

$$x_p = \frac{a_{p(r)}}{a_{1(r)} + a_{2(r)} + a_{3(r)} + \cdots + a_{n(r)}} 100 \qquad (23)$$

equation (20) becomes,

$$x_p = \frac{a_{p(r)}}{\sum_{p=1}^{p=n} a_{p(r)}} 100 \qquad (24)$$

equation (21) becomes,

$$x_p = \frac{\sqrt[r]{h_p}}{\sqrt[r]{h_1} + \sqrt[r]{h_2} + \sqrt[r]{h_3} + \cdots + \sqrt[r]{h_n}} 100 \qquad (25)$$

and equation (22) becomes,

$$x_p = \frac{\sqrt[r]{h_p}}{\sum_{p=1}^{p=n} \sqrt[r]{h_p}} 100 \qquad (26)$$

For the most part, contemporary computer chromatography software does not provide for situations where $(r \neq 1)$ so the analyst can only use peak height data from the program and make the corrections manually. Alternatively, the chromatogram can be processed completely manually in which case peak areas can be used. The magnitude of any error that can arise from not correcting for (r) when it is other than unity can only be accessed, if and when the actual value of (r) is determined in the manner discussed in chapter 2. Unfortunately very few instrument manufacturers provide a value for (r) and thus the response index will usually need to be determined experimentally by the user.

Quantitative Analysis by TLC

TLC is not the best chromatographic technique for quantitative analysis and, although it can provide quantitative results, the necessary procedure tends to be more cumbersome and tedious compared with other chromatographic methods. Furthermore, for accurate work, expensive scanning equipment is required, which, as already discussed, rather reduces the cost advantage of the technique.

In principle there are three approaches used in quantitative TLC: extraction of the spot and separate measurement by spectroscopic or other techniques, comparative techniques employing visual assessment and finally optical scanning. The first, for obvious reasons, is rarely used and, due to the difficulties involved in the extraction of the material from the spot, gives relatively poor accuracy.

Comparative Spot Assessment by Visual Estimation

The quantity of material contained in a TLC spot can not be estimated visually, as the retinic response of the human eye is not linear and varies widely with the wavelength of the reflected light. Visual estimation is further complicated by the variation of the iris with light intensity. The eye, however, is very sensitive to slight *differences* in light intensity and therefore can be used very effectively for *comparing* light intensities. In fact, in the early twentieth century, the majority of trace analyses (*e.g.,* oxygen in boiler water, heavy metal contaminants such as chromium, manganese and cadmium, alkaloids, etc.) were carried out by forming colored derivatives and visually comparing the color with that of standards.

In TLC, a similar procedure is employed. The sample is run in parallel with a series of calibration solutions each containing the solute of interest at different concentrations. The first set of calibration standards can have relatively large concentration intervals, which allows the approximate concentration of the solute of interest to be identified. A second set of standards is then made up that embraces the suspected concentration in smaller intervals and the separation is repeated. By matching the intensity of the spots, the concentration of the sample can be estimated. The TLC spots can not be rendered visible by a destructive process (*e.g.*, concentrated sulfuric acid) as such processes are not quantitative. If the solute of interest is colorless, a suitable colored derivative must be formed either before separation (in which case a special eluting solvent must be identified) or more commonly after the separation by treating the plate with an appropriate reagent. Providing the separation is clean and a suitable colored derivative is employed, the visual comparative procedure can give quantitative results with an accuracy of about +/-10 %. Today, such accuracy is generally considered inadequate and as a result, the visual comparative technique has been largely replaced by instrumental methods of spot density measurement. Scanning densitometry has already been discussed and it was clear that the method of quantitative measurement of a TLC plate would be far superior to that of color matching. Nevertheless, the necessary expense involved together with its greater accuracy makes liquid chromatography a very strong competitor.

The detector provides two services in a chromatographic analysis. Firstly, it displays the separation that has been achieved and allows the complexity of the mixture to be ascertained. Secondly, and equally important, it allows the mixture to be quantitatively analyzed with accuracy and precision. These two aspects must be foremost in the minds of the designers and manufacturers of chromatography detectors at all times. Nevertheless, versatility, ease of maintenance, size, reliability and last, but by no means least, *price,* are also very important attributes, both to the organization purchasing the equipment and the analyst who will use them. When the cost and complexity of the modern automobile are compared with those of a chromatograph there seems a serious discrepancy. As more restraints and regulations force industries to exercise even greater control over their products and the toxicity of their wastes, the need for chromatographic equipment, particularly novel detecting systems, will continue to increase. Chromatography instrumentation of the future must be more compact, reliable, and easily serviced and additionally meet all the new demands placed on the analyst. We can expect, with complete confidence, that the technical challenges will be readily met by the established scientific instrument companies. However, in addition, the price must also be competitive and truly reflect the complexity and manufacturing cost of the device. To be fair, a small number instrument companies are, indeed, pioneering the concept of low cost instrumentation without conceding to reduced quality, accuracy or reliability. These companies, I believe, will become the new leaders in the field, and ultimately, will be providing all the chromatography detectors of the future.

Appendix I

An Approximate Method for Calculating the Efficiency of a Capillary Column

The approximate efficiency of a capillary column operated at its optimum velocity (assuming the inlet/outlet pressure ratio is small) is given by the Golay equation,

$$H = \frac{2D_m}{u} + \frac{1+6k'+11k'^2}{24(1+k')^2}\frac{r^2}{D_m}u + \frac{3k'}{3(1+k')^2}\frac{df^2}{D_s}u$$

where (H) is the variance per unit length of the column,
(D_m) is the diffusivity of the solute in the mobile phase,
(D_s) is the diffusivity of the solute in the stationary phase,
(k') is the capacity factor of the solute
(r) is the radius of the column,
(df) is the film thickness of stationary phase,
and (u) is the linear velocity.

If only the dead volume peak is considered k' = 0

Thus
$$H = \frac{2D_m}{u} + \frac{1}{24}\frac{r^2}{D_m}u \qquad (1)$$

At the optimum velocity the first differential will equal zero:

Thus,
$$\frac{dH}{du} = -\frac{2D_m}{u^2} + \frac{1}{24}\frac{r^2}{D_m} = 0 \quad \text{or} \quad u = \frac{\sqrt{48D_m}}{r}$$

Substituting for (u) in (1) and simplifying,

$$H = \frac{2D_m r}{\sqrt{48D_m}} + \frac{1}{24}\frac{r^2}{D_m}\frac{\sqrt{48D_m}}{r} = 0.289\,r + 0.289\,r = 0.577\,r$$

Thus the efficiency of a capillary column of length (l) can be assessed as,

$$n = \frac{1}{0.6\,r}$$

Appendix II

An Approximate Method for Calculating the Efficiency of a Packed Column

The approximate efficiency of a packed column operated at its optimum velocity (assuming the inlet/outlet pressure ratio is small) is given by the Van Deemter equation,

$$H = 2\lambda dp + \frac{2\gamma D_m}{u} + \frac{1+6k'+11k'^2}{24(1+k')^2}\frac{dp^2}{D_m}u + \frac{8k'}{\pi^2(1+k')^2}\frac{df^2}{D_s}u$$

where (H) is the variance per unit length of the column,
(D$_m$) is the diffusivity of the solute in the mobile phase,
(D$_s$) is the diffusivity of the solute in the stationary phase,
(k') is the capacity factor of the solute
(dp) is the particle diameter of the packing,
(df) is the film thickness of stationary phase,
and (u) is the linear velocity.

(λ) and (γ) are packing constants which for a well packed column can be taken as 0.5 and 0.6 respectively. Thus, if only the dead volume peak is considered k' = 0

Then
$$H = \lambda dp + \frac{1.2 D_m}{u} + \frac{1}{24}\frac{dp^2}{D_m}u \qquad (1)$$

At the optimum velocity the first differential will equal zero

Thus, $\quad \dfrac{dH}{du} = -\dfrac{1.2 D_m}{u^2} + \dfrac{1}{24}\dfrac{dp^2}{D_m} = 0 \quad$ or $\quad u = \dfrac{5.37 D_m}{dp}$

Substituting for (u) in (1) and simplifying,

$$H = dp + \frac{2D_m dp}{5.37 D_m} + \frac{1}{24} \frac{dp^2}{D_m} \frac{5.37 D_m}{dp}$$
$$= dp + 0.372 dp + 0.224 dp = 1.6 dp$$

Thus, the efficiency of a packed column, length (l), can be assessed as

$$n = \frac{l}{1.6 dp}$$

Appendix III

Some Physical Propeties of Solvents in Common Use in Liquid Chromatography

Solvent	Cut Off (nm)	Refract. Index	Density (20°C) (g/ml)	Boil. Point (°C)	Dielect. Constant
n-pentane	205	1.358	0.6214	35.4	1.844
n-heptane	197	1.388	0.6795	98.4	1.924
cyclohexane	200	1.427	0.7739	80.7	2.023
carbon tetrachloride	265	1.466	1.5844	76.7	2.238
n-butyl chloride	220	1.402	0.8809	78.5	7.39
chloroform	295	1./443	1.4799	61.1	4.806
benzene	280	1.501	0.8737	80.1	2.284
toluene	285	1.496	0.8263	110.6	2.379
dichloroethane	232	1.424	1.3168	39.8	9.08
tetrachloroethylene	280	1.938	1.3292	87.2	3.42
1,2–dichloroethane	225	1.445	1.2458	83.5	10.65
2–nitropropane	380	1.394	0.9829	120.3	25.52
nitromethane	380	1.394	1.1313	101.2	35.87
n-propyl ether	200	1.381	0.7419	89.6	3.39
ether	215	1.353	0.706	34.5	4.34
ethyl acetate	260	1.370	0.8946	77.1	6.02
methyl acetate	260	1.362	0.9280	56.3	6.68
acetone	330	1.359	0.7844	56.5	20.7
tetrahydrofuran	225	1.408	0.8842	66.0	7.58
acetonitrile	190	1.344	0.7822	81.6	37.5
n-propanol	205	1.380	0.7998	97.2	20.3
ethanol	205	1.361	0.7850	78.3	24.6
methanol	205	1.329	0.7866	64.7	33.6
water	180	1.333	0.9971	100.0	80.3
acetic acid	210	1.329	1.0437	118.0	6.15

Index

A/D conversion 70
absolute mass detector 169
absorption cell design 55
acquisition data 67
air,dissolved,effect of 450
alkaline earth cations, separation of 232
alkyl benzenes, separation of 186
amino acids, separation of 205
amplifier
    high impedance 102
    scaling 69
angle of deviation detector 247
apparent dispersion. sensor 55
area,peak measurement of 482
argon detector 119
    mechanism 120
argon detectors, characteristics of 128
aromatic hydrocarbons, analysis of 136
atmospheric ionization LC/MS 411
atomic absorption spectrometer 429
atomic emission spectrometer 393

β ray detector 88
band width 13
base line, instability 451
base width 11
baseline correction 77
Basil oil, separation of 392
belt transport interface 401
Bieman concentrator 381
bifunctional detectors 274
bulk property detectors 6
    limiting sensitivity 259

capacity factor 39
capillary column/MS 382
carbohydrates,separation of 262
carousel transport interface 415
catecholamines, separation of 239
cell
    detector design 55
    flow through, for LC/NMR 423
chain detector 287
chiral detector 297, 310
    diode array 313
chirality, examples of 298
Christiansen effect detector 253
chromatograph control 78
chromatography nomenclature 9

chromatography resolution 475
chromatography, vacancy 460
chromic-sulfuric spray, TLC 366
circular dichroism
    apparatus for measuring 303
    spectrometer 303
circular polarized light 302
classification,detectors 6
composite resonator 307
concentration
    sensitive detectors 8
    sensitivity, of system 39
concentrator
    Bieman 381
    Ryharge 380
connecting tubes,dispersion 46
control, chromatographic 78
conversion,A/D 70
corrected retention volume 10
criteria, detectors 63
curve, Gaussian 13

data
    acquisition 67
    processing 67, 75
    reporting 75
    transmission 72
dead
    time 10
    volume 10
deconvolution, peak 476
density balance detector,LC 354
density detector 352
derivatization 464
    acylation 466
    esterification 465
    fluorescence reagents 469
    post-column 470
    procedures in GC 465
    procedures in LC 467
    UV absorbing reagents 468
derivatizing methods
    TLC 364
detectors
    bulk property 6
    classification 6
    concentration sensitive 8
    criteria 63
    differential operation 452, 453

dispersion in sensor 52
displacement effects 450
dynamic range 23
flow sensitivity 61
for GC 442
GC
   absolute mass 169
   argon detector 119
   b ray 88
   dielectric constant 167
   discharge 161
   electron capture 137
   emissivity 114
   flame emissivity 92
   flame ionization 87, 99
   gas density balance 84
   general properties 95
   helium 132
   katherometer 149
   micro-argon detector 124
   nitrogen phosphorus 110
   piezoelectric adsorption 168
   problems with 443
   pulsed dicharge electron capture 143
   pulsed helium 135
   radio frequency discharge 164
   radioactivity 157
   simple gas density bridge 155
   spark discharge 163
   surface potential 171
   thermal argon 129
   thermionic ionization 160
   triode detector 127
   ultra sound 165
general operation 440
history 3
integral operation 452, 457
introduction to 3
LC
   bifunctional detectors 274
   chain 287
   chiral 297, 310
   choice 446
   Christiansen effect 253
   density 352
   density balance 354
   dielectric constant 266
   diode array 192
   disc detector 294
   electrical conductivity 223
   electron capture 348
   evaporative light scattering 211
   fixed wavelength UV 183
   fluorescence 201
   fluorescence,single wavelength 201
   heat of adsorption 328
   interferometer 255
   liquid light scattering 215
   low angle light scattering 217
   mass 358
   modified moving wire 289
   moving wire 285
   multi-electrode array 240
   multi-functional 273
   multi-wavelength fluorescence 206
   multi-wavelength UV 188
   multiple angle scattering 219
   radioactivity 315, 325
   refractive index 247
   spray impact 343
   thermal conductivity 355
   thermal lens 264
   transport 284
   trifunctional 279
   UV 177
   UV/conductivity 276
   UV/fluorescence 274
   UV/refractive index 277
LCLproblems with 448
linear dynamic range 31
linearity 24
mass sensitive 8
mass sensitivity of system 37
minimum detectable concentration 36
noise 32
   drift 33, 452
   long term 33, 451
   measurement of 34
   short term 32, 451
non-specific 9
output 18
pressure sensitivity 60
response 31, 480
response index 25
   determination of 27
selection 441
sensitivity 36
sensor
   apparent dispersion 55
   design 54
solute property 7
specific 9
specifications 17
specifications 18
temperature sensitivity 62

thin layer 363
time constant 57
dielectric constant detector 167, 266
differential, detector operation 452
differential response 21
differential, detector operation 453
diode array
    chiral detector 313
    UV detector 192
disc detector 294
discharge detector 161
dispersion
    connecting tubes 46
    extra column 14, 42
    low,tubing 49
    peak,of system 42
    permissible 45
displacement effects, detector 450
dissociation constants 224
dissolved air, effect of 450
distance, migration 12
double beam scanning, TLC densitometry 369
dual ionization transport interface 404
dynamic range,detectors 23

electrical conductivity detector 223
electro-optic modulator 305
electrochemical detector 233
    electrode construction 236
    electrode design 235
    electronics for 237
    multi-array 240
    response 234
electrode designs,conductivity detector 228
electron capture detector 137
electron capture LC detector 348
emissivity detector 92, 114
    application of 114
equivalent conductance 226
evaporative light scattering detector 211
external standard 494
extra-column dispersion 14, 42
extraction interface,LC/IR 418

factor
    retardation 12
factor, retardation 12
fixed wavelength UV detector 183
flame emissivity detector 92
flame ionization detector 87, 99
    application of 107
    design 100

electrode configuration 101
    operation of 107
    response mechanism 103
    sensor 100
flow sensitivity 61
fluorescence
    detection 201
    multi-wavelength detector 206
    single wavelength detector 201
    theory of 201
fluorescence detection,TLC 367
fluorescence quenching,TLC 368

gas density balance 84
gas density bridge ,simple 155
gasoline,analysis of 108
Gaussian curve 13
GC
    derivatization procedures 465
    detectors for 442
    detectors, problems with 443
GC/atomic emission 393
GC/IR 389
GC/MS 380
germanium disc interface for LC/IR 419

heat of adsorption detector 328
    performance data 346
    theory 330
    trace 329
height
    of theoretical plate 15
    peak 11
height, peak, measurement of 487
helium detector 132
helium plasma,spectrometer 393
helium,analysis of 134
herbicide analysis 112, 143
heterogeneous radioactivity counter 324
high speed LC 4
history
    detector 3
    GC detectors 84
    LC detectors 177
hydrocarbons,analysis of 126
hydrogen isotope separation 154
hydrolysed cyclodextrin,separation of 263

instability, baseline 451
integral response 20
integral, detector operation 452, 457
interferometer detector 255
internal standard 491

iodine spot reagent,TLC 365
ion suppression 231
ion trap mass spectrometer 385
ionization detector
    argon 119
    b ray 90
    discharge 161
    electron capture 137
    flame 99
    helium 132
    macro argon 121
    micro argon 124
    pulsed discharge electron capture 144
    pulsed helium 135
    spark discharge 163
    thermal argon 129
    thermionic 160
    triode 127
IR trap 390

katherometer detector 149

laser desorption ionization 404
LC
    derivatization procedures 467
    detectors for 446
    detector, problems with 448
LC tandem systems 395
LC/AAS 429
    serpentine tube interface 430
LC/IR
    carousel transport interface 415
    extraction interface 418
    germanium disc 419
    micro-cell 416
LC/IR systems 414
LC/MS 395
    atmospheric ionization interface 411
    belt transport interface 401
    electro-spray interface 407
    thermospray interface 405
    wire transport interface 397
LC/NMR
    flow cell interface 423
    peak sampling unit 426
    tandem systems 422
lead compounds, separation of 432
light pipe 391
light scattering detectors
    evaporative 211
    liquid 215
    low angle 217
    multiple angle 219
linalool, IR spectrum of 392

linear dynamic range,detectors 31
linearity,detectors 24
lipids, analysis of 214
liquid light scattering detectors 215
low angle light scattering detectors 217
low dispersion tubing 49

macro-argon detector 121
    power supply 122
mass detector 358
mass sensitive detectors 8
mass sensitivity of system 37
matched element resonator 308
maximum capacity factor 39
micro-argon detector 124
micro-cell, LC/IR 416
migration distance 12
minimum detectable concentration 36
modified moving wire detector 289
moving wire detector 285
multi-electrode array detector 240
    electrode design 241
multi-functional detectors 273
multi-wavelength fluorescence detector 206
multi-wavelength UV detector 188
multiple angle light scattering detector 219

narcotic drugs, by ion trap 387
neomycin,separation of 208
neuroactive drugs, separation of 245
nitrogen phosphorus detector 110
    application of 112
noise
    detectors 32
    drift 33, 452
    long term 33, 451
    measurement of 34
    short term 32, 451
nomenclature, chromatography 9
non-specific detectors 9
normalization methods 496

parallel transmission 73
peak
    area measurements 482
    deconvolution 476
    dispersion in system 42
    height 11
    height measurement 487
    purity,diode array 196
    skimming 485

513

width 11
width at half height 11
peaks
    spurious 449
permissible system dispersion 45
pesticide analysis 142, 146
photometric detector 114
piezo-electric adsorption detector 168
piezo-optical modulators 306
Pockels cell 305
polarization modulation 304
polarized light
    circular 302
    production of 301
    properties of 301
polystyrene standards,separation of 252
post-column derivatization 470
preparative UV detector 188
pressure, sensitivity 60
priority pollutants, separation of 203
priority pollutants,fluorescence detection 210
processing,data 67
pulsed discharge electron capture detector 143
pulsed helium detector 135
purity, peak by diode array 196

quadrupole mass spectrometer,triple 383
quantitative analysis 475
    by TLC 498
    evaluation of chromatogram 481
    external standard method 494
    intenal standard 491
    methods of 490
    normalization method 496
    reference standards 490

radio-frequency discharge detector 164
radioactivity
    detection, theory 316
    detector,GC 157
    detectors, LC 315
range,dynamic,detectors 23
reference standards 490
References 16, 65, 80, 97, 118, 147, 173, 198, 222, 246, 272, 296, 314, 360, 374, 436, 472
refractive index detector 247
    angle of deviation method 247
    applications 259
    Fresnel method 249
resolution, chromatographic 475
response

    detectors 31
    differential 21
    index,detectors 25
    index,determination of 27
    integral 20
    mechanism, FID 103
response, detector 480
retardation factor 12
retention 10
    volume 10
Ryharge concentrator 380

sample solvent, effect of 450
scaling amplifer 69
scanner design,TLC 372
scanning densitometry 368
scintillation
    agents 315
        liquid 317
        solid 321
    sensor 326
Scott gas mixture,separation of 155
secondary ion ionization 404
selection 441
sensitivity
    concentration,of system 39
    flow 61
    limitations of bulk property detectors 259
    mass, of system 37
    pressure 60
    temperature 62
sensor
    dispersion in 52
    flame ionization 100
    scintillation 326
serial transmission 72
serpentine tubing 50
    LC/AAS interface 430
single beam scanning,TLC densitometry 369
skimming,peak 485
solute property detectors 6
spark discharge detector 163
specific derivatizing methods 366
specific detectors 9
specifications, detectors 17
spectrometer
    atomic 394
    atomic absorption 429
    atomic emission 393
    detectors 377
    helium plasma 393
    IR 390

mass
  Banner 382
  ion trap 385
  time of flight 387
  triple quadrupole 383
  NMR 422
spot detection,TLC 363
spray impact detector 343
spray,TLC spot 364
spurious peaks 449
sulfuric acid spray,TLC 365
sulphur compounds, detection in hydrocarbons 117
summation of variances 12
surface potential detector 171
syringe,contaminated,effect of 450
system, peak dispersion 42

tandem systems 377
  GC/AE 393
  GC/IR 389
  GC/MS 380
  LC/AAS 429
  LC/IR 414
  LC/MS 395
  LC/NMR 422
  thin layer chromatography 433
  TLC/MS 433
temperature sensitivity 62
theoretical plate height 15
thermal argon detector 129
thermal conductivity detector,LC 355
thermal lens detector 264
thermionic ionization detector 160
thermospray LC/MS interface 405
thin layer chromatography/mass spectroscopy 433
time
  constant, detectors 57
  corrected retention 11
  dead 10
  retention 10
time of flight mass spectrometer 387
TLC
  derivatizing methods 365
  derivatizing methods,specific 365
  detectors 363
  fluorescence detection 367
  fluorescence quenching 368
  quantitative analysis 499
  scanner design 372
  scanning densitometry 368
  spot detection 363
  spray 364

TLC/MS 433
  plate scanner 434
transmission
  data 72
  parallel 73
  serial 72
transport detectors 284
  chain 287
  disc 294
  modified moving wire 289
  moving wire 285
transport systems 403
  belt spray system 403
  carousel interface for LC/IR 415
  dual ionization 404
  germanium disc for LC/IR 419
  moving belt 401
  moving wire 397
trap, for IR 390
Tridet detector 279
trifunctional detectors 279
triode detector 127
triple quadrupole mass spectrometer 383
tryptic digest of lysozyme,LC/MS 408
tubing
  low dispersion 49
  serpentine 50

ultra-sound whistle detector 165
UV detectors 177
  absorption,theory 180
  diode array 192
  fixed wavelength 183
  multi-wavelength 188
  preparative 187
UV/conductivity detector 276
UV/fluorescence 274
UV/refractive index detector 277

vacancy chromatography 460
variances, summation of 12
volume
  dead 10
  retention 10

whistle,ultra-sound,detector 165
width
  at half height. 11
  band 12
  peak 11
wire transport interface 397